Katherine

Enjoy the Future

Jack Curtis

John B Mahaffie

2025

Scenarios of US and Global Society
Reshaped by Science and Technology

2025

Scenarios of US and Global Society
Reshaped by Science and Technology

by Joseph F. Coates
John B. Mahaffie
and
Andy Hines

Published for
Coates & Jarratt, Inc.

by

Oakhill Press
Greensboro

2025

Library of Congress Cataloging-in-Publication Data

Coates, Joseph F. (Joseph Francis), 1929-
   2025 : Scenarios of US and Global Society Reshaped by Science and Technology / by Joseph Coates, Andy Hines, and John Mahaffie.
   Includes index
   ISBN 1-886939-09-8
   1. Technology--Social Aspects. 2. Technology--Social Aspects--United States. 3. Twenty-first century--Forecasts. I. Hines, Andy, 1962- . II. Mahaffie, John B.   III. Title.

T14.5.C62   1997   97-18376
303.48'3'0905--dc20   CIP

Oakhill Press, Greensboro, North Carolina, (800) 32-BOOKS [800-322-6657].

                    0 9 8 7 6 5 4 3 2 1

First Printing, February 1997

PREFACE

In the modern world, science and technology are some of the strongest drivers of change. Discoveries in science and new developments in technology change the course of people's lives, their behaviors, and their attitudes to the world. The great and expanding influence of the telephone, the electric light, and the automobile are obvious examples. Some less obvious examples are the development of beneficial drugs, such as antibiotics, the proliferation of new fibers and fabrics, and the expansion of the social sciences. All of these have enabled people to live longer, have wider perspectives in their lives, and anticipate a prosperous future.

For the most part the influence of science and technology on our future is beneficial. The future is more likely to be positive if people can see how to use the developments of technology and science to improve the human condition. We believe we have the power to influence the future and, therefore, the responsibility of shaping the future in a positive direction. In this book we start with the assumption that people and organizations are capable of using emerging science and technology capabilities in ways that will be useful, profitable, and of benefit to humanity over the next generation.

Why We Wrote This Book

The scenarios of 2025 that are the major part of this book are the outcome of years of work in gathering and evaluating forecasts made in more than 50 science, technology, and engineering fields. When we evaluated all these forecasts we recognized that although collectively they implied a strong future for science and technology, they were mostly relatively narrow in their viewpoint and missed the larger picture of what the world could be like if all, or most of, these developments occurred. So we drew on our own experience as futurists as well as that of the other forecasters to create images of the world in 2025 that show what may be in store for us at home, in business, on the road, around the world, and so on.

Intended Audience

We expect that business people, entrepreneurs, people who are responsible for planning, those involved in science and technology policy, and those engaged in scientific and technical issues, will all find this book interesting and useful. Young people, whose future this is, should find ideas here about potential opportunities for themselves. Everyone who has a thoughtful interest in the implications of science and technology for our future will find things to interest them.

Overview of the Contents

This book is a set of 15 scenarios of the world in 2025. Each is written from the perspective of a business person or any well-informed adult who expects that business people will read the scenario and find in it familiar terms and a recognizable world view. Each scenario, therefore, is a history of the United States and different parts of the world to 2025. Sometimes the scenario's author looks beyond 2025 and discusses what might happen in the next few years. After reading a few of these scenarios, the reader will be able to build a mental picture of what the world will be like in several important dimensions. Although these do not make up a single unified scenario, the reader will discover that they are consistent with each other in many areas, such as the environment, for example. To help the reader catalog these consistencies, we have extracted key assumptions about the world of 2025 and listed them in Chapter 1. Introduction. In Chapter 1, we give the reader some advice on reading and getting the most from scenarios.

Acknowledgments

Our first thank you is to the sponsors, the people and organizations who made possible the original *Project 2025* from which this book was created. We mention them here with their corporate or organizational affiliation at the time of the project. Several have moved on to different activities since 1993. They are Rana Nanjappa, Richard E. Albright, and Thomas P. Egan, of AT&T, David Dragnich and Peter Van Voris, of Battelle Pacific Northwest Laboratories, Brent Welling and Tom Burns of The Chevron Corporation, Tom Sinclair of the Department of Trade and Industry, United Kingdom, Jack W. Hipple and Kerry Kelly, of Dow Chemical, Larry Harris of Dow Elanco, Charles J. Molnar and F. Peter Boettcher of Du Pont, Trueman F. Parish, and Dr. T. Flint Gray, of Eastman Chemical Company, Richard L. Wagner of Hercules, Inc., Dr. Herb Paschen of Kernforschungszentrum, Germany, Dr. William L. Reber, of Motorola, Helen Salter of the National Security Agency, Jean Woodard and Janet Joseph of New York State Energy Research Development Authority, J. Ron Cross of Northern Telecom Inc., Cynthia A. Donovan of Pacific Bell, Stagg Newman and Brent Welling of Pacific Telesis, Kyle M. Sawyer of Tenneco Gas, Lewis House, Andrew Eiseman, Bud C. Wonsiewicz and Armando Fimbres of U S WEST, Vince Roscelli and Lynn O. Michaelis of Weyerhaeuser Forest Products Company. We are grateful to them and to their colleagues who became involved in the project and gave us sound advice and criticism throughout.

Almost all the Coates & Jarratt, Inc., staff who worked with us from 1991-1993 have had a hand in one or both phases of *Project 2025*, and are collectively responsible for the research, review, and production of much of the material that went into this book. We gratefully acknowledge their contributions. Jennifer Jarratt, Kathleen Shaw, and Carol DiPaolo helped us with the production of the book.

CONTENTS

1

INTRODUCTION

This book is a picture of the world in 2025, as shaped by science and technology and based on forecasts and assumptions about that future world.

Four enabling technologies will be central to shaping the world of 2025, each introducing capabilities that will extend far beyond their immediate applications to effect a network of change throughout society, much in the way that the introduction of the electric light and the automobile at the turn of the century powerfully shaped today's world. These four drivers of change are *information technology, materials technology, genetics,* and *energy technology.*

A fifth primary driver of change, *environmentalism,* represents not new capabilities, but a changing worldwide orientation. An emerging pattern of attitudes and beliefs about sustainability and uses of the Earth will direct the shape of the future as powerfully as any of the four enabling technologies.

The world of 2025 can be broken down into three broad population groups: World 1, including the affluent advanced nations of Europe, the United States, and Japan; World 2, the middle, making up the bulk of the world's population, whose immediate needs and resources will be in relative balance; and World 3, the destitute nations, those on the brink of starvation, living with the constant threat of disaster. It is useful to categorize countries within one of these three groups, bearing in mind that by 2025 a worldwide middle class will have emerged, represented to some extent in every society. In many of the middle-income and destitute nations, an affluent few will make up a thin crust living essentially like those of World 1. World 1 countries may have a corresponding but thin bottom layer of abjectly poor.

The following columns show the projected population growth of the three worlds by 2025.

World	1994	2025
Total	5.6 billion	8.4 billion
World 1	1.0 billion	1.3 billion
World 2	3.5 billion	5.1 billion
World 3	1.1 billion	2.0 billion

The chapters that follow emphasize the character and quality of future life in the advanced World 1 nations, particularly in the United States, with some discussion of the midrange, World 2, and poorest nations, World 3. In the text, World 1 is frequently referred to as the *affluent* world, World 2, as the *middle* world and World 3 as the *destitute* world.

About Economics

Any change from old to new or any modification in the way a society behaves must have economic consequences, and also must be limited by economic considerations. Readers may be surprised that we give relatively little attention to the economics of transition from the world of today to that of 2025. Three points are important to consider:

1. The more radical a transition, the more speculative and conjectural are its economics. For instance, it is certain that genome research and its outcomes will stimulate a huge volume of biomedical, biochemical, and physiological research, but it would be difficult to make any plausible estimate of cost and return based on that research. Not all possibilities will be realized, as competition among alternatives gives the edge to some choices over others.

2. Many of the most important factors of change are caught up in intrinsic uncertainty that blunts the significance of any economic discussion and would distract the reader's attention from the powerful forces driving the future. For example, we live in a world that is a network of public and private, market and controlled, independent and subsidized productions and applications of energy. Choices in energy are driven not only by economic considerations, but also by social and political concerns, making it difficult to tease out a plausible *economic* analysis of what might happen to energy in the future.

3. The choices and futures that we describe are more strongly driven by the basic condition of the economy—recession, depression, growth, or boom—than by particular microeconomic choices.

Compound interest is a powerful factor in change. With regard to the macroeconomic state of the United States, we have chosen a midcourse assumption (rejecting both a permanent depression model of Gross Domestic Product (GDP) growth (0% to 1%), and a permanent boom model (3% - 4%

growth)). We assume that economic growth on a *per capita* basis will be about 2 to 2.5%.

That assumed rate suggests that real *per capita* income will double from 1995 to 2025 and, further, that there will be increasing discretionary funds available to individuals, families, public institutions, and government. A substantial portion of this new discretionary income, we believe, will be used to deal with accumulated problems of the past, such as decaying infrastructure and environmental lesions, and to invest in the new alternatives described in *2025*.

The table "Rapid Growth Is Possible" shows the 25-year history of several countries that, in 1996, are clear economic successes. Korea and Portugal, having started from different bases and with strikingly different growth rates, now have roughly the same *per capita* income. Japan—a stellar economic success by any measure, its 1995-1996 performance notwithstanding—has averaged only 4% growth over the past quarter century. Hong Kong, Singapore, and South Korea have run 6.2% to 7.1% annual growth.

Rapid Growth Is Possible: Countries with High GNP/Capita Growth Rates 1965-1990

Country	GNP per capita 1990 U.S. Dollars	Avg. Annual Growth Rate (%) 1965 - 1990
Peoples Republic of China	370	5.8
Indonesia	570	4.5
Thailand	1,420	4.4
South Korea	5,400	7.1
Singapore	11,160	6.5
Hong Kong	11,490	6.2
Japan	25,430	4.1
Portugal	4,900	3.0
World	4,900	1.5

Source: *World Development Report 1992*, World Bank, Oxford Univ. Press, New York, Table 1.

Ultimately, economic choices lie at the core of every discussion about the future. But it is important to realize that the explicit and implicit outcomes described in the chapters that follow do not call for unprecedented, revolutionary patterns of economic development. They are entirely plausible given only the modest 2% to 2.5% growth anticipated for the United States between 1995 and 2025.

Throughout *2025*, costs are given in 1995 U.S. dollars except where otherwise noted.

Where This Book Came From and 107 Assumptions We Make About the Future

2025 emerged from the second phase of a project conducted by Coates and Jarratt, Inc., a Washington, D.C., think tank specializing in the study of the future. The first phase of the project, completed in 1993, organized and analyzed forecasts made from 1970 through 1992 in virtually every subdiscipline of science, technology, and engineering.

> A *forecast* is a simple or complex look at the qualities and probabilities of a future event or trend. Futurists differentiate between the *forecast,* which is generally not point-specific to time or place, and the *prediction,* a specific, usually quantitative statement about some future outcome.

In the second phase of the project we created fresh forecasts, examining how diverse technologies from brain science to information technology might interact to influence every aspect of life in the year 2025. Those forecasts are the topic of this book. They appear in the subsequent chapters as scenarios of 2025. To build these scenarios we assume specific scientific, technological, demographic, social, and other changes will occur. These assumptions are included in this introduction.

Assumptions about the future are not like assumptions in a geometry exercise; they are not abstract statements from which consequences can be drawn with mathematical certitude and precision. They are highly probable statements about the future, forming a framework around which less certain ideas can be tested. We need to make assumptions about the future in order to plan it, prepare for it, and prevent undesired events from happening. Here, we built 15 scenarios of 2025 based on 107 assumptions that we made about the future.

Some of these statements are drawn from the project. Others, such as the estimates of future population, come from public or highly credible private statistical and mathematical analyses of trends. Still others result from integrating a wide range of material; one such assumption is that we will be moving toward a totally managed globe. To present the underlying arguments supporting each of these highly reliable statements (which amount to forecasts) would require a massive introductory section. We have, therefore, presented these statements about the future as simply and in as straightforward a manner as possible.

A few of these assumptions have a normative, or goals-oriented, aspect to them. The assumption, for example, that *per capita* energy consumption in the advanced nations will fall to 66% of the 1990 level is definitely not a trend extrapolation but a judgment about the confluence of social, political, economic, environmental, technological, and other concerns. Readers are urged to formulate and review alternatives that might characterize the next 30 years and test how those alternatives affect any other thoughts, concepts, beliefs, or conclusions about the future.

What follows is an inventory of high probability statements about the year 2025 in three categories:

A. Scientific discoveries and research, and technological developments and applications.

B. Contextual, that is, those factors forming the social, economic, political, military, environmental, and other factors that will shape or influence scientific and technological developments. These contextual areas form the environment for the introduction and maturation of new products, processes, and services in society.

C. Twenty-four additional high-probability statements that have slightly less probability of occurring.

These high-probability forecasts, especially the first 83, become assumptions in understanding how any particular area may develop under the influence of new scientific, technological, social, political, or economic developments. It would be nice to suggest that these developments are inevitable, but few developments are. Nonetheless, the convergence of evidence indicates that these 107 developments are of such high likelihood that they form an intellectual substructure for thinking about any aspect of the year 2025.

The reader need not accept all these assumptions or high-probability statements in order to find interest and value in the scenarios. The set is rich and robust enough that the reader may reject several or even many of them without undercutting the overall vision of the future in chapters 2-16.

Managing our World

1. Movement toward a *totally managed environment* will be substantially advanced at national and global levels. Oceans, forests, grasslands, and water supplies will make up major areas of the managed environment. Macroengineering—planetary-scale civil works—will make up another element of that managed environment. Finally, the more traditional business and industrial infrastructure—telecommunications, manufacturing facilities, and so on—will be a part of managed systems and subsystems.

Note that total management does not imply full understanding of what is managed. But expanding knowledge will make this management practical. Total management also does not imply total control over these systems.

2. Everything will be smart, that is, responsive to its external or internal environment. This will be achieved either by embedding microprocessors and associated sensors in physical devices and systems or by creating materials that are responsive to physical variables such as light, heat, noise, odors, and electromagnetic fields, or by a combination of these two strategies.

Managing Human Health

3. All human diseases and disorders will have their linkages, if any, to the human genome identified. For many diseases and disorders, the intermediate biochemical processes that lead to the expression of the disease or disorder and its interactions with a person's environment and personal history will also be thoroughly explored.

4. In several parts of the world, the understanding of human genetics will lead to explicit programs to enhance people's overall physical and mental abilities—not just to prevent diseases.

5. The chemical, physiological, and genetic bases of human behavior will be generally understood. Direct, targeted interventions for disease control and individual human enhancement will be commonplace. Brain-mind manipulation technologies to control or influence emotions, learning, sensory acuity, memory, and other psychological states will be in widespread use.

6. In-depth personal medical histories will be on record and under full control of the individual in a medical smart card or disk.

7. More people in advanced countries will be living to their mid-80s while enjoying a healthier, fuller life.

8. Custom-designed drugs such as hormones and neurotransmitters (chemicals that control nerve impulses) will be as safe and effective as those produced naturally within humans or other animals.

9. Prostheses (synthetic body parts or replacements) with more targeted drug treatments will lead to radical improvements for people who are injured, impaired, or have otherwise degraded physical or physiological capabilities.

Managing Environment and Resources

10. Scientists will work out the genome of prototypical plants and animals, including insects and microorganisms. This will lead to more-refined management, control, and manipulation of their health and propagation, or to their elimination.

11. New forms of microorganisms, plants, and animals will be commonplace due to advances in genetic engineering.

12. Foods for human consumption will be more diverse as a result of agricultural genetics. There will be substantially less animal protein in diets in advanced nations, compared with the present. A variety of factors will bring vegetarianism to the fore, including health, environmental, and ethical trends.

13. There will be synthetic and genetically manipulated foods to match each individual consumer's taste, nutritional needs, and medical status. Look for "extra-salty (artificial), low-cholesterol, cancer-busting french fries."

14. Farmers will use synthetic soils, designed to specification, for terrain restoration and to enhance indoor or outdoor agriculture.

15. Genetically engineered microorganisms will do many things. In particular, they will be used in the production of some commodity chemicals as well as highly complex chemicals and medicines, vaccines, and drugs. They will be widely used in agriculture, mining, resource up-grading, waste management, and environmental cleanup.

16. There will be routine genetic programs for enhancing animals used for food production, recreation, and even pets. In less developed countries, work animals will be improved through these techniques.

17. Remote sensing of the earth will lead to monitoring, assessment, and analysis of events and resources at and below the surface of land and sea. In many places, *in situ* sensor networks will assist in monitoring the environment. Worldwide weather reporting will be routine, detailed, and reliable.

18. Many natural disasters, such as floods, earthquakes, and landslides, will be mitigated, controlled, or prevented.

19. *Per capita* energy consumption in the advanced nations will be at 66% of *per capita* consumption in 1990.

20. *Per capita* consumption in the rest of the world will be at 160% of *per capita* consumption in 1990.

21. Resource recovery along the lines of recycling, reclamation, and remanufacturing will be routine in all advanced nations. Extraction of virgin materials through mining, logging, and drilling will be dramatically reduced, saving energy and protecting the environment.

22. Restorative agriculture (i.e., "prescription" farming) will be routine. Farmers will design crops and employ more-sophisticated techniques to optimize climate, soil treatments, and plant types.

Automation and Infotech

23. There will be a worldwide, broadband network of networks based on fiber optics; other techniques, such as communications satellites, cellular, and microwave will be ancillary. Throughout the advanced nations and the middle class and prosperous crust of the developing world, face-to-face, voice-to-voice, person-to-data, and data-to-data communication will be available to any place at any time from anywhere.

24. Robots and other automated machinery will be commonplace inside and outside the factory, in agriculture, building and construction, undersea activities, space, mining, and elsewhere.

25. There will be universal, on-line surveys and voting in all the advanced nations. In some jurisdictions, this will include voting in elections for local and national leaders.

26. Ubiquitous availability of computers will facilitate automated control and make continuous performance monitoring and evaluations of physical systems routine.

27. The ability to manipulate materials at the molecular or atomic level will allow manufacturers to customize materials for highly specific functions such as environmental sensing and information processing.

28. Totally automated factories will be common but not universal for a variety of reasons, including the cost and availability of technology and labor conflicts.

29. Virtual-reality technologies will be commonplace for training and recreation and will be a routine part of simulation for all kinds of physical planning and product design.

30. In text and—to a lesser extent—in voice-to-voice telecommunication, language translation will be effective for many practically-significant vocabularies.

31. Expert systems, a branch of artificial intelligence, will be developed to the point where the learning of machines, systems, and devices will mimic or surpass human learning. Certain low-level learning will evolve out of situations and experiences, as it does for infants. The toaster will "know" that the person who likes white bread likes it toasted darker, and the person who chooses rye likes it light.

32. The fusion of telecommunications and computation will be complete. We will use a new vocabulary of communications as we *televote, teleshop, telework,* and *tele-everything.* We'll *e-mail, tube,* or *upload* letters to Mom. We'll go *MUDing* in cyberspace and mind our *netiquette* during virtual encounters.

33. Factory-manufactured housing will be the norm in advanced nations, with prefabricated modular units making housing more flexible and more attractive, as well as more affordable.

34. In the design of many commercial products such as homes, furnishings, vehicles, and other articles of commerce, the customer will participate directly with the specialist in that product's design.

35. New infrastructures throughout the world will be self-monitoring. Already, some bridges and coliseums have "tilt" sensors to gauge structural stress; magnetic-resonance imaging used in medical testing will also be used to noninvasively examine materials for early signs of damage so preventive maintenance can be employed.

36. Interactive vehicle-highway systems will be widespread, with tens of thousands of miles of highway either so equipped or about to be. Rather than reconstruct highways, engineers may retrofit them with the new technologies.

37. Robotic devices will be a routine part of the space program, effectively integrating with people. Besides the familiar robotic arm used on space shuttles, robots will run facilities in space operating autonomously where humans are too clumsy or too vulnerable to work effectively.

38. Applied economics will lead to a greater dependency on mathematical models embodied in computers. These models will have expanded capabilities and will routinely integrate environmental and quality-of-life factors into economic calculations. One major problem will be how to measure

the economic value of information and knowledge. A Nobel Prize will be granted to the economist who develops an effective theory of the economics of information.

Population Trends

39. World population will be about 8.4 billion people.

40. Family size will be below replacement rates in most advanced nations but well above replacement rates in the less-developed world.

41. Birth control technologies will be universally accepted and widely employed, including a market for descendants of RU-486.

42. World population will divide into three tiers: at the top, World 1, made up of advanced nations and the world's middle-classes living in prosperity analogous to Germany, the United States, and Japan; at the bottom, World 3, people living in destitution; and in the middle, World 2, a vast range of people living comfortably but not extravagantly in the context of their culture. We use the terms *World 1*, *World 2*, and *World 3* for the emerging pattern of nations that moves us beyond the post-World War II nomenclature.

43. The population of World 1's advanced nations will be older, with a median age of 42.

44. The less-developed Worlds 2 and 3 will be substantially younger but will have made spotty but significant progress in reducing birth rates. However, the populations of these countries will not stop growing until sometime after 2025.

45. The majority of the world's population will be metropolitan, including people living in satellite cities clustered around metropolitan centers.

46. A worldwide middle class will emerge. Its growth in World 2 and to a lesser extent in World 3 will be a powerful force for political and economic stability and for some forms of democracy.

Worldwide Tensions

47. There will be worldwide unrest reflecting internal strife, border conflicts, and irredentist movements. But the unrest will have declined substantially after peaking between 1995 and 2010.

48. Under international pressures, the United Nations will effectively take on more peace*making* to complement its historic peace*keeping* role.

49. Supranational government will become prominent and effective, though not completely, with regard to environmental issues, war, narcotics, design and location of business facilities, regulation of global business, disease prevention, workers' rights, and business practices.

50. Widespread contamination by a nuclear device will occur either accidentally or as an act of political/military violence. On a scale of 1 to 10 (with Three Mile Island a 0.5 and Chernobyl a 3), this event will be a 5 or higher.

51. Increasing economic and political instabilities will deter business involvement in specific World 3 countries.

52. Despite technological advances, epidemics and mass starvation will be common occurrences in World 3 because of strained resources in some areas and politically motivated disruptions in others.

53. There will be substantial environmental degradation, especially in World 3. Governments will commit money to ease and correct the problem, but many will sacrifice long-term programs that could prevent the problem from happening in the first place.

54. There will be shifts in the pattern of world debtor and creditor countries. Japan's burst economic bubble, the ever-growing U.S. debt, and Germany's chronic unemployment problems are harbingers of things to come.

55. NIMBY ("Not In My Back Yard") will be a global-scale problem for a variety of issues, ranging from hazardous-waste disposal to refugees to prisons to commercial real-estate ventures.

56. Migration and conditions for citizenship throughout the world will be regulated under new international law.

57. Terrorism within and across international borders will continue to be a problem.

The Electronic Global Village

58. Global environmental management issues will be institutionalized in multinational corporations as well as through the United Nations and other supranational entities.

59. A global currency will be in use.

60. English will remain the global common language in business, science, technology, and entertainment.

61. Schooling on a worldwide basis will be at a higher level than it is today. Education may approach universality at the elementary level and will become more accessible at the university level through distance education technologies.

62. In the advanced nations, life-long learning will be effectively institutionalized in schools and businesses.

63. There will be substantial, radical changes in the U.S. government. National decisions will be influenced by electronically assisted referenda.

64. Throughout the advanced nations, people will be computer literate and computer dependent.

65. Worldwide, there will be countless virtual communities based on electronic linkages.

66. There will be a worldwide popular culture. The elements of that culture will flow in all directions from country to country. In spite of the trend toward "demassification" in both information and production, the global links of communications and trade will ensure that ideas and products will be *available* to all whether they like it or not.

67. The multinational corporation will be the world's dominant business form.

68. Economic blocs will be a prominent part of the international economy, with many products and commodities moving between these porous blocs. The principal blocs will be Europe, East Asia and the Americas.

69. Universal monitoring of business transactions on a national and international business basis will prevail.

70. Identification cards will be universal. Smart cards will contain information such as nationality, medical history (perhaps even key data from one's genome), education and employment records, financial accounts, social security, credit status, and even religious and organizational affiliations.

Public Issues and Values

71. Within the United States there will be a national, universal health care system.

72. In the United States, the likely collapse of the Social Security system will lead to a new form of old-age security such as one based on need-only criteria.

73. Genetic screening and counseling will be universally available and its use encouraged by many incentives and wide options for intervention.

74. There will be more recreation and leisure time for the middle class in the advanced nations.

75. The absolute cost of energy will rise, affecting the cost of transportation. Planners will reallocate terrain and physical space to make more-efficient use of resources. In other words, cities will be redesigned and rezoned to improve efficiencies of energy in transportation, manufacturing, housing, etc.

76. There will be a rise in secular substitutes for traditional religious beliefs, practices, institutions, and rituals for a substantial portion of the population of the advanced nations and the global middle class. The New Age movement, secular humanism, and virtual communities built on electronics networking are a few harbingers.

77. Socially-significant crime—i.e., the crimes that have the widest negative effects in the advanced nations—will be increasingly economic and computer-based. Examples include disruption of business, theft, money laundering, introduction of maliciously false information, and tampering with medical records, air traffic control or national-security systems.

78. Tax filing, reporting, and collecting will be computer-managed.

79. Quality, service, and reliability will be routine business criteria around the globe.

80. Customized products will dominate large parts of the manufacturing market. Manufacturers will offer customers unlimited variety in their products.

81. Economic health will be measured in a new way, including considerations of environment, quality of life, employment, and other activity and work. These new measures will become important factors in governmental planning.

82. GDP and other macroeconomic measures and accounts will include new variables such as environmental quality, accidents and disasters, and hours of true labor.

83. Sustainability will be the central concept and organizing principle in environmental management, while ecology will be its central science.

Additional, But Slightly Less Probable, Developments by 2025

These next 24 developments have a somewhat lower probability than the 83 basic assumptions. If they do occur, they could have long-term, extensive, startling, or disruptive effects on people and their societies.

1. Telephone communications within the United States and within Europe will be so cheap as to be effectively free.

2. Telecommunication costs will be integrated into rent or mortgage payments.

3. The greening of North Africa will begin, with megatechnologies to promote rain and build soil along the coast.

4. Antarctic icebergs will be harvested for watering the west coast of South America, Baja California, the Australian outback, Saudi Arabia, and other arid areas.

5. Going to work will be history for a large percentage of people. By 2020 or 2025, 40% of the workforce will be working outside the traditional office.

6. The home work/study center will be the centerpiece of the integrated, fully-information rich house and home. Mom and Dad will work there, the kids will reach out to the resources of the world, and the whole family will seek recreation, entertainment, and social contacts there.

7. Inorganic chemistry will rise to parity with organic chemistry in profit and importance in such areas as ceramics and composites.

8. Biomimetic materials and products that imitate natural biological materials will be common.

9. Micromachines the size of a typed period will be in widespread use. Nanotechnological devices 10,000 times smaller will have been developed and will be in use.

10. Radical cosmetics will leave no component of the body or mind beyond makeover. This will be accompanied by a melding of cosmetics, medicine, and surgery.

11. Ocean ranching and farming for food and energy will be widespread.

12. The asteroid watch will become a recognized institution. Among its most notable achievements will be several trial runs at altering an asteroid's path before it intersects Earth's orbit.

13. Moon mining and asteroid harvesting will be in their early stages.

14. Artificial intelligence devices will flower as aids to professionals, as adjuncts to ordinary workers, as doers of routine tasks, as checks on the functionality of software and complex systems, and as teaching and training tools.

15. Privatization of many highways, particularly beltways and parts of the interstate system will occur. This will be tied to the evolution of an intelligent vehicle-highway system.

16. Restoration of aquifers will be a standard technology.

17. Fuel cells will be a predominant form of electromechanical energy generation.

18. Mastodons will walk the Earth again and at least 20 other extinct species will be revived.

19. Biocomputers will be in the early stage of development and applications.

20. Squaring-off of the death curve will make substantial progress in World 1 and some progress in World 2, leading to most people living to 85 years.

21. Critical experiments in life extension to move the average lifetime of our species from 85 to 105, will begin. One hundred thousand people will be in a lifelong monitoring program. Massive numbers of other people will apply the treatments on a nonexperimental basis.

22. 120 mile-per-gallon cars will be in widespread use.

23. Hypersonic air carriers will be common.

24. Brain protheses will be one of the practical applications of brain technology.

Why the Future is Described in Scenarios

The scenario is an effective tool for presenting complex images in a coherent, integrated picture. These scenarios are presented as business reports of the state of the art in 2025, particularly in relation to developments in science and technology in their social context. Each chapter covers a basic social area, such as genetics, manufacturing, food and agriculture, the house and home, and so on. In each chapter we typically introduce a picture of the situation in the United States first, then go on to take a look at the scene in a World 2 and/or World 3 country. We assume that the United States, as a continental economy, has sufficient variation to illustrate almost all issues likely to be faced by the affluent World 1 nations in the future. We pick countries in Worlds 2 and 3 to discuss because the implications of science and technology for their futures often may be strikingly different.

Each scenario chapter ends with two tables that the reader may find interesting to think about in the context of the scenario he or she has just read. One table lists a series of events and actions that, by their occurrence between now and 2025, drive and support the images of the future described in the scenario. Each action or event is assigned a plausible date.

2025

The other table lists from the perspective of the year 2025 hopes and fears for the future that turned out to be unrealized between now and then. Some of these are hopes and fears that we have now. Others arise between now and 2025. The reader is challenged to think about how different the scenario might be if any or all of them had come to pass.

Because we are discussing events and actions that have not yet happened we have also created some organizations and institutions that do not yet exist and given them names and even acronyms. Some of their names may be familiar and only slightly different from their names today. Others may be new to the reader.

The scenarios can be read in any order. Together they form an integrated whole. Readers should feel comfortable about following their own interests in choosing the order in which they read them. The chapters are organized in a sequence that starts with the familiar, house and home, goes on to the four enabling technologies, information, genetics, energy, and materials, then covers the enabling issue—the environment. The next seven chapters are about large structures and systems, such as global management, transportation, and manufacturing. The last two chapters return the reader to personal experience, with social and cultural experience, and work, leisure, and entertainment.

Chapter 1, Introduction, this chapter, sets out and discusses our basic assumptions about the world in 2025 that underlie the scenarios. Chapter 2, Smart Living, takes the reader inside an American home in 2025, and looks at the state of shelter in countries like Egypt. Chapter 3, Information: The Global Commodity, tells how information technology has remade societies and its institutions and systems, like work and criminal justice. Chapter 4, Harvesting the Fruits of Genetics, examines the U.S. lead in genetics and comments on how, in 2025, we are beginning to integrate genetic approaches and solutions into our worldview. Chapter 5, Powering Three Worlds, shows how U.S. energy use *per capita* is declining and that of the rest of the world is rising as a result of population and economic growth. Chapter 6, The World of Things, highlights the revolutionary impacts of advanced materials on our society and the mixed impacts and availability of such materials in poorer societies. Chapter 7, Working Toward a Sustainable World, shows how concern for the environment is evolving into support for sustainability as a fundamental global value. Chapter 8, Managing the Planet, documents the role of the science and technology in enabling us to better manage the natural, built, and social environments. Chapter 9, Putting Space to Work, illustrates how technology and science in space amplifies our knowledge of, and ability to manage, our planet. Chapter 10, Our Built World, is about the developing global infrastructure and the uneven penetration of the more advanced and intelligent infrastructure and construction methods. Chapter 11, People and Things on the Move, looks at some non-automotive alternatives, as well as describing the gradual integration of all travel systems. Chapter 12, The World of Production, paints a picture of flexible, customized manufacturing systems that produce universal products for the global market and culture-specific products for local markets. Chapter 13, Food: A Quest for Variety and Sufficiency, points out that some of the world's people have a variety of new foods to eat, others get enough to eat but do not have much choice, and a third group faces periodic famine. Chapter 14, Striving for Good Health, describes affluent societies in which health is more important than medicine, and poor countries in which technology augments centrally-directed public health programs. Chapter 15, Our Days and Lifetimes, shows the reshaping of everyday life and lifecycles by physical and social sciences and technology. Chapter 16, Balancing Work and Leisure, describes less work and more leisure for the affluent sectors of the world's population.

We use the metric system in this book, but believing the reader may be more familiar with American measures, we include a conversion table below.

Metric Conversions

Metric Measure	Equivalent American Measure	American Measure	Equivalent Metric Measure
Linear Measure:			
1 centimeter (cm)	0.394 inch (in)	1 in	2.54 cm
1 meter (m)	3.281 feet (ft) 1.094 yards (yd)	1 ft (12 in) 1 yd (3 ft = 36 in)	30.480 cm 91.440 cm (32 m = 35 yds)
1 kilometer (km) (8km = 5 miles approximate)	0.621 mile (mi)	1 mi ("statute mi") = 1760 Yd. (1 nautical mi, International)	1,609.344 m
Surface Measure:			
1 square meter (m²)	1.196 square yards (sq yd)	1 sq yd (9sq ft)	0.836 m²
1 hectare (ha)= 100 Ares (a) = 10,000 m²	2.471 acres	1 acre (4,840 sq yd)	40.469 a = 0.405 ha
1 square kilometer (km²)	0.386 square miles (sq mi)	1 sq mi (640 acres)	2.590 km² - 259.0 ha
Volume and Liquid Measure:			
1 liter (l) - 1,000 cm³	1.057 U.S.quarts (U.S. qt) 0.264 U.S. gallon (U.S. gal)	1 U.S. qt 1U.S. gal (231 cu in)	0.946 l 3.785 l
1 cubic meter (m³)	1.308 cubic yd (cu yd)	1 cu yd (27 cu ft)	0.765 m³
Weights:			
1 gram (g)	0.035 avdp ounces (oz)	1 avdp. oz. (437.5 gr = 1/16 avdp lb)	28.350 g
1 kilogram (kg) kg = 1000 g	2.205 avdp pounds (lb)	1 avdp lb (16 oz)	453.592 g
1 metric ton (t)	1.102 short tons (sh tn)		
To compute Fahrenheit: multiply Centigrade by 1.8 and add 32.			
To compute Centigrade: subtract 32 from Fahrenheit and divide by 1.8.			

How to Use This Book

We expect the reader will finish this book with a sense of being better acquainted with the future—at least better informed about the potential effects of science and technology on our world in the next 25-30 years. As futurists, however, we are aware that there is no single future, that events, trends, and circumstances may lead to slightly or radically different outcomes. We suggest that the reader view this book as a challenge to his or her own assumptions and powers of futures thinking. All our conclusions and images should be tested against the reader's own knowledge, values, and ideas about the future. The great opportunity here is for the reader to integrate this picture of 2025 into a personal future vision and let that vision inform how he or she may begin to shape organizational, personal, or political decisions and actions.

2

SMART LIVING

"The very basis for my business over the past decade has been to build safe, flexible, sustainable housing with whatever look, feel, and contents my customers want."
— *Margaret Hunter-Chang, Housetex Corporation*

Earlier this year, Millennium Productions, Inc., taped footage for a new virtual reality (VR) drama in Oswego, New York. The VR production was a period piece, set in 1988 in a suburban neighborhood. The production company did not have to alter the scene for exterior shots of the houses, except for removing lawn-care robots, replacing modern vehicles with period ones, and dubbing in telephone poles via studio enhancement of the images. The same could not be said for interior shots. Those required complete studio mock-ups of period homes.

Not every modern neighborhood can pass for a 1980s or 1990s setting. But because of enduring tastes for housing in traditional styles, and because 64% of all affluent-nation houses today were built last century, many communities differ little superficially from their predecessors. The widespread differences are internal and hidden and in how people communicate and travel among communities.

In World 1 countries, shelter has been transformed, though not always visibly, by three central forces:

- *Information technology* — making everything a candidate for mechatronic enhancement and leading to the growing integration of house systems and functions. A room that knows all who enter and what their preferences are, for example, was only a concept a generation ago. At the same time, information technology enables better design and redesign of homes and control and maintenance of home systems.

- *Environmentalism* — pressing for the transformation of the home from an energy and resource waster in the 1990s to an energy conserver or even producer (in the case of active solar energy), and a resource recycler today. Environmentalism reinforces the trend toward greater integration of house systems and functions.

- *Materials technology* — changing the structure, cost, texture, and capabilities of materials of construction and decor and providing the basis for closing the loop of materials use in the home as elsewhere, with recycling, reclamation, and remanufacture.

The home as a workplace, common today, was a new concept in 1990. Information technology has been foremost in making this transition possible, as it has been in changing recreation, shopping, health care, and other domestic activities. Large-format flat video screens, such as we now commonly have throughout the home, were unavailable before 2011.

Elsewhere in the world, particularly in the destitute and lower middle-income countries, housing is visibly different in external appearance from what existed 35 years ago. That is because of changes in the materials used in shelter and new shelter designs adopted by developing countries. Polymer concrete prefabricated modular housing units have proliferated in those countries, as is illustrated by a case study of housing in Cairo, Egypt.

HOUSE AND HOME IN WORLD 1 COUNTRIES

It is in the affluent societies that people benefit most from advances in technology. In housing, affluent societies have seen a 35-year evolution toward new forms and functions of shelter, enabled most of all by information technology. The resulting homes are safer, more comfortable, more energy efficient, more flexible, more environmentally sustainable, more educational, and more fun than they were a generation ago. Newly built homes get a full complement of advanced technologies. Older homes can be retrofitted with almost any technology.

Settlement patterns reflect a hybrid of the old and the new

Transportation and information technologies make new settlement patterns possible by blurring the boundaries of space and time for work life and leisure. Still, traditional patterns lie just beneath the surface of modern communities. Those old patterns, reflecting how people worked and played in the last century, can be seen in the way communities lie as satellites around old urban cores throughout the United States and Europe. Metropolitan areas are still seen as city-centered. Infrastructures are still partly oriented to the old urban core-centered patterns of the last century.

A great change for North American life, mirrored somewhat in Europe and Asia, is the return of vitality to rural areas. Information technology and new transportation systems made this possible. The aesthetic and lifestyle appeal of rural areas have made them attractive. The share of the U.S. population that is nonmetropolitan continues to fall. However, the number of people living in rural areas began to rise again in the 2010s and is growing by 0.6% per year.

Housing stock is another instance of older patterns persisting. For example, the average lifetime of urban dwellings in the United States, which last century was 40 years, continues into the present generation. Lifetimes will get even longer with more durable construction and slowing population growth.

Environmental concerns drive much of housing change

Greenhouse warming has shaped public policy and private action in all of the advanced nations. Governments have acted on public concerns about pure water and air, and there has been a net improvement over the past generation in the quality of air and water. Solid and toxic waste issues led to a nearly universal commitment to recycling, reclamation, and remanufacturing of goods. Conservation of energy and materials is now solidly backed by legislation in World 1.

Environmental problems shape legislation and regulation on housing codes, energy use, and waste-management practices. They also shape change in building materials and design through their influence on cost and on consumer attitudes and values. The market for solar technologies illustrates the continuing demand for sustainable and renewable resource use. Today, 16% of U.S. homes have solar panels for some or all of their electrical energy needs.

Architecture and interior design are freed from technological constraint

Architecture and design respond to new technologies and materials and to cultural forces. Over the last 30 years, the most powerful driver of change in architecture has been information technology. Patterns of energy use, pressures for energy conservation, and the introduction of new materials into the design stream have also reshaped architecture's tool kit and design paradigm. In addition to reshaping the standard detached house, technology enables sharing systems and services in apartment style living, making that option increasingly attractive.

New materials, design practices, and information technologies free architects and engineers to design structures unfettered by many historic physical limitations. The use of glass as a structural material, for example, has led to more "invisible" superstructures in buildings and concomitant design changes.

Closely related to architecture are technologies of construction. Building technology has been greatly influenced by new materials used in construction, pressures for energy conservation, and the extensive application of computers and other information technologies at the construction site. The most dramatic effect in construction technology this century came from the substantial growth of manufactured housing — that is, more extensive preparation of larger modules in factories where quality control could be managed more effectively. Mobile robots generally assist with site preparation and assembly. Manufactured housing, making ever more extensive use of robots in the factory, has steadily modified designs and use of materials to take advantage of automation. Flexible manufacturing makes customization more cost effective for the consumer.

A CASE STUDY: The U. S. home - an integrated system

There is no typical U.S. home. Homes fall within a spectrum from traditional low tech to high tech, and range similarly in how much environmental technology is built into them. All U.S. homes are under the influence of a set of shaping forces, driving them to their roles in more careful management and integration of systems, environmental friendliness, and information intensity. The penetration of the latest technologies usually depends on the affluence of a household and on its attitudes toward and interests in technology. An exemplary U.S. home is depicted below. Similar technologies are available for attached houses and apartments.

A Traditional Design Modular House

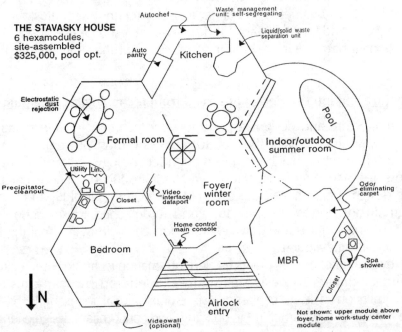

THE STAVASKY HOUSE
6 hexamodules,
site-assembled
$325,000, pool opt.

Autochef

Waste management unit; self-segregating

Liquid/solid waste separation unit

Auto pantry

Kitchen

Electrostatic dust rejection

Formal room

Indoor/outdoor summer room

Pool

Utility | Lin.

Precipitator cleanout

Closet

Video interface/ dataport

Foyer/ winter room

Odor eliminating carpet

Home control main console

Bedroom

MBR

Spa shower

Closet

Airlock entry

Videowall (optional)

Not shown: upper module above foyer, home work-study center module

N

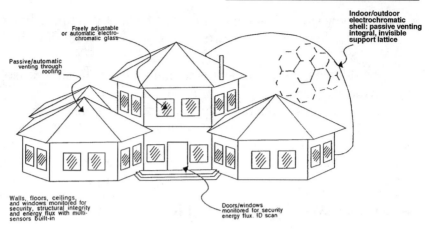

Indoor/outdoor electrochromatic shell: passive venting integral, invisible support lattice

Freely adjustable or automatic electro-chromatic glass

Passive/automatic venting through roofing

Walls, floors, ceilings, and windows monitored for security, structural integrity and energy flux with multi-sensors built-in

Doors/windows monitored for security energy flux. ID scan

Today's homes are information dense, compared with their predecessors in the 1990s. The typical household, for example, receives external information via the network for work, play, education, and household management 17 hours a day. On average, today's upscale home and car have a combined $40,000 worth of information technologies compared with $5,000 in 1990. The home has two centers of concentration of this technology — the home command center, governing home systems such as heating, ventilation, and air-conditioning (HVAC); security; and waste handling; and the home work-study center (HWSC), sometimes also called the home information center. Meanwhile, flat screen interfaces can be installed anywhere in the home. The information-dense house, manages the internal environment to satisfy divergent health and comfort needs.

The aging U.S. housing stock

The tenacity with which people hang on to old structures slows the uptake of advanced technologies in the United States. Housing sociologist Janet Jeffries identifies three reasons for the slowing turnover of the housing stock, 2005 to today:

- Aging society — With the median age of the population approaching 40, most U.S. homeowners today came of age and formed households early in this century.

- Traditionalism in the face of high tech — This more controversial supposition relates to human reactions to fast technological change. Jeffries says that people will accept advanced technologies more readily if they are packaged in familiar forms. Thus modern house designs are still sometimes reminiscent of the Cape Cod and ranch styles of late in the last century.

- <u>Economics</u> — Existing houses cost 85% of what new homes cost, making them more attractive despite the cost of retrofit technology.

In addition to the forces Jeffries identifies, the increasing ease of retrofitting and the extended lifetimes of structures and their contents avoid the need to replace structures as often. Analysts expect houses built today to last 60 to 80 years. Reconfigurable apartment complexes may last even longer, though they are rearranged and altered as often as every 10 years. Modularization of components makes alteration easier, obviating pressures to replace houses. Modular structures can also be relocated more easily than site-built houses.

Smart and smarter, integrated houses

Today's houses are smarter than their 1990s predecessors, few of which had any smart house technologies beyond programmable thermostats. Information technologies, coupled with sensors and actuators, make today's houses more fully integrated and automated.

There were two big factors in making today's houses smarter:

- The *de facto* integration and universal availability of digital, fiber optic networks, which matured only about a dozen years ago, and

- Mechatronics, making every appliance, device, and material a candidate for built-in microprocessors, sensors, or actuators.

The effects of embedding sensors and other devices are twofold:

- Devices, structures, and materials can act and react, independent of human instruction.

- Devices, structures, and materials gather information to inform central systems about their condition and use.

The net effect of this is that nearly anything in the home can become part of an integrated system. People interact by touch and voice with the system at friendly interfaces, including video screens, in most rooms and at their command centers.

Integrating the functions and systems of homes is advancing quickly, with nearly all new housing units built today having at least basic systems centrally integrated and controlled. The need to save energy and to align the household with environmental practices shape the use of information and control technologies.

Technologies at play versus technologies sitting on the drawing board

Most people have heard of Fully Automated Shelter Technology (FASTech). The automation standard and armamentarium of technologies is the descendent of smart-house standards first put forward in the late 1980s. Fully blown, a FASTech-equipped house would be the fully automated house dreamed of since sci-fi cartoons in the 1960s (*The Jetsons*).

In fact, many FASTech technologies are invented, developed, and ready to be marketed, but for cost and retrofit reasons are not finding markets in the United States. Except for a few FASTech demonstration homes and the homes of a few billionaires who insist on cutting-edge technology, the U.S. fully automated home has yet to be realized. Instead, U.S. homes combine advanced technology and automation with human labor. Robotic chefs still require some care and feeding by the user. It is far too expensive for most homeowners to have outdoor robots that remove recyclables and clean out gutters.

Integration is nearly universal in multifamily structures. Apartment, condominium, and cooperative housing complexes, as well as modern suburban developments of detached and semidetached housing have carefully managed systems including security, HVAC, wastewater, solid waste and recycling, and energy conservation. Factory construction of housing and modular housing units have made building in integration systems more cost-effective. In the most complete systems, greenspaces are also becoming integral parts.

Integration systems have been more difficult to retrofit into older housing. As a result, about 54% of U.S. housing units have incomplete integration systems, lacking either a command center or controls on the six main units of the smart house: HVAC, water, energy, solid waste, liquid waste, and security. Nevertheless, the home retrofit market for management technologies has brought tens of billions of dollars in revenues annually to the companies serving it.

Plugging into the communications network

With regulatory paths opened up since about 2005 and the standards jungles being cleared, all new building complexes and nearly all new homes have been fully wired for broadband. Yet even today, the costs of access and services keep people from using the capabilities that exist. As is typical, the technological capability is available, but developers and buyers cannot or will not afford it. The same cost problems have slowed the retrofit market.

Nevertheless, more people use more information services and networks from home than ever before. Thirty-five years ago, only 3% of the labor force worked off- site. Today, 40% work at least several hours a week off-site, often from home.

Telecommunications has meant that more people can participate in group activities without leaving home. Today's software and hardware make doing so even easier. The FolkNet software system, for example, helps users form groups who share common interests. Such systems make it easy to set up and track communications with a far-flung group, whether it be via televideo, computer messaging, or virtual conferencing.

2025

Energy use

Concern for energy costs has shown up in the materials used in home construction and in home design and management. The most widespread development has been smart windows, which respond to light and heat flux. The market for them has grown rapidly since around 2000 with an added boost from public/private conservation incentive programs.

The new helps preserve the old

Today's capability for tight control on interior home environments makes it possible to protect antiques, musical instruments, fine fabrics, heirlooms, and works of art. Humidity, light, temperature, and vibration control mechanisms can eliminate much of the damage caused over time to such family treasures. Security, meanwhile can be established not only for the whole home, but also for particular items such as antiques.

In homes with central command modules, all systems are coordinated for energy efficiency. Residents can assess their practices and monitor the outcomes of opening doors, windows, and skylights, as well as how they use their passive ventilation systems and smart windows. They can also establish preferences and let the house manage itself. EnergyMonitor systems, available since the 2010s, allow the home to automatically buy energy during low-use times when rates are lowest. This is especially attractive for charging electric vehicles and recharging heat and cooling storage systems.

Safety and security systems

An area where the smart home concept has moved rapidly is in community safety systems. Fire, flood, storm, and accident losses have been reduced to nearly half the rates of the 1990s, and insurance rates have dropped. This concept wasn't even around a few decades ago, although its seeds can be seen 40 years ago in burglar alarms linked to police stations. Since then, information technology has allowed the maturation of comprehensive monitoring and alert systems that encompass health, accident, crime, fire, and weather warnings. Condominium complexes and lifestyle communities led the deployment of community safety systems where the housing manufacturing industry has worked closely with developers, governments, and service providers to build integrated community-scale systems. This individualized, coordinated, community-scale approach has been the most effective way of deploying advanced technology while still allowing for diversity and flexibility in regulation, codes, and construction.

Safe home, healthy home

"Safe home, healthy home" is the American Home and Building Technology Association (AHBTA) motto. By 2011, the AHBTA had established standards, revised triennially, for home safety in materials and devices.

Overall, only 39% of U.S. homes are AHBTA-approved Safe Homes. AHBTA-approved homes have higher resale values. Among other successes, Safe Homes, with medical monitors and complete voice-actuated Safe Fixtures, have helped hundreds of thousands of elders maintain their independence.

Any individual home or workplace information equipment now is usually secured against tampering and theft. Traditionally, codes and key locks protected computers and other information consoles. The difficulty was in protecting information as it was being transmitted through the network. One universal information security retrofit device, for example, is Nab-A-Thief™. This device combines user-tracking technology with a thumbprint scan identity checker. The device can be installed on nearly any information device. Similar tracking and ID devices are widely used to secure appliances and structures.

Mixing outside and inside

Integration and control technologies would seem to create closed, artificial environments. In fact, they also allow a blending of the outdoors with the indoors, to almost any degree people want. Indoor greenspaces, invisible barriers to the outdoors, and home ecospaces are examples of how people manipulate the aesthetics of their indoor environments. A burgeoning hobby is the art of bringing ecospaces into balance and keeping them that way. Among the most popular things to do today is to create a tropical climate in a living area of the home for year-round enjoyment. At the same time, 18% of detached and semidetached homes have a vegetable garden or greenhouse.

Managing the home sustainably

Environmental sustainability is a central goal of home management. Federal and local regulation, as well as public attitudes about living sustainably where possible, have transformed two critical aspects of home life: energy use and waste handling.

Because much of the housing stock in the United States is 40 or more years old, retrofit systems are essential for converting houses to sustainable practices. The home environmental audit, now a standard practice countrywide, identifies where changes in construction, technology, and practices can make a home more environmentally friendly. Although people did energy audits starting in the 1990s, more careful systems models of homes and the integration of environmental technologies with other systems make it possible to make much greater gains in energy and resource efficiency today. It is educational for family members to see an energy and resources flow model (in 3D and color) of their home as a dynamic system. They can have routine access to a model in the home televideo system.

Information technology and the modern home

All new houses and buildings are fully fibered for broadband communications. Technologies have long been available for retrofitting nearly any older home with broadband communications technologies and a complement of integration technologies. At a minimum, installers can give older homes full access to communications, information and image services, and two-way utility control.

The Eames Bed

Named for Charles Eames (1907-1978) who developed the concept, and patented in 2019, the Eames bed is a home medical device. Built into an ordinary bed, the $1,500 device monitors the occupant's health. Users can get a report each morning, or have all data or emergency signals sent to medical professionals automatically. For the bed's 89,000 users, heart attack deaths have dropped by 16% measured against people of similar health status who do not use the bed.

Monitoring technologies can go too far

Sometimes capabilities can lead to overuse or ill-conceived uses of technologies. One example was the EntryWatch system installed on a voluntary basis by 16 electrical utilities in the United States in the 2010s. The systems were a marketing failure for every one of the utility companies. These systems incorporated devices that measured heat loss or gain when a person opened a door or window. Digital readouts at the door or window announced the exact, to the penny cost of each opening and closing.

According to surveys done on homes equipped with the devices, people began to exhibit "prisoner behavior." Some restricted their trips outdoors. Others covered the readouts with black tape, preferring not to know how much they were spending. On the other hand, others praised the system for helping them teach spouses and children not to waste energy.

Sensing and reacting

The advent of multisensor packages for home control, safety, and security is at the heart of home integration. In 2005, nearly all commercially available home sensors did only one thing, such as burglarproofing or fire safety. By 2015, the sensor market took off with affordable packages that monitored indoor air pollution, health conditions, and aesthetics such as noise and light. Today, embedded sensor arrays are standard in structural members and finishes. The latest sensor arrays now in development can track over 200 environmental compounds and conditions; and automatically control appropriate filters, scrubbers, and ventilators; and monitor structural members for stress.

The home work-study center

Part of any affluent and most middle-class homes is the home work-study center (HWSC). A HWSC combines the information technology needs of the worker, the student, and the leisure user in one integrated computing and telecommunications package. The ergonomics of the HWSC are nearly as critical as the information technologies included, and designers continually modify their designs for optimal use by children, adults, the aged, and for mixed purposes. A typical HWSC is depicted schematically below.

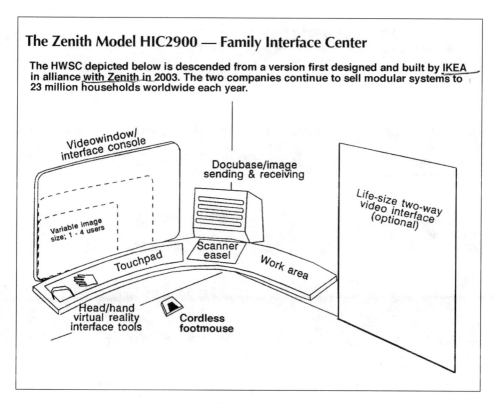

The Zenith Model HIC2900 — Family Interface Center

The HWSC depicted below is descended from a version first designed and built by IKEA in alliance with Zenith in 2003. The two companies continue to sell modular systems to 23 million households worldwide each year.

Work life patterns depend on the home as a workplace

Work life shapes life at home. In 1995, about 3.5% of the workforce were involved in some kind of off-site or distributed work. By 2005 that rose to 20%. Today it is 40%. In that pattern of off-site work, the majority of off-site workers still appear at a traditional central workplace once or twice each week. Another large group of workers perform their work at home or at local satellite centers within two miles of their residences. The most striking development is the number of professional, managerial, clerical, and support functions carried on at home. Over the past generation there has been a widespread development of small professional services businesses, involving one to three people, and dependent on a handful customers or clients. This pattern evolved in the late 1990s and the first decade of this century. Over 66% of these ventures, by a recent estimate, use the proprietor's home part or all of the time. This powerful trend led to changes in the use of space in the home and to the design of new homes and home workplace technologies.

The materials revolution in house and home

Traditional and advanced materials shape home construction, furnishing, and finishing. The box below shows the outcome of shifts in materials use in a typical house, from 1990 to 2025. Note that the average home size has shrunk from 158 to 139 square meters.

The Changing Materials Makeup of a U.S. House

Typical 1990 158 m² home		Typical 2025 139 m² home	
Amount of material	% of total cost	Amount of material	% of total cost
23 m³ lumber	18%	22 m³ lumber	17%
42 m³ concrete	14%	18 m³ duracrete	12%
451 m² sheathing (metal >>> wood)	10%	502 m² sheathing/sealer (composite >metal)	12%
208 liters paint	2%	200 liters coatings	2%
136 kg nails	1%	127 kg nails or adhesive	<1%
12 windows	4%	15 windows	6%
604 m² gypsum wallboard	7%	516 m² wall panel (gypsum > composite)	10%
232 m² siding (alum. > wood > brick)	8%	195 m² siding (alum/brick/wood >poly)	9%
232 m² insulation	2%		
185 m² asphalt shingles	2%	185 m² durashingle	2%
27 m HVAC ducting	2%	30 m CEC ducting	1%
137 m plumbing pipe (copper >plastic)	2%	143 m plumbing pipe (composite > copper)	1%
229 m copper wire	<1%	305 m wire (copper ≈ fiberoptic)	1%
fixtures/appliances	22%	fixtures/appliances	20%
other materials	5%	other materials	7%

The lingering importance of traditional materials

Traditional materials, sometimes in new forms, coexist with advanced materials in home construction. For example, wood is still a favored material for interior finishes. Usually it is genetically engineered plantation wood, made rot resistant, fireproof, and given exotic grain patterns and colors. For eleven years, extrudable plasticized wood has been on the market for custom and standard finishes, furnishings, and interior decor.

The biggest developments have been the many new biochemical treatments for wood. These have extended the lifetime of the home and reduced the need for maintenance and repair, and they have made lumber and construction waste safe and environmentally benign when burned or disposed of. Lumber from genetically engineered plantation trees is proving its superiority in resistance to rot, pests, stress, and water. Its longevity remains to be seen.

Framing and structural members, on the other hand, are most often made of extruded polymers, metals, and composites, in the case of roof trusses and joists. Because 86% of housing units are factory built in modules, designers choose materials that are readily manipulated by automation. Extrudable, moldable, and formable materials make structures more economical and flexible.

Mass customization with coextrusion

Mass-produced housing can be customized through surface finish coextrusion where a color/texture finish is applied in the factory to standard interior or exterior sheathing. First used to build standard model varieties of housing modules, increasingly coextrusion and other automated mass-customization mechanisms are used to let the buyer specify combinations of finishes and options at the sales console.

Getting the house you want

More people today are customizing their new homes in the factory. Among the tools making this possible are architectural design systems. A premier version of these is the latest system from EXPERTS, Inc., described below.

A top-of-the-line automated design system

EXPERTS, Inc., has just announced the franchising of its new architectural design system intended to be used by private citizens, although it is also applicable to the professional architect, in laying out and practicing alternative designs and patterns for homes and apartments. The machines will first be made available through franchised mall shops operated by Space Consultants, Inc., which has 243 shops in shopping malls throughout the United States.

EXPERTS claims that even a nine-year-old can effectively operate the system. It purports to combine information from the work style and psychological inventory to augment and shape the users' judgments about the design, layout, furnishings, and decor of house and home. It provides alternatives within that framework as a function of income, ethnic background, and 13 other cultural and social variables. The system, operating through a Sony MARK6 Virtual Reality Suit allows the operator to walk through and visually and kinesthetically experience alternative designs. One purported weakness is that they have not yet satisfactorily perfected the experience of flopping into furniture.

EXPERTS is seeking a commercial alliance with BigW, the totally customized, wood-plastic-based home fabricator.

Can you build one yourself?

Although home assembly and construction experts recommend against build-it-yourself houses, this practice appeals to a certain sector of the market, and as many as 20,000 modular houses are assembled by hand each year. About 12,000 traditional, site-built houses are built each year. Fabrihomes, Inc., and other factory housing companies, recognizing the small but significant market for do-it-yourself (DIY) housing, sell kits to the DIY market. Most difficult and problematic, however, is the use of adhesives and sealants applied on-site. Most kits include a video that offers special instruction on the more difficult techniques.

Automated construction sweeps the housing market

Factory construction brought automation to housing starting about 25 years ago. It was not until the 2010s that outdoor robotics made widespread use of construction robots possible. Their earliest uses were in straight-line or laser-guided functions such as grading and paving. By 2008, robotic excavators were in use to dig foundations. Remote operators were then and are generally still on hand to guide the systems around problems. Today, machines such as the Caterpillar Model X34 Robostruction and assembly trailer, described below, make site assembly cheaper and faster.

Caterpillar Model X34 Robostruction and assembly trailer

This device's 2025 model comes with a robo-operated, laser-guided ground preparer. It is usable in all conventional applications. It includes automated assembly capabilities usable with manufactured structures involving metals and all construction-grade plastics for buildings up to three stories high.

Home life in the smart, integrated home

The result of 35 years of smart-house evolution is safer, more efficient, and more pleasant homes. People are free to do more work and to play more, because their houses can manage themselves. That is the ideal. In reality, every home owner has different practices and experiences with automation and integration, and housework and management loads vary.

Getting food on the table — finally getting easier

Only 10 years ago, people still engaged in the hassles of meal preparation or were regular subscribers to meal delivery services. In quick succession, there was the development of the tunable microwave (2016), robot chef (2019), and automated pantry (2020). Today, about 80% of homes have the latest microwave technology. Only 22% have robotic chefs and automated pantries, which are usually used together.

Dodging maintenance chores

U.S. households are locked in an endless struggle to reduce and remove chores. Technophiles make a hobby of automating household chores and programming appliances and domestic robots to do more. Many spend more time tinkering with their systems than they would doing the chores the old fashioned way — using dumb tools and appliances. New homes built or reconfigured since 2009 have a video and data record of their histories. These records are effectively owner's manuals for homeowners.

Attitudes about household chores

The 87th Annual Gallup Poll of American Values (2024) looked at U.S. household chores and people's attitudes about them. Leading the list of the most hateful chores were doing the dishes, servicing the building diagnostics system, cleaning the HVAC precipitators, and reprogramming the home security to accommodate house guests. Doing the dishes is no surprise. That was also high on the list back in 1990. The home automation industry has been promising foolproof automated dish handling since the early 2000s.

Other 1990s chores that no longer concern Americans are vacuuming, ironing clothes, and taking out the garbage. How many kids understand what "taking out the garbage" is?

The same Gallup Poll looked at people's favorite home maintenance chores. For the fifth straight year, the favorites were testing the health monitoring module and View Vision shopping.

Managing sleep, waking, and circadian rhythms

Even circadian-rhythm-management can be automated. For example, Hitachi has the Waking and Sleeping Health Center, adaptable to most homes, that provides induced sleep, round-the-clock monitoring of vital signs and health, stimuli for resting, thinking, or physical activity. It can be used simultaneously by up to three people, if they are all within 40 feet of the unit. These are still expensive technologies. The average American that tries to manage sleeping and waking still uses melatonin treatments, caffeine, and bedroom environmental controls to regulate perceived noise levels, light, and temperature.

The home replaces clinics and hospitals as the key health maintenance center

With the growing emphasis on self-care in the United States, the home has become the center of health care for most people. Routine health and medical monitoring is done daily through health-monitoring systems in the home. Where needed, external information capabilities and human medical advice are available through televideo links.

HOUSE AND HOME IN WORLD 2 COUNTRIES

The typical middle-income country, like a paler version of World 1, has a mix of the old and new in housing. Income levels usually dictate the uptake of advanced technologies for households. New housing units nearly always make use of advanced building materials, like those used in World 1.

A runaway bestseller across World 2 is home security. These are not the built-in, integrated systems of the affluent home; they are add-on devices that detect entry, tampering, and unwanted trespass. Rising violent crime and theft in the developing economies has made security devices an annual $13.2 billion market.

There is much less of a market for retrofit information technologies and automation. The table below summarizes the housing situation in World 2 countries.

House and Home in World 2 Countries

Housing stock	In most countries, as much as 75% of the housing stock was built before 2015.
Settlement patterns	Continuing metropolitanization means more people live within 40 kilometers of the center of a major city than ever before. Most urban people live in multifamily structures, whereas rural people live in detached houses.
Design and construction	Designs are traditional, as are most construction practices. Innovative design comes with prefabricated housing, which is growing as a share of new housing units.
Materials use	Traditional materials, such as wood, stone, mud, adobe, concrete, and brick, are used along with new materials, which include extruded structural polymers and sheeting.
Energy use	The widespread adoption of insulation practices has reduced energy use for home heating and cooling, but net energy use has increased because of the increase in use of electrical appliances and central HVAC.
Environmental technologies	Reduced emissions furnaces and hearths are growing in prevalence.
Waste management	Recycling and waste-handling systems are taking off, starting with demonstration projects in the 2010s.
Work life at home	Most workers still work outside the home. However, many ply a home-based trade, run a shop at home, or provide a service from home. Few professionals and nonindustrial workers as yet work at home.

Home life and leisure	Advanced video technologies and home computers are restricted to the more affluent households of the middle-income countries. Poorer people may have access to information technologies for education and leisure at community centers.
Information technology	Home computers are still rare. Radio, cable, and broadcast television provide most connections to the outside world of information.
Safety and security	Home security systems are among the first homeowner investments. Sensing systems and perimeter penetration detectors are used by most households that can afford them.
Food preparation and consumption	Traditional food preparation practices, including hearth, stove, and microwave cooking continue to be most common. More food preparation work is done by hand than in the affluent countries, though packaged convenience foods have grown to be an annual $600 billion market worldwide.

HOUSE AND HOME IN WORLD 3 COUNTRIES

In the world's destitute countries, housing is traditionally an ad hoc technology. People work with the limited resources at hand to build and furnish their houses. By some estimates, over a billion people in the world live in hand-built shacks, typically made of waste materials found in municipal landfills. The conditions in these communities are unhealthy and desperate, despite decades of international aid efforts to improve them. Those efforts continue, as new planned communities for the poor take shape.

The big change is prefabricated housing. Modular prefabricated housing units are now a staple of development. The practice began with experiments last century, and took off in the beginning of this century. The table below outlines a series of such schemes.

A Comparison of Housing Schemes
in Developing Countries
2000 - 2025

Years	Model	Locations	Material	Method	5	6	7	8
2017	Concept-7	Al-Uqsur, Egypt	Reinforced polymer composite	Cast/layup	P	H	Y	$990
2015	Earthtech	Mexico City, Laredo	Rammed earth	Earth compacted in reusable frame	S	H	N	$210
2023	Sakura Model-D	Port Moresby, Papua New Guinea	Fiberglass	Spun/cast	P	L	N	$550
2015-2019	Casa Fabrica	Lima, Belo Horizonte, São Paulo	Thermally compressed plastic	Panels and frame members	S/P	H	N	$423
1998-2014	Jambo Nairobi	Nairobi, Kinshasa	Advanced concrete	Cast	P	M	N	$352
1997-2003	Desai Complex	Bangalore, Colombo	Polymer concrete	Cast	P	M	Y	$809

Key: Column 5 — P=prefabricated, S=site fabricated
Column 6 — L, M, and H for low, medium, or high alterability (by the owner)
Column 7 — Y for stackable or N for not stackable
Column 8 — Cost to dwelling resident at time of project in 2025 dollars

In the next section, a case study set in Egypt depicts shelter in a typical World 3 society.

A CASE STUDY: Shelter in Egypt — building communities for the poorest

Egypt is a country with a mix of middle-income people and destitute people. The country's 98 million people have a *per capita* annual income of $2,425. That average masks the widening gap between rich and poor in Egypt. People in the poor communities average half that amount.

About 22 million of Egypt's people are destitute, living in shelters built of waste and surplus materials, often in squatter settlements surrounding large cities. Several million such people in Cairo, for example, are Zarrabs — people who make their livings scavenging garbage from the city's waste heaps. By one recent estimate 170,000 Egyptians are homeless in Cairo alone.

This case study examines house and home in Cairo, Egypt, with a special emphasis on its shanty towns, which are typical of housing in World 3.

Shelter and living conditions in Cairo, Egypt

Egypt's destitute live in violent, unsanitary conditions. Although outright hunger is rare, malnutrition is widespread, affecting at least half of all children in squatter communities and shanty towns. People suffer from preventable infectious diseases, due to poor sanitation systems and practices.

Most dwellings are of materials scavenged from waste dumps, or from traditional materials such as stone, mud brick, and concrete. Recent additions to this, sponsored by the Government Housing Ministry, is the use of recycled plastic materials, fiber-strengthened, and fireproofed. Rising fast in squatter communities around Egypt are new prefabricated housing complexes. These offer the destitute a path to better, more healthful lives.

Sprawling poor communities surround core cities in Egypt. Most people live in single-story shacks, 12% in high- or low-rise apartment buildings. The communities are crisscrossed by foot and bicycle paths. Buses and trucks, mostly privately run, offer cheap transportation around cities.

A typical prefabricated housing complex

The Al-Uqşur Project, begun in 2017, is a successful example of a planned community for destitute urban peoples. Al-Uqşur is a sprawling squatter settlement southeast of Cairo. The project involves selling families basic housing modules (the Concept-7 Dwelling Module) developed for Al-Uqşur and subsidized, costing families the equivalent of nine month's income. Each family could install and improve a base structure through sweat equity, perhaps buying add-on units later. Already about 2,300 families live in Concept-7 houses.

The Concept-7 Module, depicted below, is of reinforced polymer composite prepared by a combination casting/layup process. The modules may be made partly of recycled polymers. The Concept-7 module is configured according to traditional designs found in Cairo dwellings. The module conforms to Cairo's building codes and standards. The units are readily joined side by side. The ideal home is 2 to 4 units, depending on family size and home-enterprise work. Al-Uqşur residents continue to modify their units by adding lean-to structures and other rooms using found materials such as metal sheeting and scrap lumber.

Schematic of Concept-7 Module

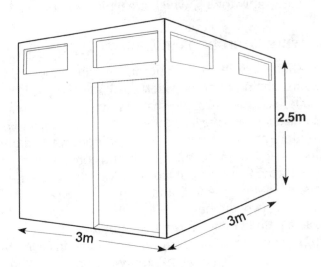

Work life in and out of the home

The home is typically a workplace for the poor of Egypt. In urban areas, 28% of dwellings have workshops in them. The work done ranges from fabrication of shoes, cooking utensils, and other handicrafts, to appliance repair, tailoring, and food preparation. A third of residents work nearby in workshops, stores, and enterprises in other people's homes and yards. Another third work elsewhere in the community. Most people tend gardens, often on their roofs, as a sideline.

Waste, water, and environmental management systems

Largely based on international assistance and central government action, urban Egypt has begun to conquer its waste, water, and other sanitation problems. The secret to its emerging success has been to design systems that depend little on compliance with regulations by individuals. For example, people would not comply with rules about keeping sewage out of storm sewers, so the government decided to separate the two centrally using membrane technology.

Recycling, on the other hand, invented itself in Egypt. The Zarrab scavengers found economic incentives to collect and sort waste and to sell it to recyclers and remanufacturers. Thus, widespread recycling and reclamation took off in Egypt before they did in World 1.

Sewage treatment is decentralized in Cairo, and Al-Uqşur has its own microbial treatment system. The effluent is made usable for fertilizing home gardens and is given free to Al-Uqşur residents. Some is sold to the Upper Nile Farm Cooperative.

Energy remains a low technology at Al-Uqṣur. The worldwide warming trend of nearly 1.5°C so far this century has added to the discomfort during Cairo's summers.

*   Most homes have a solar cooker for the yard or rooftop; some have solar water heaters.

*   Temperatures rarely drop below 9°C in winter, so home heating is not a problem. Some homes have Goldstar photovoltaic air conditioners. The units cost four months' wages. Other residents use electric fans.

*   A program of tree planting in the community has served well for providing shade. The new date palm variety *Arecacea nishidae* survives in urban conditions on much less water and provides additional food for residents, from its annual date crop.

Home life and leisure are shaped by poverty

The average poor Cairo household has electricity, typically powering a light or two, a television, a radio, and perhaps a small refrigerator. Hundreds of thousands use portable photovoltaic panels for their electricity. About 22% in the Cairo area also have cellular telephones. Despite their poverty, the poor of urban areas in Egypt typically have community centers, cafés, educational centers, and mosques where they can have access to things they are too poor to have at home, such as video technology, air-conditioning, and health diagnostic equipment.

Rooftop gardening mitigates poverty for most households. Gardens are a highly promoted option for urban Egyptians. The Concept-7 structure used at Al-Uqṣur, for example, can sustain an intensively worked, containerized roof garden. About half of Cairo's residents have vegetable gardens. Making garden soil lighter has enabled roof gardening for many. Inexpensive, lightweight synthetic soils made of waste polystyrene foam are a stock item at dry goods stores in Cairo. Genetically altered crop plants such as the pygmy vegetable varieties are successful rooftop crops.

About a third of poor Egyptian households have 0.4 cubic meter sulfur dioxide refrigerators from China. Fresh food is important as these refrigerators have limited capacity. These refrigerators helped solve the need for alternatives after the international agreement on CFC elimination by 2013.

Critical Developments 1990-2025

Year	Development	Effect
1998	Environmental Audit Certification Act passed (United States)	Made home energy and environmental auditing respectable and reliable and led to the growth in that service in the United States.
1998	Sustainability principles become part of school curricula worldwide	Beginnings of the movement to sustainability, especially influential in the home through schoolchildren
2002	Selling Sustainability campaign (United States)	Heightened awareness of how people can save energy and resources at home
2002	U.S. Energy Transition Act passed; massive conservation campaign	Helped inspire manufacturers and builders to implement more conservation technologies
2007	Full-fledged home command/integration systems become commercially available	Beginning of the information technology revolution in home management still underway today
2008	The nonmetropolitan United States population resumes growth	Reflect the growing ability of people to use information technologies for work and leisure, making central location less necessary
2009	EnergyMonitor systems come on the market	Allow households to make more energy consumption decisions and control energy use more effectively
2011	International Agreement on Standards in Construction	Developed standard measures of performance and capability for materials, including their energy efficiency and insulative capacity
2013	Recycling and Reclamation Act (United States)	Regularized and extended recycling/reclamation practices, including those done at home
2014	Factory manufactured homes surpass site-built homes in housing starts annually	Led to the recognition of efficiencies and economies to be had in factory building, and the greater flexibility and customization options available to the buyer

Unrealized Hopes and Fears

Event	Potential Effects
Extensive turnover of the housing stock built in the 1900s	Most houses could be built with a full complement of command and integration technologies, obviating the need for partial-solution retrofitting
A rash of accidents, including fatalities, caused by home automation technologies	Could slow or destroy the market for home automation technologies and smart-house systems
Widely available cheap energy	Remove economic incentives for home energy conservation, including materials use and control technologies
Cheap, renewable, easily disposed of building materials	Little development of recycling, reclamation, and remanufacturing practices, faster turnover of housing stock
Extensive ozone destruction and consequent high rates of ultraviolet solar radiation	Ultraviolet light would lead to surface damage. This or some other large-scale environmental threat could make necessary complex protective systems for houses and other structures and for the people in them

3

INFORMATION: THE GLOBAL COMMODITY

Energy technology powers the human enterprise, but information technology empowers people, things, institutions, and nations. Nothing since the beginning of the Industrial Revolution has affected the human condition as positively as information technology. The only technology to approach it in significance is electricity, which turned night into day and gradually reshaped most social and personal activities. Information technology is everywhere and in everything.

The first phase of the Information Age drew to a close in the late 1980s and early 1990s. Three fields we recognize as basic today — telecommunications, computation, and imaging — had by then progressed to the point that technologies under those rubrics had developed into relatively complex technical, business, personal, and social systems. The transition to the current phase of the Information Age is marked by massive integration of all three of those systems as well as the convergence and interchangeability of their components. By 2005, television, broadcast, cable, satellite, and various forms of wireless communication had become so thoroughly meshed that there is no significant difference to the user, but only to those responsible for their technological and social management.

By 1990, the early mainstay of modern computation, the so-called mainframe, had been largely superseded by broad-scale applications of smaller computers of all sizes and by the linkage of smaller computers to massive parallel processing machines. By then, the computer had also burst out of the confines of fixed facilities. The computer by the year 2000 became a component of virtually every product and every technological system. This concept was labeled "mechatronics," implying the combination of mechanical and electronic capabilities.

The launching of the current phase in terms of image technology is even more striking. Photography, cinema, video, bar codes, facsimile, and numerous other areas had developed as technically and economically independent enterprises. With the advent of universal digitization of data and low-cost, high-capacity telecommunications and high-capacity computers, all those technologies fused into telecommunications as we know it today. It is useful today to talk about information technology in the somewhat obsolete categories of the last century, because it causes us to tease apart the conceptual components of our information world, which are so entwined that we often fail to see their distinct elements.

Data to wisdom

Information technology in the last half century has affected information in four stages. The early use of information technology was as a mechanism for collecting, processing, and presenting data. The next significant stage was the interpretation of that data with the assistance of machines, to convert it into information. That stage spanned from 1975 to 1995. Emerging out of that presentation-of-information phase was the development of systems that went beyond information to generate knowledge. For example, in scientific laboratories machines create new models, new pictures, new images, and new ways of designing and building molecules. In quite different ways, expert systems have radically altered financial services, medicine, design, and scores of other activities by creating knowledge from information.

The fourth stage, wisdom — the understanding of how one should use information and knowledge — is the goal still sought. The acute awareness of our lack of wisdom is highlighted by the fact that the Harvard Business School, the University of Chicago, the University of California-Berkeley, and Iowa State University have shared in a $26-million grant to set up programs of "Wisdom in the Information Age." Because the grants were made in 2021, it is premature to expect or describe significant results. The grants have spawned imitators in 75 universities and 14 commercial and private think tanks.

The emergence of the three broad families of information technology and their interchangeability led to what had been called "systematization" of information technology, the linking of everything to everything else. By the 1980s, it was common to talk about networks. There was even nomenclature for local area networks (LANs), wide area networks, and metropolitan area networks. These networks evolved under different business auspices for specific purposes.

It is difficult to believe today that as recently as a generation ago, large blocks of the populations of World 1, the most advanced nations, were put off by information technology. They even had a terminology about the preferred interfaces being user-friendly. Computer users then sought systems that were interactive, forgiving, and helpful. Today, it is difficult to imagine systems not being robust in the face of our errors and ignorance and unable to teach and train us in their own use.

A U.S. CASE STUDY — Information technology remakes a society

While the last generation's basic changes in information technology have already been mentioned in passing, it is worth noting that the reason we have such difficulty in appreciating the ubiquitous effects and power of information technology today is that, to a large extent, they are invisible. Virtually all communications networks have become a matter of total indifference to the user, because systems are intelligent and take care of integration on their own. Although the devices we use are visible, the rest of the technology is not.

Soft intelligence

At the end of the last century, a great deal of empty hurly-burly occurred around the concept of artificial intelligence and the related category of expert systems. Many people who believed that intelligence was an exclusively human characteristic were offended by the thought that a machine could be intelligent. Eventually, however, artificial intelligence software became common, and folks began to advertise it as "soft intelligence," which was contracted to "SI." SI has become a virtually universal characteristic of all organizational enterprises. Not only is SI used to introduce vision and speech into systems, but also it is extensively used as an augmentation of virtually all professional work and is a common tool in all kinds of network management, whether electronic or other. Also, SI has become a primary factor in educational technology, in allowing the teaching system to respond to the level of knowledge of the student, to the material to be transmitted to the student, and to the preferred learning strategies of the student. Although the nomenclature battle has been lost, the substantive war has been won. The result is that SI is virtually everywhere and is the key to many of the new innovations such as the application of virtual reality to simulation, training, and recreation.

The effect is even more striking with computational tools. The vast number of embedded computers (5.6 billion in 2022) and the associated sensors and actuators in devices everywhere has made the bulk of computations not only invisible but a matter of complete indifference to us. We are aware of the devices that we choose to interact with, but a generation ago the keyboard and the awkwardly manipulated, so-called mouse were almost the only means of access. Today eye, voice, sound, and pointing finger are the primary interfacial tools.

Imaging technology is universally visible, but again, how it works and the technologies for controlling it are invisibly embedded in devices and systems.

Let us look at some of the high points of information technology today.

Screens, early known as flat screens, have revolutionized architecture and design. The typical business has 2.6 screens per worker. The typical upper-middle-class home in the United States now has 37 screens. Screen technology took off with the development of the high-definition television at the turn of the century, which made it practical to have aesthetically satisfying screens of any size from the four-square-centimeter lapel screen to two-meter wall screens. At home the screens are no longer furniture — bulky boxes taking up floor space — but have become wall and surface decoratives, both aesthetic and entertainment sources as well as functioning work tools of the household.

Knowbots boomed when three things came together: (1) the ability to commission a search of an information system with relatively simple oral commands, (2) the assignment of knowbots to individual users by name and confirmatory identification number, (3) the knowbot was given a humanlike form.

My favorite knowbot

Some people choose to have a knowbot look like Santa Claus; others, a cartoon of their grandfathers; others choose arbitrary, privately designed figures or copies of commercial figures. The most popular ones are George (Washington), since that knowbot never lies, and Data, which is a play on an old television series character. The third most popular is Nemo, the ship captain in Jules Verne's *Twenty Thousand Leagues Under the Sea. Nemo* is Latin for *nobody*.

Talk-listen-read, the ability of devices to both listen to standard language and to talk, is now universal. Other devices have the capability to read the written word. One of the interesting side effects was the reintroduction of penmanship into schoolrooms in 1998. The talk-listen-read capability has practically eliminated the keyboard except for the Infoquarians, a club whose members use obsolete equipment for nostalgia and fun. The favorite is the Macintosh II used in 1987. Extremists use quill pens.

Mechatronics has already been mentioned. The embedment of computer chips coupled with sensors and actuators has made everything smart. The smart house has on average 241 embedded mechatronic devices. The most striking application of the mechatronics concept around 2010 was to water and sewer systems. Another striking example was the application to building large, relatively featherweight structures in which the whole building dynamically responds to forces and pressures operating on it. The Mechatronic Society of America estimates that, as of 2023, 16,000 products in commerce have mechatronic capabilities.

Compact interactive video discs are a mainstay of the operation, maintenance, repair, and service sector. All complex equipment comes with its own video CD explaining how to make repairs and demonstrating repairs from the point of view, vision, and angle of the person making the repairs. They have been a great success and are available in multilingual form. General Motors, a 20th century automobile firm, introduced instructional CDs in 1997 as a sales stimulus. Many people prefer direct broadcast to the cumbersome CDs.

Photonics, beginning about 2002, made deep inroads into silica-based electronics dominating computers for a quarter century. By 2020, 90% of all microprocessors were photonic. The bulk of the traditional silicon microprocessors were used for replacement parts. Photonics offered higher speed and simpler integration with fiber-optic communication.

Personal communicators—personal identifiers

The personal communicators were the precursor to a system of communication that depended not on the location of a telelink but on following the individual wherever he or she went. The complete integration characteristic of The Network made this practical. As one might expect, however, all good things have their abuse. Rock stars, actors, many public political figures, were ceaselessly harassed by people who could buy their personal identifier numbers. Many other people found that the personal number facilitated harassment sometimes in the form of repeated calls and other times amounted to cutting a person off by a group overloading the system by repeatedly calling the number.

Things began to settle down after about 2010 so that today 57% of the population have personal identifiers, and 36% of those have unlisted numbers. It is estimated that another 7% work with a pseudonym. Apart from the high abuse potential for prominent figures and victims of harassment, it is a widely accepted and much-welcomed technology. It has spawned a number of subsidiary activities, such as renting a personal identifier in connection with travel to obscure or dangerous places, whereby the personal identifier is merely the sponsoring company's code number.

Encryption has been a thorny issue for over half a century. The waning of the East-West conflict in the late 1980s and early 1990s led some to believe that the need for national security would be diminished. However, the rise of the half-dozen mid-size military powers and the last seven wars involving United Nations intervention have made it clear that encryption for military security purposes must still be with us.

Business and industrial security has become the bone of contention in encryption. The Supreme Court found in 2003 that the National Security Agency (NSA)-White House plan for limiting encryption codes was a violation of First Amendment freedom of speech. There has been a rapid proliferation of private encryption technologies since then.

Digitization of all data has made text, numeric, audio, and visual images fully manipulable in storage, transmission, reconversion, and, most significantly, intermixing. Digitization effectively has made all the databases of the United States into The Database. Digitization has also led to the proliferation of image entertainment in which the purchaser substitutes for·the original stars.

Digitization has made computer monitoring and control practical in increasingly broad areas, since systems can now feed their data into a central controller more readily. Digitization has also led to the creation of many counterfeit, false, or artificial documents; anything that is graphic or acoustic can now be thoroughly manipulated to any size, shape, format, or apparent age, with any other visual, acoustic, or numeric materials.

The long-time pattern of computer capacity doubling every 18 months held true for a surprisingly long time. The first teraflop computer was introduced in 1998. Many people saw that as the end of that pattern. Computer capacity continued to grow as forecast through 2017, when the rate of improvement began to substantially fall off. Computers today are 5,000 to 10,000 times faster than they were in 1990.

2025

With each tenfold or hundredfold increase in computer speed, new practical problems were embraced in direct real-time control or manipulation. For example, the dynamic management of street traffic became practical around 2004. The ability to control and manage a street network of 1 square kilometer to optimize traffic flow and to interface with adjacent square kilometers with only somewhat less reliability, was dependent on computer capacity, models, and sensors.

How we got where we are

Throughout the second half of the 20th century, the automobile accounted for 10% to 11% of the GDP, counting original sale, servicing, maintenance, repair, and collateral industries. The information-technology-based industries far and away swamp that modest contribution to the GDP. It has remained for the Department of Commerce to tease out the exact contribution of information technology to the economy, since so much of it is built in, embedded, and hidden in everything from construction to services. The most conservative estimate is that information technology now accounts for 40% of the GDP. That is easy to accept when one realizes that of the total workforce, 64% are information workers and 14% of those are full-time information workers. A more generous estimate puts information technology as 53% of the GDP.

Tools of the Information Age

- Flat screens
- The Network
- Knowbots
- Universal telecommunications
- Talk-listen-read
- ID cards
- Personal communication numbers
- Geographic information systems
- Chemical tags
- Simulation

- Universal digitization
- Smart cards
- Mechatronics
- Photonics
- Robotics and automation
- Fuzzy software
- Satellite and wireless systems
- Image manipulation
- Soft intelligence

With a family of technologies so universal in application, it would be unrealistic to attribute development and evolution to any single or small set of variables. Complex situations have simulated the perpetual need for a means of dealing with complexity, which often amounts to information management. On the other hand, each new capability creates a fresh complexity. The expanding spiral of complexity in our technologically dependent world lies at the core of the evolution of the Information Age. The continual pressure for ever-higher market performance implying ever-lower costs seems to be endless and led to the commoditization; that is, making commodities of virtually all of the information technologies and services common from 1980 to 1997.

Factors driving the Information Age are:

- Continuing pressures for improved productivity.

  — Today manufacturing in the United States, which is the world's second largest manufacturing nation, accounts for 18% of the GDP, and employs only 4.1% of the workforce. In this century, manufacturing has made the same shift that agriculture did in the middle of the 20th century, replacing human labor with information largely in the form of automation.

- Pressures for speedier, faster, and more reliable delivery of goods led to extensive networking in all logistics systems for management and control.

- New technology has continually opened up new opportunities for further productivity improvement and new products and services.

- Demands for convenience, efficiency, time, mobility, quality, entertainment, and ease of use, as well as distributed work (that is, work outside the traditional office), were powerful incentives for cheaper, more rugged, more mobile, and highly individualized systems and subsystems. Mobility and convenience are a regenerative cycle: each step stimulated the other's next step.

- The rise of knowledge work and the associated information tools makes much of knowledge work independent of time and place.

- Information overload was a much-talked-about problem late in the last century and early in the present century, but that problem disappeared with the advent of the multidimensional, dynamic, and interactive devices. The users of today's systems have unprecedented control over information flow. Many people choose to use the built-in stress sensors that determine whether they are psychologically moving into overload.

- In the second half of the 20th century, the beginning of mass data collection by old methods created issues of access. The automation of record-keeping was the two-edged sword that provided both safeguards and more potential opportunities for abuse by both private and public organizations. The situation described under "Issues of the Information Age and Their Resolution," below, seems to have come to a relatively successful resolution of those pressures, at least for the present.

- The globalizing of all the world economies, which evolved rapidly in the late 1990s through 2005, created pressures on corporations to be able to operate in all places at all times at all distances with minimum delay and minimum movement of people.

- Government was the primary and most obviously significant respondent to the growing complexities in the economy. Fifteen to 30 years ago it found itself constantly pressured by a rapidly changing pace of business and by globalization and it had to resort to increasingly sophisticated technological tools for monitoring, regulation, and control.

- Environmentalism and its associated concerns for energy and resource conservation has been a factor in the promotion of both visual and traditional data collection to promote a sustainable society and environmentally benign business practices, products, and services. Information technology has become the primary instrument of the environmental movement.

- Military and government R&D, reflecting a split interest in national security, has been a great stimulus to the Information Age. Traditional and military concerns in the 174 countries and sovereign areas of the world have led to much more complex planning for a much larger number of military and quasimilitary activities. Government has consciously recognized since 1999 that economic security was a key element of national interest. That has been a stimulus to unprecedented developments of national and international statistics by the federal government.

- Unlike population growth in the other advanced nations, the U.S. population grew substantially since 1990, adding some 65 million people to the 255 million then. This provided a powerful incentive for internal economic growth, innovation, and the adoption of new technologies.

A generation of technological integration and economic expansion

The simplest way to capture the scope of information technology today and the radical changes it has brought about in the last third of a century is to look at some of the quantitative and institutional changes in that interval.

- Internet became prominent in the 1990s and was superseded by gigabit networks around the turn of the century, and by 2007 terabit networks were common. Broadband interactive services digital networks

(ISDNs) were laid in place by 2005 and continue their expansion. The ISDN was critical to the development of The Network, since digitization of all data is one of the core developments in the Information Age.

- In 2005, optical fiber to every new home, building, structure, and new facility, was mandated by federal law. Optical fiber to already constructed homes had reached 72% in urban areas, and 14% in smaller and rural communities. Universal fiber to the home (97%) was achieved by 2022.

- High-definition TV, the first breakthrough in large-size screens, occurred in the late 1990s. It was given tremendous stimulation by the integration of the telephone and cable systems and led to the nearly universal availability of video on demand by 2006. Three-dimensional television became common (22% penetration) by 2015.

- Universal wireless coverage reached all communities of 100,000 or more by 2003, and all counties of 250,000 or more by 2004. Mobile video telephony achieved the first practical application in 2001. The associated surge in automobile accidents, however, led to legal restraints with slow growth, particularly in buses and certified work vehicles, through 2017.

- The increase in microelectronic chip capacity slowed from 2015 to today. From 1995 through 2012, doubling continued roughly every 18 months. Doubling time currently is about 4 years.

- In 2008, chip density reached 250 million transistors per chip. In 2017, one billion transistors per chip were achieved, but that never reached the market in significant numbers.

- The Network, which can be looked at as an integration of smaller telephone, cable, wireless, and computer networks, has grown tremendously. In 1988 there were 3.7 million LANs; today there are 11.6 million LAN equivalents in the United States.

- In 1995, there were 70 billion minutes of international telecommunications traffic. By 2010, that had reached 700 billion, and today (2025) it is 1.3 trillion. Thirty-five percent is machine-to-machine communication, 20% is person-to-machine, and 45% is person-to-person.

- Lassettre made an interesting forecast in 1991 of computer power growth. His forecast, reproduced below, is overlaid with the historic data from 1995 to the present (dots).

Computer Power Growth

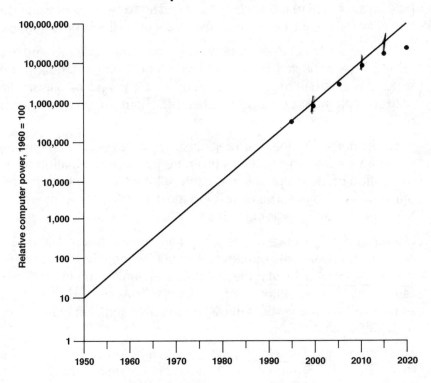

Source: E.R. Lassettre, National Engineering Consortium, *2021 AD: Visions of the Future*, 1991, p. 49.

Some unexpected developments in the Information Age

Before turning to specific changes in various areas of society that accompanied the movement into our Information Age, it is worth noting some unexpected consequences.

New modes of thinking. Advanced computer technology and imaging by the late 1990s resulted in the widespread use of computers that were interactive with the user. These computers were dynamic insofar as things moved across the screen and were three-dimensional, giving a depth to imagery that was surprising. Furthermore, the screens were no longer merely screens but were multimedia with acoustics and voice to accompany images. The most important consequence of multimedia machines was the new modes of thinking. By 2003, it was clear that many people over age 30 were not able to work with the new technology to its fullest capacity. This put a high premium on younger workers, and it was a stimulus for introducing computer-assisted and other high technologies into colleges. It also quickly propagated down into secondary schools.

Language has been reshaped in the Information Age.

Psychiatric consequences. The development of technologies for consciously manipulating brain function, enhancing creativity, increasing stamina and performance, influencing attention, giving what was, in the old terminology, a "psychedelic experience" and feelings of exultation are all well known. The effects, however, on the morbid mind were also interesting. The first case of paranoid schizophrenia reflecting the Information Era was in 2007, where patient X reported that in his delusion, "nothing talks to me." There has also been a group of schizophrenics who seek to "release the person from the system." Some believe that their knowbots are real and that they are the agents of alien galaxies here to take over our world. Others believe the knowbots are their captured souls.

The partial or total rejection of the information era is reflected in the clubs and associations that celebrate earlier times using old technology such as word processors and even typewriters and pens. They take great pride in mastering obsolete and purportedly more humane technology. It is a hobby or commitment on the part of different people.

Strong resistance has been met in seeing art, health, literature, and political interactions as mere information. There is great resistance on the part of many people to the handling of that material in the same categories in which we handle more traditional and routine information.

Affinity groups, that is, groups of people voluntarily coming together because of some common interests, have flourished in the last quarter century, belying the early belief that information technology was isolating and alienating. The overwhelming number of people in the advanced nations have expanded their orbit of connections, although research shows that these new contacts are generally topic-specific. By concentrating on the intense interest of the participants, they prove on balance to be most satisfying. In 2015, the Massachusetts Institute of Technology (MIT) Laboratory for Telesocial Studies showed that in 2002, the 2,000 contacts that the average middle-class citizen had over a lifetime in terms of personal face and name recognition had been augmented by 68 GlobeNet contacts, 80% domestic and 20% overseas. In the course of a lifetime, the total number of GlobeNet contacts will run to approximately 1,600. GlobeNet has for billions of people meant the end of boredom.

Farming is an example of continuity and change as a result of the movement to the Information Era. Growth of the large industrial farms slowed during the 1990s as they went for more intensive agriculture and the transition to a less chemical- and pesticide-intensive agriculture. The greatest effects, however, were on the small- business farm, capitalized at $200,000 to $500,000. They became information-driven just like the rest of business, with information and electronic controls managing feeding, grain stock, planting, and soil treatments.

The most dramatic effect was the resurgence of what have come to be called "hobby farms," usually run by people who had small land holdings and where both the man and woman of the family work off the farm for primary income. The other kind of hobby farm is run by exurbanites who have moved to rural areas or to the metropolitan fringes to farm a few hectares and have struck a felicitous balance between old-fashioned agriculture and the use of modern information as their key planning tool.

Effects of the Information Age

Information technology is the only truly universally enabling technology. From 1980 to today there have been four primary enabling technologies: information, genetics, energy, and materials, but only the first has been universal in its applications and consequences. It is referred to frequently in the other chapters.

The home

Home life has been radically altered by information technology. The home has become safer and more secure, both from human and natural intervention, from accidents, emergencies, and disasters. Eighty percent of midpriced and 100% of all high-priced homes are smart, and over half of the less-expensive homes have some smart features.

The spread of distributed work, particularly work-at-home, has led to the rapid evolution of the home work-study center (HWSC) center, sometimes called the work-study-entertainment center, as the centerpiece of domestic life. An *ad hoc* consortium, "The Home for Millennium Three," in 2005 held a nationwide contest for the design of an electronic home work-study-entertainment center. The top five winning designs were built as prototypes. That launched work-at-home in a new, positive direction. The HWSC is the place where the adults in the family do a portion of their work, the children reach out to the resources of the world, and the family seeks entertainment, recreation, the pursuit of hobbies, and a wide range of social contacts.

By 2005, 19% of the workforce was in distributed work. By 2024, 39% was in distributed work, that is, working at some location other than the traditional central workplace at least two days a week. The average time is 4.5 days per week for those in distributed work.

Recreation and entertainment

One effect of the Information Age has been to promote all kinds of scorekeeping, for personal performance, group performance, and team performance. The information era has also seen the rise of new kinds of entertainment and recreation as well as massive public competitions. One of the more interesting elements of the information era has been the rationing of limited public recreational space such as access to the Appalachian Trail and access for climbing the most popular mountains. The nationwide system of certification provides access to those who can make the best and fullest use of the restricted facilities. The rationing operates not by money but by skill and competence. One consequence has been the spinoff of many kinds of training programs to qualify for those new steps-up. For mountain climbers, there are six certification grades, each about one-tenth the size of the grade beneath it. For example, only 1,000 people worldwide are certified to climb Everest and Annapurna.

Work

As early as 1993, 60% of the workforce was in information. By 2006, this peaked out at 66% using the obsolescent nomenclature. The method of tallying work categories that was developed in the 1940s proved unreliable by 1980. The new organization of work that was proposed in 1997 and partially implemented by the government in 2001 had become standard by 2005. Consequently, the transition of work categories from the Industrial to the Information Era has not been smooth or clear. Today, well over 98.5% of all work involves some use of information technology, whether direct or embedded.

The displacement of manufacturing workers from 1980 to 2005 created a number of social proposals. One proposal that was put forward in the 1980s for distributing the benefits of manufacturing automation was given a try around 2007. The concept was to have individual citizens, nominally displaced workers from manufacturing, own robots and lease them to manufacturers as a source of income. The National Federation of Robot Leasers was formed, and the system worked within limits. But by 2015, the system effectively withered away for lack of interest.

Science and technology

Almost all scientific research is pure knowledge work. Even empirical field work now involves extensive use of space satellites. The collection of ground data by video, audio, and other sensors, night-and-day, round-the-clock, is routine. The development of miniature and full-scale artificial animals from insects to hippos has added new understanding to ethology and animal behavior and has been a primary component of ecological management.

In more traditional laboratory sciences, the universal availability of information technology and multidimensional modeling have drastically reduced the amount of hands-on experimentation and greatly increased the amount of experimentation and design on computers. In many of the most advanced laboratories, the experiments themselves are automatically set up and operated by labots. The old smell of the lab and the touch of the equipment is alien to most working scientists.

Health and medicine

The health care debate of the late 1990s gave unexpected impetus to the automated medical center. Based on the work of Project Caduceus, there was the now-popular proliferation of shopping mall and work-site health centers. These are expert systems tied to physical equipment. Typically, the way the earliest versions worked, a person walked into the center, was greeted electronically, asked to urinate into a bottle, spit into a test tube, puncture the skin to get a blood droplet for the slide, and place each of the containers into the appropriate receptacles. While the samples were being processed, the person sat at the Caduceus and engaged in a medical dialogue. Effectively Caduceus did everything that a physician at that time did on the first two visits, including presenting alternative diagnoses, the need for further tests, and the subsequent diagnosis on the completion of the tests.

The early version of Caduceus could diagnose 614 diseases. The system has measurably reduced health-care costs in the United States. At the same time, it has given people great confidence in the services that they are receiving. The current system Pasteur is the fourth-generation descendant of Caduceus.

In the last few years, typically 74% of the U.S. adults have used Pasteur, and 84% of urban families with children have used Pasteur.

Education

The effects on education have been three-fold, representing the effective meeting of the 100-year objectives of public education in the United States:

- Universal Education

- Lifelong Education

- Individually-tailored Education

Educational tools and apparatus are everywhere. Although nearly all children (96.3%) go to a public institution for schooling, the experience is hardly like what was provided in the schools of the 20th century. The in-classroom learning time has shrunk greatly, and the schools are directed at physical, social/interpersonal, and artistic development. The educational components of traditional reading, writing, and arithmetic are split 60/40 between school and home. For high school students, the shift has been even more striking to a 40/60 split. High school is primarily for interpersonal development, hands-on activities, and group activities such as teams, theater, and song.

College just is not what it used to be. The typical student now enters college with one year of advanced placement, and it is not unusual for extremely bright students to earn two-and-a-half years of advanced placement. The college is primarily a social acculturation institution for youth and young adults. It has also become a site for continuing education by people of all ages.

The sea change in education in the United States has been the shift from primary, secondary, and tertiary (college) to quaternary education, that is, lifelong, individualized education. The sites of quaternary education are 50% at home, 15% at work, and 35% elsewhere.

Video University has one to 16 alternative versions of every college and university course offered in the United States and Canada. It also has selected courses in 16 foreign languages for most of the 4 years of undergraduate work and a more limited amount for postgraduate training. These are routinely available in the classroom for subsequent collateral exercises with human instructors and professors or for rental at corporate facilities or for private rental use at home.

The major award in the field, the Athena, is given to the best instructor or instructor team in the arts, sciences, and social sciences and humanities, on a three-year cycle.

The classroom in the Information Age

As a child grows older, school becomes increasingly a socialization institution, and more and more of the cognitive content of school is conducted off-site, primarily at home. Even social-ization in the physical and collective activities in school has a strong information component. Fourth grade students collectively will build a building and subject it to a storm or hurricane. This is simulated on screens. In the seventh grade, 13-year-olds design and operate some kind of organization. It may be a Red Cross chapter, a business selling toys, whatever the group of four or five chooses to do. Perhaps the most interesting simulation is in the last year of high school. Typically, 16- and 17-year-olds team up to simulate a marriage and household, learn-ing the ins and outs of income, expenses, planning, furnishing a home, confronting acts of God, and so on.

At each level, the traditional basic training in reading and writing, arithmetic, geography, and history is heavily augmented by work on computers and screens. Children early become vir-tuosos in both traditional reading of books and publications and contemporary reading on screens. Younger children particularly love the old-fashioned books because they can carry them, take them to bed and on trips, and hide them away. They enjoy the kinesthetic feel and smell of the paper. The individualization of education from kindergarten on allows each child to progress at his or her own pace, eliminating much of the sometimes adverse effects of one child being far ahead of his or her classmates in an open-classroom situation.

For very bright children, it is standard to begin research projects in second grade. For those in the 99.9th percentile, the research can be quite provocative and stimulating – anything from the origin of buttons to the history of Nepal. They find it particularly attractive working on The Network, where they are linked not only with their classmates but to children throughout the nation and in many cases outside the country, who share their special research interests. In the education of extremely bright children, a major component of socialization is to learn to toler-antly deal with less capable people. Socialization is often carried on through interpersonal simulations. One of the dramatic effects of the Information Age has been the flowback of college-level material into high school, and in turn the flow from high school into elementary school as children are able to progress intellectually at their own pace. The result is that a large percentage of students move into higher education with advanced placements.

Preparation for work life has always been a challenge for schools. Currently there are computer video games which present some 34,000 different occupations and crafts and are designed to acquaint the student with not only the superficial aspects of those occupations but the day-to-day nitty-gritty elements. One consequence has been a vast proliferation of vocational training, and diversification of childhood expectations for future employment.

Criminal justice

Several components of the criminal justice system, police, corrections, parole, and probation, have been augmented in the last 15 years by crime prevention. All of those components depend heavily on information, with the most effective application being at the prevention and the parole/probation stages. Personal and family profile behavior analysis has gone far to identify crime-prone children and adults. It has been more successful with those prone to physical violence than with those open to commercial, financial, and infor-mation-related crimes. Similar results apply to parole, probation and correc-tions. As discussed below under "Issues of the Information Age and Their

Resolution," the applications of information technology in these areas are constantly interacting with the concerns for privacy and other civil rights.

Testimony and forensic evidence have been radically changed in the Information Age. The use of testimony from remote sites, the universal availability of video recordings, and the capability to take interactive testimony at distances, have greatly improved the efficiency of criminal justice, and, incidentally, of civil justice, proceedings.

Business

The first effect has been to virtually wipe out the notion that a firm is in the shoe business or the food business or the lumber business or the hotel business. All of these enterprises are mere information machines that incidentally provide shoes, food, lumber, or hotel space. Information management has become the central focus of successful business.

The second effect on business in the information era has been to force organizations to not deliver a single product or a single service but to deliver total service packages. This is embodied in the concept of the integrated performance systems (IPSs), in which more and more companies sell outcomes, not products. For example, one firm sells energy, one sells health, one sells transportation, one sells housing, one sells food. The way this is done is, in many cases, to provide a total package that reduces costs, increases diversity, and most interestingly, has had profoundly positive effects on the environment. An IPS, on average, uses 34% less total energy and the system lasts 117% longer, with 80% less maintenance and repair.

Rural America

Rural America was a primary beneficiary of the information era. Twenty-five years of rural decline, outmigration, job loss, and stagnation were reversed when fiber optics linked rural areas to metropolitan business centers and to global business centers. An equally dramatic effect was the expansion of metropolitan areas into sprawling suburbs. The changing patterns of movement to work and the large-scale growth of distributed work made it attractive, for reasons of time, flexibility, and the amenities of the open landscape, to move to the metropolitan fringes.

Disaster and risk management

As part of the general trend toward a totally managed globe, the handling of both natural and human disasters has been improved by information technology. The most dramatic of these interventions is the San Andreas project, which is directed at relieving the stresses to prevent San Francisco Quake 3. The 7.8 earthquake that hit San Francisco in 2011 was enough to put the fear of the future into both the state and federal government. The present projects involve a 2,000-kilometer stretch of faults and a complex daily and weekly schedule of lubrication (water injections) into the fault to steadily relieve the tensions that can lead to a big quake. The goal is to drop the quake by 5 orders of magnitude from one Richter 8 to thousands of Richter 3s. Injection, which has been in use since 2016, seems to be working. There is steady and measurable slippage. An average of 23 quakes per day occur in the 2.5-3.5 Richter range along the fault.

The effectiveness of emergency police, fire, and ambulance service, hinges on the fact that emergency vehicles have a built-in database that fully informs the crew chief during transit about the site of the emergency. In the last 20 years, deaths from fires due to toxic smoke inhalation have dropped 47%. Traffic accidents have had a 21% increase in survival rate in the same period.

Government

Activities in government that parallel those in the private sector such as purchasing, hiring, and service delivery, have all changed in the same ways that they have in the private sector. The special effect of information technology on government has been to stimulate democratic processes. Scientifically impeccable wide-scale surveys are a routine part of local, state, and federal government. Much of government's personal services under the broad category of health and welfare are now thoroughly interactive. Close attention is paid to the quality and effectiveness in delivering services. This has enhanced services to the elderly, the indigent, and the handicapped. A study in *Political Science Review* in June 2014 compared the quality of legislative debate in terms of the detail of options considered, and found striking improvement between 1990 and 2014. The effect of the Information Age on the judiciary has been to accelerate the processes at state and federal levels leading to much more coherent pretrial preparation of material and to higher levels of interaction between jurists and contenders. The effects on the criminal justice system have already been noted. It is also worth noting, however, that the average time to final resolution of criminal trials has dropped 34% in the last 20 years. The effects are more dramatic in civil suits, in which the time to resolution has dropped 84%, not counting the large numbers that have been shunted out of the legal system to arbitration, conciliation, and mediation.

From 1998 to 2010, the postal service received a shot in the arm as it became the general electronic point of contact with government for large numbers of people. As The Network grew, however, that role of the postal service sharply declined, and it is now primarily effective in some rural areas.

Quality comes to religion and goes

One basic effect of the Information Age on organized religion has been to open the system to the expression of the members in regard to shaping policies and practices. The second effect has been to widely expand pastoral services. Pastoral services do not solely involve clergy in one-on-one relationships to the membership, but often involve many-to-one, and particularly interesting has been the development of religious-based affinity groups under church auspices, which discuss various special interests or affinity group concerns.

In the early teens, with the great movement toward quality in the service sector, a number of churches saw that many of the religious functions are closely analogous to private-sector services. Because the quality movement had been so successful in the corporate world, the attempt was carried over to religious institutions. Lutherans and Roman Catholics, with their commitment to the confessional, albeit even then a declining enterprise, attempted to use quality control measures by assigning penitent code numbers, coding the penalties, and making a continuous record of the behavior of the penitent. All were maintained at a high degree of security by the pastor. After three years of experimentation in seven churches (two Lutheran and five Roman Catholic), it was concluded that the pastoral functions as reflected in the confessional had no significant effects on behavior, but did raise spirits.

Issues of the Information Age and their resolution

The issues associated with the Information Age crop up at different places along the spectrum — from data to information to knowledge to wisdom. As is often the case, one resolution later leads to a new issue. Many would argue that if we did not collect certain data we would not be able to abuse it. Others argue that we need the data, and the problem is to safeguard it. Others argue that well-developed knowledge bases will provide their own unique safeguards. Lying in the background is the plea for us all to be wiser and more anticipatory in what we do.

Radiation health effects from the universal availability of information technology intensified as an important public policy issue in the late 1980s. In spite of increasing research, it never was clearly established that radiation from information technology was a health risk. High-voltage transmission lines, however, were established as a risk, especially to children, by 2001. The effect on information technology was to stimulate a round of designs to keep people at a greater and presumably safer distance from the radiators. The flood of innovative designs also introduced a new generation of enhanced technological capabilities.

Local confrontational challenges and codes limiting telecommunications had become the bane of the several industries involved. Fifty-four state codes and thousands of local variations had to be rationalized. The Supreme Court found that local restraints on rules and regulations governing information technology were in violation of the Constitution's Commerce Clause, and that all such codes and regulations were voided. Responsibility was totally with the federal government or through agreements with international organizations.

The age of the image

Many folks believe that we have passed through the Age of Information and are emerging into the Age of the Image. Digitization of images has had dramatic effects on the way we do our recreation and entertainment. It has also shaped the way we think. One interesting commercial application is to pop a customer into a product advertisement. Food ads are particularly effective with this. A customer-of-the-week is selected by lottery and integrated into a food ad, with results often that are just cute, but sometimes strikingly hilarious. The incongruities between the character and the ad have drawn a great deal of public attention.

In home entertainment, personalized videos are now available. One hundred thousand were sold in 2019. Five million were sold in 2024. The technology is relatively straightforward, although complex in detail. A favorite film is the classic Charlie Chaplin, *The Great Dictator*. Another favorite is *Snow White and the Seven Dwarfs*. In the first case, typically a purchaser will have members of his or her family take the lead roles, and in the latter case, there is an intermediate process in which the family is cartooned before taking the roles.

Imaging has affected the way we process data. The ability to picture things in three dimensions and dynamically moving has long ago been used in the laboratory in genetic engineering and the design of proteins. That capability has moved into the general economy and into business. Flow models literally showing the cash flowing are now common.

The explosive development in imaging technology from 1995 to 2012 created widespread concern about the legitimacy of images. The early and relatively benign abuses came about in the cinema. Quickly, broadcast news, historic programs, and entertainment moved into creating synthetic events showing real characters, living and dead, in circumstances that had never occurred. The Authentication and Certification Act of 2006 required that all images be certified as to authenticity, and that four classes of false, synthetic, doctored, or modified images required suitable warnings and descriptions. It also clarified intellectual property rights to altered images.

Privacy Safeguard Act of 2003

The primary irritations had come from the business sector in the constant push for the next edge, the next marketing opportunity, the next understanding of microdemographics. Businesses have pushed toward more finely tuned data collection and analysis. The Privacy Safeguard Act of 2003 has worked reasonably well. It prohibits commercial enterprises from basing any advertising on populations smaller than 1,500, 2,000, or 10,000 people, depending on the nature of the product or the service. In other words, one cannot market to individuals on the basis of individual knowledge about that person or a small group like that person. An escape hatch has proved to be quite popular, in that release from that restraint can be granted by individuals at both a personal level and household level. As it stands today, 63% of the U.S. public have waived those commercial privacy rights. The remaining 37% have chosen to use them.

Privacy is another outstanding issue in relation to information technology. It has been a concern since the late 1960s and its intensity skyrocketed around 2002. The privacy issues frame themselves in the United States around the Fourth Amendment, which makes us secure from search and seizure. Search and seizure originally assumed physical property and physical intrusion. Information technology allows us to acquire information about people and their affairs with no physical intrusion into their private space. Consequently, the question of expanding privacy rights under the Fourth Amendment has been a nettlesome issue.

The ultimate risk to privacy is government. Courts, corrections, and probation systems are the main troublesome actors in the United States. The crime prevention issue centers around the ability to identify highly crime-prone groups and individual members of those groups and the limited surveillance and intervention now permitted. There are three levels of probationary and parole surveillance that have caused great concern.

Infoterrorism has flourished as information became central to the economy. After a terrorist group effectively shut down Cleveland, Ohio, from all external communication for seven hours, Congress was strongly motivated to relax the constraints on monitoring and surveillance groups and to severely upgrade the penalties.

Terrorist acts by individual groups, vengeful workers or former workers, and psychotics are chronic. There seems to be no clear way of preventing them as the cost of both information and transportation continues to fall.

The economics of information. The Nobel prize awarded in 2016 for the theory of the economics of information altered the tax structure and the payment structure for information. It has not, however, removed all of the problems associated with the cost and value of information. Laureates Lee and Richovsky recognized the limitations on Claude Shannon's information theory as the key to the economics of information. But working with that as a base and combining some thermodynamic concepts and work early in the century on time and human values have led to their prize-winning work. The primary consequences of their work, which began to have its effect around 2006, were:

- The monetization of information on a national and international scale opened up tax incentives to shaping information-rich products and services more in the public interest.

- The United Kingdom from 2007 to 2011 tried an experiment in the economics of information by making all network services free. The five-year experiment proved a smashing success, and now with minor exceptions, notably the interlacing with overseas information nets, all information networks in the United Kingdom are free.

- In the United States, an alternative tack was taken. AT&T led the way in conjunction with US West in the reformulation of payment for information services. Research had shown that late in the 1990s in steady state there was generally less than a 2% fluctuation in the annual use of information by a typical family or business. The great experiment in 2016 that proved successful was to annualize network use costs and to build them into mortgage payments or rents. The effect was to submerge awareness of the cost of the network. This led to a rapid expansion in its use. The cost of the networks had fallen so low relative to the cost and value of end uses that this new formulation of price structure was both realistic and a great user incentive.

Systems vulnerability. Aside from the interference of terrorists and psychos, the simple complexity of information systems has left them vulnerable to system failure and to the effects of natural hazards. The insurance industry of the United States has made redundancy the universal model for business and for those private individuals who wish their databases to be safeguarded.

Many systems exceed the intellectual capabilities, particularly the temporal capabilities of the human being. The tension is between turning operations over completely to machines or having human beings as backup. In the late 1990s and through the first decade of the present century, problems of this sort came up with regard to aircraft for which the information capabilities of managing the aircraft made the pilot redundant and gave him or her little opportunity to practice. No one was fully satisfied with virtual reality training.

Safeguarding intellectual property. In a world increasingly drenched with information, the ownership of intellectual property has turned into a chronic issue. The development of the global patent system in 2009 helped, but the abuse in terms of products sold primarily on a regional basis has not been dealt with. Strictly intellectual property involving no significant physical embodiment has proved to be extremely troublesome. The fifth triennial meeting of the Global Intellectual Properties Commission is scheduled for June 2026.

Informationally impoverished people who have limited access, skills, and understanding of the information world are purported by many to be suffering serious disadvantages. In spite of extensive research, it has not been established that there are significant adverse consequences for those of limited ability. The usual situation seems to be that people with limited capabilities rely on friends, relatives, and associates for the information that they cannot quite manage or control themselves. The issue, however, is chronic.

The world has watched as a half-dozen times in the last quarter century, distant, rebellious people have risen up against their oppressive governments. The pressures created by the worldwide visibility of those events could only have been possible because these people had the competence and skill to deliver their message and the accounts of events in real time. The argument, therefore, is that other oppressed people around the world, by being isolated from the information systems, are likely to remain perpetual victims of their oppressors.

Global applications

The Information Age has not uniformly penetrated all parts of the world. It has, however, to some extent penetrated every part of the world. Some of the highlights of that follow.

GlobeNet, 2003, sponsored by the United Nations Educational, Scientific, and Cultural Organization (UNESCO), provides continuous video around the world. Because uplinks are not available everywhere, it provides a minimum of 10% of the time, 17 hours a week, of uplink. It has 11 roving uplink teams that travel to different parts of the world. GlobeNet was launched after the undeniable success of the Motorola Iridium project (Motorola was a firm that was prominent in information technology toward the turn of the century). The favorite content is health, child education, adult education, agronomic practices, animal husbandry, and kit assembly and use. GlobeNet has been extremely successful in promoting various kits applicable in specific regions of the world — electrical, household plumbing, and six auto kits are the most popular. Many programs are cartoons with culturally specific formats. They are designed to be informative and entertaining.

The environment has grown to be of such dramatic importance since the confirmation of global warming in 2013 that the *Annual Condition of the Planet Report* has become a landmark document. The quantitative scaling of individual nations and regions, and for larger countries, provincial and state conditions, has been a great stimulus to positive response to environmental issues. For many of the 174 independent countries of the world, the annual report has become both a recognition and a stimulus. The *Condition of the Planet* has as its long-term objective stimulating sustainability, particularly in agriculture, mining, and water use. The *Condition of the Planet* has also become a tool for reporting on national as well as corporate abuse and misuse of the environment. The 2024 report earmarked 1,600 facilities polluting at level one (critical risk to people and the environment), and 3,600 at level two (degrading the environment), and 14,000 at level three (important and readily correctable). The annual report depends heavily on space satellite data and extensive networks of traditional on-the-ground database generation. The annual planetary report takes up one-quarter of the GlobeNet time for two weeks every year.

Regional telelinks have become important as conflicts have grown among European, North American, and Japanese trade blocs. Telelinks have expanded on a highly subsidized basis. For example, the United States, Canada, Mexico, and the Caribbean islands all enjoy the same telelink user fees irrespective of the actual costs. The same applies to Europe, including Turkey and the Commonwealth of Independent States (CIS). Their latest expansion is the complete CIS in 2016, Russia having been included in 2011. The Japanese telelinks have been much more competitive. The United States and Japan are in intense conflict for telelinks to the Philippines and to Australia. On the other hand, Japan has been successful in mainland China and the Chinese peninsula. There has been for the last seven years a raging conflict to establish regional telelinks throughout Latin America with the United States and Japan being the primary contenders and Europe a secondary contender.

Arms control and disarmament have evolved slowly, control more slowly than monitoring, though it did receive a powerful stimulus with the nuclear disaster in 2011. The general flow of arms has declined dramatically since the peak of international conflicts and the last of the seven wars in 2015. The problem is still far from resolved.

One of the benefits appropriate to global information has been large-scale conciliation as conflict resolution. It is not uncommon now for as many as 30,000 people in an internal or transborder conflict to participate over a period of eight to 15 months in broadscale discussion. There have been several successes as well as some regrettable failures.

International standards, especially in information technology, have become a successful global enterprise, building on 100-year history. Today 99.7% of all information technology is fully compatible. The remainder is associated with the intended introduction of new technologies and the exploration of new bandwidth incorporations and allocations.

Information terrorism and crime have been a steadily evolving concern of the global community. The knotty issue of dealing with different international standards, practices, and customs has been resolved at the international level; it was ruled that the law of the country in which the alleged crime is committed is the law that prevails in the pursuit, apprehension, trial, punishment, and incarceration of the criminal. The amount of contraband material—drugs, stolen goods, illegal materials—has over the last 12 years declined by 63%.

Financial services. The most complete and earliest globally integrated sector has been financial services. The centuries-old institutions of transborder finance burgeoned as global financial transactions became commonplace in the early 1980s and reached an unprecedented level by 2003. The market collapse of 2007 and the earlier collapse of derivatives in 2002 were largely due to the suborning of public officials, in the former case, and the overly zealous application of computer programs in the latter. They have led to stringent global financial flow controls. As a crime control measure, the United States in 1999 called in all currency and reissued it in machine-readable form. The ability to literally trace the flow of individual $50 and $100 bills had a radically depressing effect on organized criminal activities. During the call-in which occurred over a period of seven years, there was a 12% depreciation in the value of U.S. currency for each year after year one in which it was turned in. Only 40% of currency was turned in during year one, and by the end of year seven, 13% of currency had ceased to have any value. There was an inevitable and necessary escape hatch from the regulation for found money, lost money, buried resources, and small holdings. A minor industry flourished in that seven-year period, in helping foreigners exchange money. Surprisingly, only 0.002% of turned-in currency was counterfeit.

One of the goals of calling in currency was to draw in the large caches of criminal- and drug-related funds outside the United States. The program was extremely successful. The major flow of illegally gotten funds into the United States occurred in years two and three, when presumably the holders of those funds realized that they had no choice but to cut their losses and turn in their money. As a result of the call-in, 1,034 arrests were made worldwide, with 930 prosecutions and 812 convictions. All of these were big fish.

INFORMATION TECHNOLOGY IN WORLD 2 AND WORLD 3 SOCIETIES — The world's largest information technology market

Worlds 2 and 3, the middle and destitute societies, together underwent a population growth from 4.6 billion in 1994 to 7.1 billion in 2025. They have proven to be the world's largest market for information technology. However, they are by no means the dominant market for high-end, and highly integrated systems. Worlds 2 and 3 each have some big cities that are modeled on Western cities, and large portions of them are completely modern in their information technology. However, as one moves into the countryside, the situation becomes more spartan.

None of the 174 countries and sovereign states in the world are out of touch with the rest of the world, although it still is the case that, particularly in World 3, large portions of their indigenous populations have little or no outgoing communication. As discussed under "Global Applications," below, World 2 and World 3 have been primary beneficiaries of GlobeNet.

Satellite delivery has proliferated in three forms: (1) GlobeNet, sponsored by UNESCO; (2) commercial broadcasting, which in its early stages, through 1999, was largely sponsored on a user-fee basis (this proved not to be feasible in Worlds 2 and 3) with the bulk of the costs borne by advertising; and (3) individual public networks of countries or regions. Six regional satellites operate over Africa, four over South America, and six over Southeast Asia.

In metropolitan areas, which are now the dominant domicile for people throughout the world, wireless was popular from 1989 through about 2011, when the transition to fiber optics began on a country-by-country basis. The metropolitan areas of the world, including Worlds 1, 2, and 3, are now fiber-optic wired, with wireless being the feed-in and feed-out for many people.

With regard to public networks, the World Bank has been a force in promoting more effective government planning. In particular, it has insisted on the use of information technology as a form of public participation with central government planning authorities. It has also provided teams — there are 75 now in the field — doing video recordings and in many cases video broadcasting of the status of World Bank-funded programs in World 2 and World 3 countries. This unprecedented feedback has added measurably to the sophistication and effectiveness of international programs and has had the net effect of raising the productivity and income of several World 2 and World 3 countries.

The Peace Corps network has become extremely popular as the United States Peace Corps became internationalized. The United States is now the training ground for Peace Corps participants from 13 advanced nations and 22 World 2 countries. PeaceNet has been so popular that it has raised the prestige of the Corps. PeaceNet, although it has some visual capabilities, is primarily an information networking system that brings analytical capabilities and databases to 6,000 locations throughout Worlds 2 and 3.

Today the world's largest corporation is Universal Technology (UT), formerly known as AT&T. It changed its name in 2013, although it maintains the obsolescent name in North America. Currently, UT has 267 primary world business alliances, average duration 23 years, covering 137 countries and indirectly covering 26 more states and sovereign areas. For countries within its global alliance, UT provides manufacturing, distribution, broadcasting, and wire and wireless services.

The global division of labor applies to information technology as well as everything else. One of the most striking outcomes was the dominance that India came to enjoy around 2005 in software design.

Information Technology in Worlds 2 and 3

Number per 100 Households

The Device	World-2	World-3
Screens	226	11
Knowbots	10	2
ID Cards	400	300
Smart Cards	4	0.2
Personal Computers	6	0.7
Telephone—Basic	50	4
Telephone—Advanced	7	1
Embedded Computers	500	60
Fiber to Home	26	3
Wireless and Other Personal Communications	80	7

Critical Developments 1990-2025

Year	Development	Effect
Late 1990s	Flat screens introduced.	Changes the use of computer screens from office or domestic furniture to decorative and work/entertainment tool.
1998	Knowbots become prominent.	This leads to widespread public approval of the systems and greatly expands use by those semiliterate with regard to computers.
1999	The U.S. government calls in all currency and reissues in machine-readable form.	Crimp in organized crime and international terrorism.
2000	Environmentalists see information technology as a primary tool for promoting sustainability.	Broad pressure for widespread global and national surveillance by satellite and by collection of ground-truth data.
2002	Collapse of derivatives market.	Securities and Exchange Commission (SEC) intervention to severely restrict derivatives.
2003	GlobeNet initiated by UNESCO.	Continuous around-the-clock video information primarily to countries in Worlds 2 and 3.
2003	Supreme Court drops limitations on encryption codes.	Widespread industrial and business applications expand; spying in disarray.
2003	Privacy Safeguard Act.	Severely limits use of databases for intrusive exploration of customer base.
2005	Contest held on "Home for Millennium III."	The contest and building of prototypes began the radical change in domestic architecture, design, and furnishings.
2006	Telecommunications Computer and Business Stimulation Act.	Relieves industry of monopoly constraints.
2006	Authentication and Certification Act.	Requires certification of images with regard to authenticity or extent of doctoring.
2006	AT&T and US WEST annualize telecommunications fees and integrate into mortgage payments.	Dramatic expansion in use of information technology.
2007	Market collapses.	Tighter controls on public officials connected with financial services.
2009	Adoption of global patent system.	Greatly reduces abuse of intellectual property.
2011	San Francisco earthquake 7.8 on the Richter scale.	Large-scale civil works to prevent San Francisco Quake 3.

2011	Supreme Court finds state and local regulations governing information violate the Commerce Clause.	All responsibility for regulation is in the hands of the federal government or through treaties with international organizations.
2011	Second nuclear disaster in Pakistan.	Stimulates international attention to disarmament.
2011	First triennial meeting of Global Intellectual Properties Commission.	Much improved control over intellectual property abuses by World 2 and World 3 countries.
2013	Global warming confirmed as real and significant.	Pressure for effective information gathering, including initiation of the annual *Condition of the Planet* report.
2016	Nobel prize to Lee and Richovsky for economics of information.	Radically alters public policy toward information, including making it a practically taxable asset.
2018	Nobel Prize to Smith and Garcia for their research on the structure of the workforce.	New understanding of work and its flow and organization in the Information Age.

Unrealized Hopes and Fears

Events	Potential Effects
Health effects from electronic-device radiation.	Severe cutback in the use of such devices.
Split of nations into electronic haves and have-nots, with powerful political disparities in power.	Division within countries into two hostile groups of exploiters and exploited, rich and poor.
Massive invasions of privacy by both commercial and government data collection and use.	Arbitrary intrusions on people's well-being and their ability to engage in private and commercial activities, and government political intervention in citizens' private lives.
Collapse of the public school system, replaced by a dual system of fee-for-service for the prosperous and public schools for the impoverished.	A two-class society, information-rich and information-poor.
Identity cards promote fascistic governmental implications.	The end of free movement, democracy, the closing down of the open society.

4

HARVESTING THE FRUITS OF GENETICS

We have entered the age of the gene. Watson and Crick's discovery of the structure of DNA in the last century and mapping the human and other genomes in this one have provided the keys to applied genetics. We have spent much of the last 30 years deciphering raw genetic data and converting it into information and knowledge. Today, that conversion accounts for almost one fifth of the U.S. gross domestic product (GDP).

Applications are everywhere, including health, behavior, forensics, livestock, fisheries, pest management, crops, food, forestry, microorganisms, chemical engineering, environmental engineering, materials, manufacturing, energy production, and information technology. Genetics improves on nature, enhancing human health, economic performance, and the environment. It alleviates arthritis, creates new foods such as the beetato, and made history of many crop pests such as the fungal rice blast. It creates new techniques, approaches, and even organisms, such as the toxagen, for breaking down many hazardous wastes. The table below highlights some of the important capabilities and consequences that genetics has brought to business and society over the last 35 years.

Application Area	Capability from Genetics	Examples of Consequences
Health	Identify, treat, and prevent genetic diseases and disorders	Elimination of almost 2,000 single gene diseases, such Huntington's Chorea; 50% reduction in diseases with genetic predispositions, including dozens of cancers
Behavior	Understand chemical, physiological, and genetic bases of human behavior	85% reduction of schizophrenia; overhaul of education curricula to tailor learning based on individual genetic/ cognitive profiles

2025

Identification/ forensics	Unprecedented accuracy contributes to a decline in many crimes, including kidnaping and fraud	27% reduction in auto theft attributed to deoxyribonucleic acid (DNA) security locks
Livestock	Produce transgenic custom-designed livestock by calling up and obtaining genes from databases along the lines of cookbook recipes	Revival of the once-flagging pork industry with the advent of 37 popular varieties such as the ultra-lean "Pig-No-More"
Fisheries	Create and combine transgenic fish with different tastes and textures, doubling production from last century	Overwhelmed natural fisheries supplemented with aquafarms raising hearty new strains such as the "octosquid"
Pest management	Target specific pest species and behavior patterns and genetically engineer resistance to pests in crops	Crop loss due to pests reduced 63% in United States; elimination of Lyme disease and elephantiasis by genetic alteration of the vectors
Crops	Increase yields, growth rates, resistance to disease, longevity, and reduce the need for water and fertilizer	Irish potatoes, Kansas wheat, and Japanese rice continue to set yield records due to elimination of intermittent blights
Food	Add more foods to the human diet and customize foods to particular needs and requirements, such as tastes, preparation, and preservation	The number of foods making up the bulk (90%) of the human diet increases from 6 to 37; all 11 principal grains are dietarily balanced for protein.
Forestry	Develop superior and faster-growing strains through disease resistance, herbicide resistance, and artificial seeds	Worldwide tree/forest coverage doubles
Microorganisms	Widespread use in medicine, food and agriculture, chemicals, mining, waste management, and environmental cleanup	Use in bioreactors to produce commodity and specialty chemicals, medicines, and foods
Chemical engineering	Use databases of molecules and chemical reactions in more biologized industrial processes	Random hit-or-miss approach gives way to rational chemical design, contributing to the industry's 54% reduction in design time
Environmental engineering	Monitor, remediate, and enhance the environment, such as breaking down solid and toxic wastes to restore degraded ecosystems	Bioremediation now used as the primary cleanup mechanism in 40% of hazardous waste sites in the United States

Materials	Manipulate materials at the molecular or atomic level so manufacturers can customize materials for highly specific functions	35% of people in affluent nations use some form of biosensor to monitor their health
Manufacturing	Use more biological processes at smaller scales, approaching the nanoscale	Bioreactors in widespread use in the nondurable goods sector or affluent nation manufacturing
Energy	Expand the use of biomass and aid enhanced oil recovery	Conversion efficiency of biomass could triple with recent feedstock breakthrough
Information technology	Apply genetic algorithms in software programming	Neural network computers mimic the intelligence level of chimpanzees

For convenience, the term genetics, as used in this book, incorporates applied genetics, aspects of molecular biology, and biotechnology. Genetics is the key enabling technology of the 21st century, rivaling information technology, materials technology, and energy technology in importance. The effects of all of the enabling technologies are far-reaching across business and society, but advances in genetics in particular are fundamental to many science and technology areas and societal functions, including health and medicine, food and agriculture, nanotechnology, and manufacturing. Unlocking the secrets of genetics has required understanding the underlying physics, biophysics, and chemistry of DNA, such as how and why it conducts electricity.

Through genetics, we understand and manipulate:

- the structure and functions of complex biological molecules, particularly proteins, and employ rational design for synthesizing complex molecular assemblies and cells;

- the functions of many tissues, organs, and systems;

- gene expression to influence an organism's development, growth, and aging; and

- novel traits and novel organisms.

Rising public interest in genetics is tied to the growing realization that humanity is capable of directly shaping our and other species' evolution. We no longer have to wait for nature's relatively slow natural selection. Genetics brings the capability of speeding and redirecting evolution along paths of our choice. Adaptations that once took generations can now be made in months or years through genetic manipulation. For example, it was inevitable that rare genetic conditions such as autosomal dominant polycystic kidney disease, in which fluid-filled cysts form in the kidneys and in half the cases ultimately required dialysis or transplant, would be eliminated through natural selection in hundreds of years. With genetic therapy, however, that process was accelerated and accomplished in just 15 years from the first therapies to the ultimately successful one.

We not only use more biological and natural processes today, but we improve on them. Many biosynthetics are more suitable for human purposes than their natural analogs. For example, peptide nucleic acids (PNAs) that are used to block the message of colon cancer genes from being carried out can actually improve natural antibodies. Indeed with a-life (artificial life), we are rethinking the definition of life itself (see box below).

Artificial life enhances our understanding of natural life

Artificial life is devoted to the creation and study of lifelike organisms and systems built by humans. Computer-generated simulations synthesize evolutionary processes. The strong a-life school argues that these simulations are living things whose essence is information embodied in machines or robots. They point to computer viruses, which evolve strategies to prolong their existence, and suggest long-term issues of our responsibilities to new life forms.

The so-called weak a-life school avoids the living/nonliving question and focuses on their lessons for our and other species' evolution. Practical applications include the growing capability to evolve, rather than design, new products ranging from pharmaceuticals to robots. For example, thousands of tiny robots without advanced programming are simply set on a task. Those who perform best naturally are selected. After many iterations, a best candidate emerges, and the reasons why are probed and applied for future generations.

Harvard's first doctorate in a-life was granted in 2015.

This power has inspired a profound global debate about how genetics should and should not be used. Net forums, electronic town meetings, newspapers, and journals are filled with discussions of what behavior society should encourage, given our growing capability to influence it.

We are still evolving a philosophy of how to incorporate genetics into our world view. Over the years, the nature/nurture debate often split along ideological lines, with conservatives tending to favor the gene-determinism notion, and the liberals favoring environmental determinism. Today, the ideological slant is largely relegated to the fringes, as consensus is growing that the ratio between gene determinism (nature) and environmental determinism (nurture) slightly favors the former—roughly 60% to 40%. A side effect of the debate is a great increase in the research demand for twins, as subjects for experiments aimed at proving the case of one side or the other.

On the economic front, genetics is rewarding those who invest in it for the long haul. It has been an industry for patient capital. Its spread over so many industries has made it an increasingly important factor of the national and global economy. There are several boom regions: San Francisco with its large number of genetics companies and its proximity to Silicon Valley; Boston with its universities; and Houston, which has diversified away from oil and energy to health-related genetics. There has also been a niche industry based in Salt Lake City that has grown up around the reliable genealogies kept by Mormons.

Genetics is not a typical industry, in that it is not measured as a separate entity. It is a part of, or embedded in, so many industries that government statisticians do not attempt such a measure. Economists' best guesses are that genetics accounts for about 20% of gross domestic product, or roughly $2 trillion in 2024. The only segment of the economy contributing more is information technology, which is estimated at 40% to 53% of GDP.

R&D Spending on Genetics: 1992, 2025

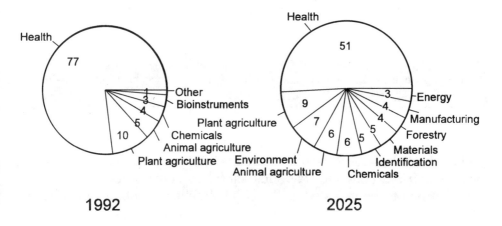

1992   2025

Genetic history: a new national pastime?

Newsgroups, videos, books, videoconferences, and courses have sprung around genetic history. Many people are interested in tracing their family's roots genetically. Thousands have gone so far as to obtain genetic samples of ancestors—even when it entails exhuming bodies.

A second thread is interest in the genetic history of other species. The restoration of the Mastodon in 2014 sparked interest in species restoration. The classic film *Jurassic Park* enjoyed a huge revival that year, especially when the interactive element was added to it. Plans to revive other species are under consideration. Politicians and the scientists involved are proceeding cautiously, but surely species restoration is an industry poised to explode in the 2030s.

Last century's emphases on using genetics for improving human health and battling human disease are being supplemented with more exotic applications, e.g., manufacturing and materials, human enhancement, energy, environmental engineering, and species restoration and management. The food and agriculture industries, for example, have been steadily expanding their use of genetics for decades. Advances have often come from applying what seemed like isolated breakthroughs into a systems framework. In 2020, for example, researchers in Egypt working on eradicating a species of locust engineered a microorganism that later proved useful in converting crop wastes into biomass energy.

Some false paths in genetics

Genetics involves a complex system of predispositions, biochemistry, and environmental factors. As a result, there were many wrong turns on the road to prosperity. There were many Interferons—a 1980s drug once hailed as a cure-all that ended up with limited use in cancer therapy. It turns out that finding the gene for a disease or disorder is one thing, and devising a safe, affordable test for it is quite another.

Lots of the failures came from petty schemes. The Burma government, for example, spent lots of money trying to develop and breed in a gene for submissiveness to authority, without success.

Some companies took a limited view of genetics as a manufacturing tool using recombinant DNA for making proteins. Their products often failed, as the synthetic proteins were difficult to administer as therapeutics, because they were big and fragile.

And, of course, there is the now infamous attempt to mimic human brain cells by trying to induce them to grow on a silicon chip.

Researchers in these early days often did not understand the underlying principles behind a purported breakthrough. Many discoveries or processes had to be worked out by trial and error. Gene therapy, for example, began with *ex vivo* injections into cells and their subsequent replacements—a process that had to be repeated as the original cells died off. It took another decade to work out the retroviral approach widely used today that is far less invasive, and much less expensive.

The pace of genetics developments

The speed at which developments have occurred has depended largely on four factors:

- the pace of technical progress (development of new technologies, such as for automating genome mapping and sequencing)

- the extent of unexpected complexities (for instance, changing a gene or two can alter more than just one function)

- economic considerations (costs, for example, including gene therapy treatments in standard insurance packages)

- social and ethical issues (for example, social resistance to genetic testing as a condition of employment)

Many times, progress in a particular area stalled and then a new finding reinvigorated the R&D community. Often, old ideas were rejuvenated as the supporting knowledge base that developed in the interim led to new insights.

Engineered biomass, for example, lay largely dormant for about two decades until the Cropagen treatment came along in 2020.

Genetics knowledge and capabilities are largely in the hands of World 1 nations, which have the knowledge base and infrastructure required to support the R&D fundamental to success in genetics. World 1 nations have so far realized their greatest share of profits from their home markets. Over the last 10 years, that gap has begun to narrow as burgeoning middle and destitute nation markets are tapped through alliance, acquisition, and licensing arrangements. Strategies for affluent markets differ considerably from those in Worlds 2 and 3. Customized genetic therapies, which are increasingly common in World 1, have barely found a market in the middle nations. Similarly, in food and agriculture, genetics has greatly increased the diversity of affluent nation markets, while its primary effect in middle-income and especially in destitute nations has been to increase the supply of staples.

THE WORLD 1 NATIONS—GENETICS: A LEADING KNOWLEDGE INDUSTRY

Genetics has been dominated by the affluent nations. It is knowledge work in every sense of the term. The technological and intellectual infrastructure required to be a major player has kept the affluent nations ahead of World 2 societies who are eager to catch up. The footsteps of nations like Thailand and India are getting louder, especially in plant genetics.

Health and food have been the big drivers of genetics in World 1. The United States sells more genetics-related products and services than anyone else, with Japan close behind. Europe continues to lag. The United States has been particularly strong in health and food, whereas Japan has been strong in environmental applications. Europe has been hampered by social objections to genetics. Last century, Germany passed one of the world's most comprehensive laws designed to regulate the use of recombinant organisms in laboratories and industry. This handcuffed genetics developments throughout Europe due to Germany's central role in the EC. Despite the removal of these restraints by 2009, it has proven difficult to catch up.

Cooperation among affluent nations regarding genetics has been strong. The cooperation has its origins in the various international genome projects. It arose largely from practical necessity, given the rapid dissemination of new findings since the establishment of the GlobeNet last century. The difficulty in keeping secrets and the growing web of global alliances made it more sensible to accommodate rather than fight. Of course, cooperation worked because the commercial interests were carefully negotiated. It continues with the move toward global management and regulation. The International

Organization for Standardization (ISO) 46000 standards for genetics were a capstone achievement in 2024.

A U.S. CASE STUDY: Still the leader in genetics

The U.S. R&D and science and technology base, supplemented with talent drawn from a global labor pool, remains first class. The large internal market for health spending was the early driver for genetics, but the emerging markets of the middle nations have become increasingly important. At the same time, these World 2 nations are becoming stronger competitors in the genetics marketplace.

Human health continues to drive genetics

Genetics is enabling health professionals to identify, treat, and prevent genetic diseases and disorders. The centrality of genetics in diagnosis and treatment is clear, particularly in early diagnosis, in testing for predispositions, and in therapies. There are thousands of diagnostic procedures and treatments for genetic conditions.

Diagnostics detect specific diseases such as Down's syndrome and behavioral predispositions such as depression. Treatments included gene-based pharmaceuticals, such as those using antisense DNA to block the body's process of transmitting genetic instructions for a disease process. In preventive therapies, harmful genes are removed, turned off, or blocked. In some cases, healthy replacement genes are directly inserted into fetuses, via injection, inhalation, retroviruses, and sometimes pills, to alter traits and prevent diseases. For those already sick or impaired, genetic therapies using similar techniques have an outstanding record of reversing or correcting conditions.

The evolution of health-related genetics

- Identifying single-gene diseases and disorders

- Identifying multiple-gene predispositions

- Developing tests for genetics flaws

- Mapping biochemical pathways and developing therapies and treatments

- Developing preventive methods

- Enhancement

Our understanding of genetic diseases is along a continuum. Three areas are under particularly good control. The best understood are the single-locus genes that are strictly heritable from parents. Next are the genes associated with the transition of normal cells to cancerous ones. This is particularly important because most cancer patients do not die of their initial tumors, but from metastasizing cells that manage to escape the tumor and grow into secondary tumors in other organs. The breakthrough was finding and developing a monoclonal antibody (MAB) treatment for the genes that code for cellular adhesion proteins central to a metastasis. A third area of solid understanding is in autoimmune disease genes. Defects in these genes cause the immune system to destroy healthy cells to which they accidentally bind. Understanding of these mechanisms continues to grow, and here again, MAB treatments are proving effective.

Although genetics will be the greatest driver of advances in human health this century, it is not a panacea for all human health problems. Health is a complex of interacting systems. The benefits of genetics are also weighted more heavily to future generations, because prevention is such an important component. Genetic therapies are ameliorating conditions in middle-aged and older Americans today that will not even exist in future generations. For example, psoriasis has been brought under control for many via gene therapy. The recent development of an effective prenatal diagnosis, however, means that no future child need be born with the condition.

Genetics and reproduction

Some of the uses of genetics in reproduction are:

- in utero testing for traits and predispositions

- in utero alteration of traits through genetic manipulation, including baldness, schizophrenia, diabetes, sexual predisposition, and some mental abilities

- gene replacement therapy

- intrauterine karyotyping by examining fetal cells in maternal blood

- replacement of defective fetal proteins in utero

- DNA-mediated gene transfer using retroviral vectors, including during in vitro fertilization

- enhancement therapy, such as for height; elucidation of the mechanisms for nearly all genetic diseases and the onset of most cancers and all autoimmune diseases

- automated genetic diagnosis; genetic testing of embryos for traits; DNA probes for diagnosis; identification of the genetic locus for every disease-related gene

---

First step: single-gene diseases and disorders

Almost half of the approximately 4,400 known genetically-based diseases and disorders (up from 4,000 last century)—caused by one or more genes—are now under effective control, that is, there is a preventive measure or therapy available. All human diseases and disorders have had their linkages, where they exist, to the human genome identified. The mapping of the genome has driven a move to more customized health care by enabling advances in diagnosis, prevention, treatment, and enhancement. Some diseases and disorders have been wiped out, in cases where access to care was widespread and affordable. Many rare diseases have been eliminated, as they were easier targets for researchers. More mainstream diseases such as cystic fibrosis and eczema, and conditions such as near-sightedness, have practically disappeared as well.

In some cases, a large percentage, but not all, of a disease's occurrences have been eliminated. Many Alzheimer's disease cases are linked to a single errant gene and can be repaired with gene therapy. Fifteen percent of arthritis cases have been eliminated over the last 15 years with genetic diagnostics and therapy. Researchers project that another 15% to 25% of arthritis cases can also be eliminated if the testing is more widely disseminated and the cost of the treatment is reduced, or insurance coverage of it expanded. Some cases, however, still elude researchers. Similarly, a small percentage of breast cancer cases have long been eliminated while others persist.

The mysterious role of "junk DNA"

As the human genome project found, most DNA does not code for genes. Of the three billion chemical bases that make up human DNA, only 3% code for proteins. The role of the rest of the 97% has long mystified researchers, but the pieces of the puzzle are slowly falling into place.

Experiments in which some of the junk DNA is removed, have shown that it plays a role in normal genome function, because complications develop in its absence. For instance, it appears to mediate and control gene expression, such as when and where they are turned on and off.

Some junk sequences, such as the Alu sequence, have remained the same for millions of years. Natural selection would preserve such a sequence this long only if it served some important function. Still, the details of its role and how it works remain to be discovered.

Sometimes, even if a single gene is found to be responsible for a disease or disorder, there can be many ways it can mutate. For instance, one study found 23 types of mutations that could occur on a gene linked to Type II diabetes. Similarly, although cancers arise from genetic regulation of cell growth going awry, they can be initiated many ways—from viruses, radiation, environmental poisons, defective genes, or combinations of these factors. Another common set of mutations are the repeats, which are like stammers in the genetic message. Triplet repeats, for example, have turned up in many places, ranging from cerebrospinal ataxia and speech impediments. And some disease genes have been identified without an effective treatment coming with it, as is the case with Lou Gehrig's disease.

The economic impacts of these findings have been significant. In the case of osteoporosis, for example, fixing a single gene that significantly increased the risk of getting the disease later in life saved almost $1 billion a year in the 1990s and 2000s—when $10 billion a year in medical bills were attributed to it. Of course, the procedure for fixing the gene cost money too, but far less than the care required when an elderly woman fell and broke her hip. A positive social impact was the addition of an average of three months to women's life expectancy. It has also improved the quality of life of older women, enabling them to stay active in political, social, community, and business life longer.

Identifying genetic predispositions

The chemical, physiological, and genetic bases of human behavior are generally understood. Direct, targeted interventions for disease control and individual human enhancement are commonplace. Brain-mind manipulation technologies to control or influence emotions, learning, sensory acuity, memory, and other psychological states are available and in widespread use. The incidence and severity of conditions, such as hyperactivity in children, have been substantially reduced. Each sensory activity can now be enhanced by genetic treatments where necessary. Near-sightedness, poor senses of taste and smell, and some forms of deafness can now be prevented or treated genetically, assuming that the appropriate prenatal diagnostics are performed.

Genetic predispositions have been identified for thousands of diseases, disorders, and behaviors. Some genes have been found to be responsible for a variety of disorders, such as the P16 and P53 genes for many cancers. P16s are housekeeping genes, which when functioning normally spot nucleotide mismatches and orchestrate repair enzymes. P53s are tumor suppressor agents that serve as brakes on abnormal cell growth. Both are susceptible to mutation leading to multiple types of cancers. Some forms of rarely fatal conditions, such as hay fever, have been discovered to have important genetic components.

Genetics has been used as ammunition in the nature versus nurture debate. Many on the nature side see genetics explaining everything, whereas the nurture side downplays the explanatory power of genetics. Most people fall between these two poles and see genetics as a powerful, but not an all-inclusive, explanatory tool. Current research is emphasizing the linkages, rather than an either-or approach. A team at the Institute for Psychiatric Genetics, for example, found that two organisms can have the identical genotype (the physical genes), but different phenotypes (how the genes are expressed) based on the environment in which they are in. If one version of a gene provides a better protein than another in a given environment, then that version will prosper. The implication is that with our growing genetics capabilities, we cannot only influence the phenotype but the genotype as well. For example, a person allergic to cats only finds out about his or her condition in the animal's presence. In addition to the obvious step of simply staying away from cats, he or she now has the ability to act on this information and use genetic therapy to alleviate the condition—and even enjoy the company of cats.

Questions of whether conditions have genetic predispositions have drawn the social and biological sciences closer together and stimulated joint R&D. A practical experiment underway in Australia—where it is very sunny, without many trees, and with a relatively thin ozone layer—is attempting to genetically predispose people to sun-avoidance.

There have been genetic links found for many behavioral conditions, such as drug and other addictions, alcoholism, sexual orientation, aggression, neuroses and manic depression. Great strides have been made in understanding many genetic origins. Unfortunately, the reality has often turned out less than the promise. For example, great expectations about genetic links to hyperactivity have fallen well below expectations—it turns out to be a complex condition with less than 10% of cases tied to a gene defect. Similarly, the magic genes for alcoholism, obesity, and other conditions have not eliminated these problems, but they have greatly increased the effectiveness of treatments.

There has been a 20% reduction in the crime rate this century. Although the degree to which this is related to genetics is arguable, most criminal justice experts attribute an important role to genetic testing. For example, genetic testing is often used for early intervention to influence the behavior of those predisposed to commit crime. In a finding with important implications for criminal justice, knockout mouse experiments, in which the gene for the serotonin receptor is "knocked out" or replaced by an inactive copy that holds its place, leads to more aggressive behavior in the mice. This research was important evidence for the concept that aggressive behavior in people could be linked to some dysfunctional serotonin systems. Treatments to balance serotonin levels continue to improve. Longitudinal studies with prison inmates are ongoing efforts to fine-tune these therapies. At the same time, studies linking genetics and criminal behavior have often had to tread carefully. It has just been since about 2013 that it has been politically safe to publish results that could be interpreted as having adverse implications for a social group. Even now such results are carefully couched, stating that genetic and behavioral links are primarily tendencies rather than immutable facts.

Researchers are investigating, for example, whether there is a genetic predisposition to accident-proneness. Similar research has been investigating the more general questions about predispositions to certain emotions, cognitive styles, and behaviors. Research indicates that most people change their behavior or alter their environment—such as avoiding smoking or milk or dairy products—after a genetic test reveals a predisposition.

Genetics has contributed to the understanding of personality, intelligence, behavior, psychoses, and neuroses. This understanding has driven as well as been driven by society's greater emphasis on mental and spiritual health and well-being vis-a-vis the physical aspects. There have been many defendants claiming in courts that "the genes made me do it," but the courts have not yet accepted this as a defense—although genetic information is factored into sentencing.

The most positive outcomes have been in education. Some progress comes from the correction of genetic disorders, such as locating and testing for a gene associated with dyslexia and devising the appropriate therapy. More recently, some teachers are designing classes for individual children based on their genetic makeup. A pioneering program at Penn State University found genetic markers that correlate with an individual's intelligence quotient (IQ). Subsequent research has been looking for and finding more and more genetic links to dozens of components of intelligence. The combined applications of this research are enabling custom-made education for some—mostly those who can afford to pay for it.

The links between genetics and brain science continue to solidify. The popularity of psychiatric genetics, first as a field of inquiry, and now one of the hottest majors on campuses across the country, portends strengthening of this link.

Genetics' contribution to brain science

Molecular biology, genetics, and immunology are providing a rich tool kit of new molecular tools for investigating the brain:

- *Antibodies* permit the visualization and localization of specific neurochemicals and cells.

- *Tissue cultures* allow brain cells to be cultured and studied in the lab.

- *Mapping* the human genome has generated new knowledge about the genetic basis of brain structure, development, and function.

- *Genetic engineering* creates tools for research, and treatments for mental and neurological disorders, such as growing neurons in culture and biochemically altering their development potential or function.

Brain science is evolving into a new enabling technology, on a par with genetics. Its rapid advances in the 21st century have carved a central role for it as a shaper of other sciences and technologies, as well as business and society. Geneticists are playing primary roles in brain science, helping to unravel brain structure and function at the level of genetic regulation, brain development, and neurotransmitter-receptor communications. Researchers have developed a genetic map for nearly all recognized brain disorders, along with a rough guide to the type and extent of environmental and psychosocial influences that interact with the genetic template. One of the painstaking successes is progress against the rare brain-related psychiatric disorder known as Wilson's disease, which had schizophrenic-like symptoms and sometimes killed people. The gene responsible for the toxic buildup of copper in this disease was identified last century, but it mutated in many ways and only persistence in identifying the biochemical pathways from gene to brain have enabled the present 55% success rate in reversing its effects.

Bioelectronics researchers have learned how to place embryonic brain cells on silicon chips and induce them to grow along desired paths. The next step, which has been researched without a breakthrough for decades, is to get the brain cells to grow connections to one another—in order to crudely mimic the circuitry of the brain. Enthusiasts believe it may be soon be possible to make biochips that drug makers could use to test new compounds or that may enhance human functions like memory or learning.

Developing tests for genetic flaws

There are thousands of diagnostics commercially available, ranging from those that identify specific diseases or disorders to cognitive and intelligence tests. Some are generic across the population, whereas others aim at specific racial, ethnic, or other groups. Tests have long been available for some diseases such as Charcot-Marie-Tooth disease, myotonic dystrophy, hereditary breast and ovarian cancer, and Alzheimer's Types 1 and 3.

Testing for flawed genes is now a $15 billion industry in the United States, including carrier screening, prenatal diagnostic testing, predisposition testing, and confirmatory diagnostic testing. A basic diagnostic test for the most common and harmful conditions is required for all pregnancies and is covered by all insurance plans. Mandatory diagnostics account for 70% of the industry today, due to their numbers. But the fastest growth is in more sophisticated tests for would-be parents, especially for those couples interested in enhancing their children's traits.

The mandatory diagnostic tests required for children to attend public schools screen for 253 diseases and disorders. Some of the tests are public health measures, such as susceptibility to the resistant forms of tuberculosis. Others are educationally oriented, such as tests for dyslexia, incidences of which, with the help of genetic therapy, have been reduced by 85%. Of course, other diagnostics are available in the schools, but they are administered on the basis of family history, or as a consequence of a student's symptoms and behavior.

Genetic diagnostics are also mandatory for admission to the military and many public service fields, such as police, fire, and education. Many insurance companies cover diagnostics, since detecting a potential condition or predisposition before it becomes a problem can save them money. Additionally, if a person applying for insurance is discovered to have a serious condition, such as a 90% likelihood of lung cancer, the insurance company can apply to the federal government to have the person placed in the national high-risk pool.

The insurance industry has been slow to get into genetic testing. Until 2005, they were using only the results of tests done by others. But as the costs came down, accuracy improved, and social objections lessened, the industry began doing some testing of its own. A key justification by the industry and its supporters for getting into testing was that insurers were willing to develop and pay for some sophisticated tests that would not have been developed at all or for a long time. The industry scored some important successes in collaboration with the big genetics companies, such as Genentech. For example, in 2017, a test for an elusive form of autism was developed that enabled parents to decide whether to risk having children with the condition. Isolating its genetic location also catalyzed research that has just recently led to the first promising therapy. The use of testing by insurance companies would not have passed muster, however, if the arrangement to cover high-risk applicants in a government-sponsored pool was not created by federal legislation in 2011.

A consequence of the boom in genetic testing has been a corollary boom in genetic counseling. The shortage of counselors in the late 1990s and 2000s has been addressed—some say a bit excessively if you take a look at the Network ads. At any rate, there has been tremendous development in the field of counseling over the last 20 years. The fundamental principle is that the person doing the science is not the person doing the counseling. It was learned that the skills required for the former were much different from the latter—and vice versa—and it was best to keep them separate.

Some decisions, such as what to do if one possesses a gene identified as directly responsible for a disease or disorder, have been fairly straightforward for most people. Few people choose to have a child that is genetically doomed. The exceptions are typically for one of two reasons. People are afraid of the results and avoid the tests, or they have religious or moral objections to tinkering with what is seen as God's plan. Even more difficult are cases where tests reveal a condition, but no treatment is available.

Public policy has often lagged the science and technology. Before the High Risk Insurance Act of 2011, people who refused a genetic test were often denied employment. There are still policy vacuums regarding some important genetics' issues. For instance, there is no formal public policy on parents who do plan to have children that diagnostics reveal will have serious genetic defects.

Many analysts regret the lost opportunity to design a proactive public policy covering such questions at the turn of the century, before developments took place that limited policymakers' flexibility. Now, people can assess whether or not they as individuals would be hurt by policy changes regarding genetics. Twenty-five years ago, when genetics was still fairly new and its eventual consequences unclear, there was a chance to make policy based on all people being treated as equally at risk, and subsequently being more willing to share that risk. Now, the federal high-risk insurance pool budget is strained, given its commitment to cover those people who insurance companies are not legally required to insure.

Mapping biochemical pathways and developing therapies and treatments

For many diseases and disorders, the intermediate biochemical processes that lead to the expression of the condition have been clarified—a recent estimate from Roberta Lee, head of the President's Advisory Board on Genetics, was that about 30% of the pathways are well understood. This information is particularly useful when combined with an individual's environmental, medical, and genetic histories. These histories are on record and under full control of the individual. They are protected on the network by cryptography or biometric identification of the user. Of course, some people still swear by their health smart cards, not trusting information only available through the network—despite the overwhelming evidence that you are more likely to misplace a card than lose information on the network.

The focus on biochemistry has created a boom for pharmaceutical companies. Today's pharmaceuticals have evolved generationally.

- The first generation consisted of recombinant DNA (rDNA) and MAB drugs based on making proteins.

- The next generation was antisense technology (the synthetic molecules that were easier to deliver than bulkier proteins) using compounds to block genetic instructions.

- The third generation involved the direct transfer of genes through hazardous and expensive bone marrow transplant or other implantation or injection.

A new generation of techniques is evolving from improvements in these three areas. For example, therapeutic gene pills are being tested now in the labs.

Biochemistry, and designing delivery systems appropriate to individual biochemistries, are the pharmaceutical industry's core competencies. Access to personal information has been essential to their success, as it enables the customization of treatments and dosages to the individual, rather than to a hypothetical average person.

Custom-designed drugs such as hormones and neurotransmitters are as safe and effective as those produced naturally within humans or other animals. Drugs are designed to act on individual cells rather than on organ systems as in the past. Drug manufacturing via genetic engineering offers greater precision, purity, and reproducibility of products, complete process control, the ability to improve or modify products, and the ability to create products that affect only the target site. Drugs made of genetic material, nucleic acids, carbohydrates, and nonexclusively protein-based synthetic molecules, such as PNA analogues of DNA or ribonucleic acid (RNA), are in widespread use.

Although the move to genetics-based treatments forced pharmaceutical companies to reevaluate their inventories to serve World 1 markets, it also opened opportunities in middle-income and destitute nation markets. The customized treatments in the affluent nations have remained beyond the means of most people in the middle-income and destitute nations. Mass-market-designed drugs, however, are affordable. Although some in these nations complain that they have financed the pharmacogenetic revolution in the affluent nations by buying these older generation products, there is little doubt that people's status in the World 2 and 3 societies has been significantly improved. In fact, an unanticipated consequence of the pharmaceutical boom in some less well-off nations has been a resurgence in population growth, as people live longer with the benefit of more sophisticated treatments. In Peru, for example, average life expectancy has increased four years so far this century.

The evolution of antisense

Antisense, or left-handed DNA, targets messenger RNA, binds to it, and keeps an unwanted message, such as triggering cancerous growth, from being transcribed. Antisense was invaluable to alleviating extreme obesity and many other metabolic conditions. Antisense is also used in many products, such as foods that people cannot metabolize, and therefore, do not gain weight from eating.

The path to success for antisense has taken decades. In retrospect, it was found that the following criteria had to be met for success in the marketplace. The criteria included:

(1) synthesized easily and in bulk
(2) stable in vivo
(3) able to enter the target cell
(4) retained by the target cell
(5) be able to interact with their cellular targets
(6) not interact in a non-sequence-specific manner with other macromolecules

---

An important area of progress coming from mapping the biochemical pathways is in alleviating common conditions. Arthritis, for example, has been reduced in severity by genetics-based therapies intervening in biochemical processes involved in the swelling of the joints. The greatest single success in treating cancers to date has been with suicide cells, which are targeted to the cancer via MABs and kill all cells that they sit in—including some healthy ones as well. Delivery mechanisms have also been a heavily researched area, with the key advance being the use of smart materials able to sense and respond to conditions in the body.

Genetic therapies are now routinely preventing and reversing conditions. One of the early diseases that effective genetic therapies were developed for was cystic fibrosis, which once afflicted one in 2,500 newborns. Most victims died before age 30. A therapeutic gene is spliced into a virus and inhaled into the lungs, where it prevents the growth of the deadly cysts.

By 2035 or so, nanomachines (one-billionth of a meter) in the bloodstream will monitor and in some cases compensate for chemical deficiencies. These self-assembled nanovessels are expected to float in the bloodstream and search for and destroy harmful bacteria, fat, or cancer cells.

Developing preventive methods

Prevention today is centered in reproduction. We have complete control of human reproduction, including when and how conception occurs, and what gender traits we want in the child. Based on techniques pioneered with cows, it is possible to detect the genetic defects in fertilized cells just days old via in vitro fertilization. A sample cell is screened for diseases and disorders, and if a harmful gene is found, the cells are discarded. These embryonic treatments and improved birth-control technologies, including the descendants of the RU486 morning-after pill have gradually lessened the once potent political issue of abortion—by avoiding it. Harris polls find that 73% of Americans now favor using gene therapy to improve babies' physical characteristics or intelligence, up from just 43% in 1992.

There was a great outcry in the late 1990s over reports of human embryos being secretly cloned in laboratories. Legislation banning human cloning was passed in 2003, although organs may be cloned. Developing other preventive methods required that legislation prohibiting tampering with genes be repealed, making it possible to alter genes linked to undesirable conditions—so that they need not be passed on to children. Most such laws were repealed by 2010, when legislation establishing a fixed list of genes that could be manipulated was passed. A panel meeting every three years reviews and adjusts the list. It considers the unanticipated effects of manipulating one function of multipurpose genes when making decisions on whether to add a gene to the list.

Genetic prevention has meant great progress over some cruder preventive measures of the past, such as removing breasts or thyroids to prevent future cancers. These procedures are still sometimes necessary, mostly for older people, who were born after prenatal preventive techniques were developed. The selecting out of harmful genes will eventually reduce them drastically, but mutation will always replace them to some low incidence.

Challenge for the next decade: enhancement

The human species is the first species to influence its own evolution. One of the first enhancements was the use of human growth hormone. There was a great deal of debate over whether this treatment was really necessary, because in many instances its use is more cosmetic than vital to health. Today, most insurance plans cover hormone treatment for dwarfism, but not to make short people taller—this coverage is only included in more expensive plans. An interesting social consequence is that height has become a measure of social status for some. Research has historically shown a correlation between income and height. With genetic enhancement, the correlation has been strengthened, as the enhancement therapy runs into thousands of dollars. People in households with incomes above $100,000 are on average roughly seven centimeters taller than people in households with incomes below $100,000.

Basic forms of physical enhancement come through transplanting organs including the heart, kidney, lung, eyes, ears, skin, endocrine glands, nerves, bowels, and pancreas. The organs come from other humans as transplants or from synthetics cloned in the lab. Organs grown in other animals for use in humans have been in R&D for decades without being able to get beyond the trial stage, except for skin and hair.

The trend is drawing attention to mental enhancement—primarily intellectual. Parents lacking math skills, for example, can now shop for genes predisposed to mathematical excellence, and have them inserted prenatally in their children. Other parents are selecting traits such as artistic ability, musical talent, or athletic prowess for their children. Of course, some challenging social questions are bound to arise as genetics leads to increasingly talented and intelligent children growing up in a society in which they are in many ways superior to their parents, teachers, and government authorities. Optimists anticipate a more informed and enlightened society. Pessimists worry about older people being warehoused in communities or homes for the genetically impaired.

In several parts of the world, particularly Asia, the understanding of human genetics has lead to explicit programs to enhance people's overall physical and mental abilities. Some countries, for example China, have adopted eugenics under the cover of enhancement. China long ago banned marriages that might produce children with mental or physical defects. It now forbids couples from having children whose genetic tests indicate having any of several hundred diseases or disorders. Despite the policy of strict control—including jail time for violators—in practice the controls are often evaded. A lesson many Western nations have learned is that public policy fails when it tries to reach 100% compliance. A combination of information and incentives generally proves more effective than mandates.

Slow but steady progress: aging and life extension

More people in World 1 societies are living to their mid-80s while enjoying a healthier, fuller life thanks in part to genetics' role in changing people's behavior regarding their health. Biologists now manipulate the genes involved in normal and abnormal development, growth, and aging, although no fountain of youth has been discovered. It is evident that aging is a complex set of processes and is not likely to be significantly influenced by any one discovery.

Genetics' greatest role in life extension has yet to arrive. Today, the average life expectancy is 85. Genetics promises to extend that average life expectancy to 100 by end of the century, when the first generation of children benefiting from genetics' advances will be reaching the end of their life. Advances include the use of prenatal gene tests that can indicate embryos likely to result in a fatal disease or serious disorder.

Two developments boosting life expectancy are genetics-based advances in reducing free radicals in the body, which have been found to hasten aging, and by increasing the body's production of free radical scavengers. Another promising area for life extension is continuing advances in programmed cell death, in which cells activate intrinsic death programs. Links to genetic activation have been found for a limited number of cells. Research is ongoing in developing genetic programs that cause cancer cells to die and immune cells to live longer.

Identification: the DNA does not lie

DNA identification has been the single most important advance for criminology in the last 50 years. It has contributed to declines in violent crime, the identification of deadbeat parents, and the prevention of fraud. For example, paternity suits in which the accused denied being the parent practically disappeared by 2006, because the threat of the test guaranteed that the guilty party would be identified.

Back in the 1990s, forensic application of DNA became popular in cases of rape, murder, and mayhem where there were blood samples. By the year 2000, the sharp decline in the cost of doing an identity match had led to massive DNA sampling. By 2002, legislation established a program to set up a DNA databank of children. An early application enabled parents to identify their children in cases of kidnapping or accidents.

Fortunately, the legislation involving access to genetic information was taken out of the hands of state governments, where the variations from one to another were quite extraordinary. One consequence was a small population boom in the Midwest that was attributed to people attracted by the generally strict privacy laws in these states. The Genetic Recording Act of 2004 built in substantial safeguards at the national level. One goal was to provide important biological data for individuals. For instance, the act addressed an anachronism of 20th century medicine, in which medical records were owned by doctors or hospitals rather than individuals, and placed medical histories under the full control of individuals.

There were also the epidemiological elements of genetic information that were quite important. Under the act, no use or application of genetic information was permitted that allowed a personal identifier, except at the request or with the permission of the individual. Epidemiology still obtained vital public health information without tying it to individuals, where it could possibly be used against them. Guaranteeing the anonymity of people's records boosted the supply and quality of public health data. In cases where data has been spotty, small royalties have been offered as an incentive.

A number of striking applications developed out of DNA databasing. One is tracking population flows and migrations. By 2012, it was established that the gene pool of self-reported black Americans consisted of about 37%

white genes, and the gene pool of self-reported white, non-Hispanics consisted of about 1.7% black genes. In the so-called "old confederacy" that rose to 13%.

DNA: the ultimate personal identifier

DNA is useful for identification because it is relatively inert, but it can be copied. This inertness is what makes it useful for identification purposes and why it can be recovered intact from ancient times. Controversies over the admissibility of DNA evidence in trials in the 1990s have long since blown over. Every state has developed DNA databases to varying degrees of sophistication. They started with taking blood samples of prisoners. Key data now are routinely noted on birth certificates.

The DNA identifiers code in an interesting way regarding older people. It became voguish around the turn of the century, when the great expectations for DNA identification were becoming real, for older Americans to leave a bequest of their tissue for analysis on behalf of their descendants. To some extent, a still popular process is exhumation sampling. Currently, about 19,000 graves are opened in the United States each year to exhume genetic samples. Both the cost and the legal constraints to prevent frivolous exhumation have kept the number relatively low. Of course, cell banks, in which people can leave a portion of their cells for descendants, should eventually eliminate the need for exhumation.

Unanticipated consequences sometimes arose from the forensic applications of DNA identification in violent crimes. For example, it led to a decline in instances of rape, but a simultaneous rise in murder associated with rape. This presumably was a response to reduce the likelihood of being identified as a suspect. The forensic applications also extended effectively to military injuries, deaths, and accidents, such as plane crashes.

Genetics helps manage the animal, insect, and microorganism kingdoms

The genomes of prototypical animals, fish, insects, and microorganisms, including goats, pigs, fruit flies, and locusts have been worked out. Knowledge of the genome has led to more refined management, control, and manipulation of their health, propagation, or elimination. Other species serve the needs of people while at the same time, the sustainability principle has led people toward a stewardship role for the planet, in which a balance is sought among the needs of people, other species, and the planet itself.

The biological functions and behavior of many animals and insects are manipulated through biochemical or genetic means. In agriculture and plant genetics, as with human health, working out the biochemical pathways dominated genetics at the turn of the century.

There are routine genetic programs for enhancing animals used for food production, recreation, and even pets. There has been a global boom in the goat population, as it turns out to be especially well suited to genetic manipulation. In the affluent nations, goats often produce pharmaceutical compounds, whereas goats in the middle and destitute nations produce high-protein milk. Enhancements in affluent nations are generally aimed at improving people's quality of life, whereas in the less-developed countries, more basic necessities are targeted. For example, work animals are the targets of massive investments in genetics applications and R&D.

Customized livestock

Genetic engineering of animals has enabled total control over livestock reproduction, leading to increased growth, shortened gestation, and higher nutritional value. Farmers can draw on a network cookbook of recipes for custom-designed livestock. They simply call up and obtain the genes they want from databases, transmit them to the local biofactories, and the animals with the desired characteristics are produced and shipped.

All transgenics must pass muster with the International Commission on Animal Care in Agriculture established in 2017. They are subject to an extensive physical examination, typically by virtual reality. Field visits occur only when something is suspected to be amiss. Transgenics must not threaten other species, and must not be in a condition of undue suffering as a result of the engineering—even where it might suit human purposes. The latter condition arose after horror stories of the treatment of transgenics. Network videos of hogs too musclebound to even stand led to public pressure to provide a reasonable standard of comfort for transgenics. The global network of animal rights groups has been effective in identifying and reporting scofflaws.

Transgenic animals, usually cows, sheep, or pigs, are used as living factories to produce needed proteins and other compounds internally or in their milk. These sites were once dubbed "gene pharms." The animals are bioreactors, producing, for example in the case of goats, a vital protein called AAT for fighting emphysema. Farmers also benefit from transgenics and other livestock by using engineered drugs and hormones to increase feed efficiency. Sometimes, the animal's digestion is modified to accommodate available feedstocks. Veterinary pharmaceuticals have boomed alongside human pharmaceuticals.

More varied menus from livestock genetics

Relatively few animals are grown on farms that grow human foods. They are generally handled as separate activities. The somewhat mislabeled fat-free swine came into substantial commerce in 2001. Swine have been among the most popular animals for genetic manipulation in agriculture because their size can be easily controlled from 15 to 100 kilos. Their genetic manipulation has been so fine-tuned that since 2006, at least 85 genetic varieties of swine have been introduced. A brief fad for duck-tasting pork settled down into a fairly steady but small seller. The introduction of chicken swine could not compete with the low cost of poultry itself. The growing popularity in venison in the early decades of the century has been even surpassed by the transgenic beefison, which has all the richness and flavor of venison, but the desirable physical and bulk characteristics of beef.

The economic ramifications have been impressive. Instead of, or in addition to, building a multimillion-dollar factory, pharmaceutical companies can either buy or lease farmland and raise a herd of livestock to breed the needed compounds, saving millions of dollars. The field has come a long way since the controversy over bovine somatotropin, the growth hormone supplementing the cow's natural hormones. In the 1990s, this case crystallized the fears of many people about genetics. A movement started to use milk only from "natural" cows. The agriculture industry was much more careful about subsequent introductions of genetically manipulated products, and over time the introductions engendered less and less opposition, to the point where they are routine and unnoticed today.

Strange bedfellows: animal rights and genetics

Animal rights activists have long protested against the use of animals for experimentation. They were successful decades ago in eliminating testing for nonessential experiments, such as for human cosmetics, by tapping the public's conscience. They were not successful in eliminating animal experimentation where human disease was the subject.

Ironically, given the perception that the activists were antiscience, genetics along with information technology have significantly reduced the need for animal testing. Animal testing has been increasingly replaced with toxicological models using genetics, experts systems, and computer simulations. Some genetic therapies prevent the onset of diseases and disorders, which took away the need for a drug treatment that may have used animal testing. Also, the use of technologies such as polymerase chain reaction (PCR) and cloning enabled the creation of synthetic substitutes for animal testing in some cases.

Transgenics are also adapted to withstand rough environments. Genes from the hearty llama in South America, for example, were introduced into their Middle Eastern relatives, the camel, and vice-versa, to greatly expand the range of each. Some species have been introduced into entirely new areas. The modification of parrots to withstand cold North American temperatures, for example, has been a boon to bird-watchers across the United States. Transgenic pets continue to be popular. Genes from mild-mannered Labrador retrievers, for example, have been bred into the fierce Pit Bull Terrier, and led to a resurgence of this breed once outlawed in many communities. Astromals[tm], four-legged 4-kilo fuzzy creatures with wings that can fly for short bursts, have been a big favorite of kids since 2019.

Experiments to use animals as breeding grounds for human organs continue on a small scale. It turns out that pig organs are the best candidates, but this research has been superseded by advances in synthetic organs. Animal experiments have been more useful in providing insight into their human counterparts. For example, experiments with growth and fatness in pigs in the 2000s vastly improved our understanding of human obesity. Cases of extreme obesity—where people weigh more than 200 kilos—are now successfully treated with a combination of gene and behavioral therapy. Unfortunately, those who could be helped do not necessarily come forward or cannot afford the treatment. Tests on other animals have led to sunless tans (certainly important, given the fragile ozone layer), 20-20 vision and better for the visually impaired, and most welcome for many, a full head of hair for those previously marked by their genes to be bald.

Fisheries and the coming aquaculture revolution

Genetic engineering of seafood has resulted in fish with different tastes and textures. In some cases, single species are directly modified, and in others, desirable qualities from different species are combined. For example, the tunasword, in which abundant tuna fish have been modified to taste like the less abundant swordfish, has brought the once-expensive taste of swordfish into the mainstream diets. A strain of oysters was modified to give it a less squishy texture, which was found to be a turnoff to many people squeamish about seafood. At the same time, another modified strain of nonedible oysters serve as natural filters for pollutants in many bays along the U.S. coastline.

Genetics advances have been helped along with a complementary growth in aquaculture. As fisheries across the globe had to be shut down in order to prevent their complete depletion from overfishing, aquaculture and genetics teamed to fill the gap, boosting output twofold over last century's average production. An unfortunate byproduct was a massive decline in employment for fisherman, and the resultant social problems of trying to retrain fisherman for other work—many were third, fourth, and fifth generation practitioners and strongly resisted the demise of their occupation. Although many natural fisheries have been restored, their careful management and advanced technologies have kept the demand for fishermen low.

Selective breeding techniques used in agriculture are now routinely applied to fish. Hearty species such as the tuna and catfish have become favored due to their suitability for fish farming and their receptivity to engineering. Fish farms prosper at sea as well as on land. Ocean ranches manage fish populations in their natural habitat. On land, tanks extending thousands of square meters house fish in environments designed to maximize their growth for human consumption—as with transgenic livestock, the fish farmers must provide a reasonable standard of comfort for their fish. Another genetics-led boom for fisheries came in krill and kelp, when they were modified to suit people's tastes. Krill dip became a national fad 10 years ago.

Pest management: outsmarting nature's evolutionary wonders

Genetics plays a central role in pest management, which has been evolving toward targeting specific pest species, in some cases targeting particular behavior patterns. Insects have proven remarkably adaptable to human efforts to eliminate them. An arms race between insects and pesticides has been marked by humans winning battles, but insects winning the war. Genetics is turning the tide.

Breeding pheromones into surrounding plants to lure pests away from their intended prey has been gaining favor over the last decade. There have been several course corrections, such as redirecting resources away from breeding in pesticide and herbicide resistance, as sustainability phased out the use of these chemical agents. Almost 75% of the year 2000 market for herbicides and pesticides has been replaced by biological or genetic alternatives. Sustainabilty advocates are pushing for total replacement in a decade.

Pests are now routinely sterilized through genetic engineering to disrupt their populations. Genetically engineered resistance to pests in crops is now common, through techniques such as inducing the plants to produce their own protective compounds. Genetics has enabled advances such as:

- crop plants altered one gene at a time to build in pest resistance

- insecticidal crops (inbred pest killing mechanism)

- crops that better tolerate negative environmental factors such as salinity and wind

- crops that store better

- products highly selective for pest species

- products that can be patented

Insect disease vectors are being targeted through genetic engineering to control their populations and thus the diseases they carry. The virtual elimination of malaria in World 3 took advantage of mixing in mosquitos bred without the capacity to carry malaria with their disease-carrying counterparts. Several generations later, fewer and fewer carried the disease. In combination with other public health measures, malaria cases are down 93% from last century. Other biological control methods used today include the synthesis of effective cost-competitive, pest-specific, safe substances that regulate growth, cause molting failures, accelerate aging, or otherwise breed in disinterest in destroying crops.

Boosting nature's bounty with plant genetics

Scientists have worked out the genome of prototypical plants, such as corn and wheat. This has led to more refined management, control, and manipulation of their health and propagation. The overarching goals are to reduce the time of the breeding cycle, speed up plant evolution, and develop patentable varieties.

Expanding diets through plant genetics

Genetics has expanded the number of foods in the human diet. Back in the 1980s and 1990s, there were about 3,000 edible plants. Three hundred of these were actually consumed around the world, 30 were significant in commerce, and 6 provided 90% of human nutrition. By 2010, the number of foods drawn into commerce had risen to 111, and today it is estimated that there are 212 in commerce in the United States, that number being made up of both modified and transgenic species.

The primary factors holding back many of these foods from entering into commerce was the presence of an unpleasant odor, a poor shelf life, or great difficultly in preparation time and energy. Those factors have been designed out. Other foods are from transgenic plants, not merely natural species that have been improved be genetic manipulation, but species that represent crossovers from distinctly different plant and animal genetic lines.

A second Green Revolution in agriculture

Farmers have near total control over plant genetics. Today's plants are more productive, more disease-, frost-, drought-, and stress-resistant, balanced and higher in protein content, lower in oils, and with more efficient photosynthesis rates or feed conversion rates than their 20th century predecessors. Natural processes such as ripening are enhanced through stimulatory microorganisms.

Genetics customizes and fine-tunes crops, building in flavor, sweeteners, and preservatives, while increasing nutritional value. Most crops are precisely suited to the resources at hand, to climate and other conditions, and to market needs. Many crops have gained 25% to 50% productivity over the last 25 years.

Some crops have been engineered for three decades now, including squashes, cotton, tomatoes, potatoes, sugarcane, soybeans, corn, peppers, peas, wheat, and rice. The money saved through these innovations has been quite formidable. For instance, last century, fungal rice-blast disease was costing rice farmers in Southeast Asia about $5 billion a year—with engineering, it is no longer a problem.

Linking farm and food

The close systems linkage between farm production and food preparation continues to grow. The integration of those two, while far along by 1997, really reached an unprecedented level by 2014. The optimization of food size, color taste, strength, cooking characteristics, and flavor to go through the food preservations processes (freeze-drying, frozen food preparation, open-stand sales, irradiating) have basically transformed the food delivery agency.

Five field tests conducted in all of the 1980s rose to hundreds in the 1990s and to thousands by 2020. The first step in agrogenetics centered around the cloning of known disease-resistance genes. Next was the identification of novel resistance genes and breeding them into populations. A more recent approach has been to genetically engineer plants to produce specific antibodies against their likely disease invaders. An example of this evolution was that crops were once, and in some cases still are, inoculated with nitrogen-fixing bacteria, but now they possess the characteristic through genetic engineering.

Genetics also plays a role in developing or enhancing soils for agriculture as well as environmental restoration. About 15% of the world's farms employ some form of restorative agriculture. A prime beneficiary has been Papua New Guinea, where thousands of square kilometers of deforested areas are now a healthy ecosphere once again and are producing valuable export crops such as mangos and kiwis. Farmers design crops and employ more sophisticated techniques to optimize climate, soil treatments, and plants. Crops are sometimes engineered to fit soils, or soil is engineered to fit crops.

Transgenics are not just in agriculture

Although agriculture has been the prime beneficiary of genetics, horticulture has also received a boost. Transgenic flowers, for example, combine heartiness, fragile beauty, colors, and textures in thousands of new varieties. A 2023 Net survey of the Flower Forum and its subgroups found that 76% of participants had at least one transgenic nonfood plant in their home.

Genetics also accounts for a large share of the millions of dollars spent on seeds, not only for crops, but also grass and flowers. The global seed market is in the tens of billions of dollars and continues to be a rich source of export income for the nation.

Expanding food choices

Foods for human consumption are more diverse as a result of agricultural genetics. Grocery shoppers sitting in front of their screen and ordering today have choices that would have bewildered people last century. Older people in the United States can barely conceal their amazement at how young people not only freely try new foods, but also seem to be in a continual search for the latest novel food. So many plants that once were not candidates for consumption have become standard fare, due to the removal of a noxious trait. For example, seaweed, always a part of some diets, became a widespread choice as its texture was made more palatable to a wider range of people and tastes.

Old fashioned foods linger on...barely

It has been recently estimated that 7% of the population prefer old-fashioned, that is pre-year 2000 plant and animal foods for a variety of reasons, and 3.5% exclusively consume them. The rest prefer them but consume them to varying degrees.

There is far less animal protein in diets in advanced nations today compared with the 1990s. Twenty-three percent of the U.S. population are vegetarians. Protein substitutes or enhancements, such as super-rich fish, are in widespread circulation. Even cultural staples, such as rice and beans in Mexico, are routinely altered nutritionally to have balanced proteins. There are more vegetables carrying proteins, for example, eggplant, potatoes, turnips, and radishes. Health, environmental, and ethical factors in tandem with better-tasting and more-convenient nonmeat foods through genetics have powered this change.

Synthetic and genetically manipulated foods can be matched to an individual consumer's taste, nutritional needs, and medical status. Many older people decry—most with good humor—the guilt- and consequence-free consumption of previously unhealthy foods, such as cake, cookies, and potato chips. A recent Net advertisement extolled the virtues of "extra-salty (artificial), low-cholesterol, cancer-busting french fries."

Food scientists have characterized molecular and structural properties of foods, and have identified structural-functional relationships and defined how these properties affect processing, storage, and acceptance of foods. They have determined the molecular and cellular bases of biological activities in food. Preservation methods have undergone a radical overhaul since the 1990s. Toxicity testing is much more sophisticated.

Genetic probes monitor safety and quality in food processing, such as the ability to detect single organisms, and more accurately set acceptable levels of microbes. This has been especially helpful to risk-benefit analysis routinely used in government risk-management programs.

Fine-tuning forestry

Forestry has drawn on genetic engineering and tissue culture to improve tree species, in the process becoming steadily more like agriculture. Genetic manipulation has resulted in superior tree strains through improved disease resistance, herbicide resistance, and artificial seeds. There have been substantial gains in productivity based on faster-growing species, better-quality lumber, and fine-tuned wood characteristics. The relatively long turnover time for gene testing in the field (5 to 15 years) was sped up by somatic embryogenesis, which enables the rapid multiplication of desirable genotypes.

Species are adapted to specific conditions and environments, including ones previously thought of as poor, or not even considered. Trees are routinely engineered for use in paper, with the important environmental characteristic of allowing nonchemical pulping. The introduction of the nitrogen-fixing-from-air capability into tree crops has been a boon to productivity and cost savings.

Forestry management has been driven by a doubling in global demand for forest products since 2005 and by environmental considerations, such as reforestation, biodiversity, and global warming. Genetics has also been instrumental in the global restoration of many denuded areas. The U.S. expertise has been critical in helping World 2 and 3 societies to get their environments into shape. Many of these nations, influenced by sustainabilty as well as economic opportunity, have come to view their forests and associated biodiversity as national assets.

Microorganisms are everywhere

Engineered microorganisms are used in the production of commodity and specialty chemicals as well as medicines, vaccines, and drugs. Groups of microorganisms, often working in sequence as living factories, produce useful compounds. They are also widely used in agriculture, mining, resource upgrading, waste management, and environmental cleanup. Oil- and chemical-spill cleanups have been a high-profile application that remains important today, even with double-hulled ships and other improved safety measures.

A boost to the application of engineered microorganisms came from the development of so-called suicidal microorganisms. They were developed in response to fears of runaways, particularly in bioremediation of solid and hazardous waste sites and agricultural applications such as fertilizers. Engineered microorganisms can self-destruct by expressing a suicide gene after their task is accomplished. An additional control element came with a detonator mechanism that can be activated in the event of a malfunction. Although they are not 100% effective, with redundant approaches there has never been a problem with large-scale releases.

Newer applications of genetics

Genetics has long been a force in human health, food, and agriculture. Over the last few decades, however, genetics has been having a greater impact across diverse industries, such as chemical engineering, environmental engineering, materials, manufacturing, energy, and information technology. It also contributes to the burgeoning field of artificial life.

The biologizing of chemical engineering

The various genome projects have created databases of molecules and chemical reactions. Chemical engineers have gained greater capabilities over processes at the industrial scale. Chemicals are now routinely devised to mimic and in some cases improve on those in nature.

Chemical engineering has been biologizing over the last 25 years. The challenge for chemical engineers today is to improve their understanding of more complex, weaker biological interactions and apply them to the chemical knowledge base. The hot research areas are in larger molecules, molecular recognition, attraction, and evolution and self-assembly in substances, and self-replicating systems. An expanding branch of chemistry known as combinatorics creates many variations of an existing molecule and then uses the ones that are interesting. Large libraries of molecules based on this approach are screened for effectiveness as drugs or other useful compounds.

Modified genes and biologically-active membranes are routinely synthesized in the lab. Chemical labs also synthesize gene-derived enzyme catalysts and other microorganisms not found in nature. Chemical processes are based more frequently on biocatalysts and their mimics and antibodies, as well as immobilized enzymes and their analogues. Synthetic bioactive agents inhibit or activate enzymatic or receptor functions.

Genetics has been critical to the chemical industry's shift away from bulk chemicals to higher value-added products. Genetics is especially useful for fermentation that produces amino acids, which are in turn used as food additives or as animal feed supplements. Dozens of industrial enzymes are used as biocatalysts in applications, such as producing simple sugars from more complex ones and breaking down cellulose in cotton to soften new blue jeans. Engineered enzymes offer the advantages of working at mild temperatures, producing chiral compounds, being biodegradable, not requiring organic solvents, and producing very specific reactions. Artificial DNA, synthetics not mimicking natural compounds, have brought new capabilities such as DNA chlorination and bromination that eliminate the troublesome byproducts characteristic of previous approaches.

Helping to clean the environment

Genetic engineering is used for a vast range of environmental services—from breaking down toxic wastes to restoring degraded ecosystems. Environmental applications of genetics have been gaining momentum since 2010. Of course, 13% of GDP growth in the United States over the last 35 years has been devoted to environmental cleanup or preservation.

Ecologists and environmentalists use genetics as a source of greater understanding of organisms' interactions in ecosystems and as a key to reclamation or bioremediation. Current research is looking into applications of genetics for totally artificial environments, as in space and seabed stations. More and more proposals for terraforming Mars are presented each year. The Senate held hearings on the subject in October 2023.

Bioremediation now accounts for almost 40% of the hazardous waste cleanup market. Engineered organisms speed cleanup, which would occur naturally but at nature's much slower pace. An important breakthrough for engineered organisms was developing the ability in 2003 to cost-effectively break down polychlorinated biphenyls (PCBs), which were once thought to be virtually indestructible. Microorganisms biologically convert or consume pollutants. Oil spill and hazardous waste cleanup were the first major commercial activities for biotechnology in environmental applications. Typically, the toxins and wastes are broken down to subatomic particles, reassembled, and converted into harmless or sometimes useful byproducts. An interesting application of DNA identification in forensics is determining who and what contaminated a soil. The ability to genetically manipulate organisms has introduced more transgenic species into the open environment, increasing the numbers of fish in the ocean and the survivability of vertebrates on land. There continues to be great interest in transgenic organisms as either threatening invaders or as with the goal of enhancing species or creating new balances among species. There have also been successful, small-scale projects of adapting animals to better suit their environment than the other way around. In cases where a food source was disappearing, genetic modifications have enabled species to adapt to a new food source. Genetic manipulation has presented fewer threats to the environment than many environmental groups had forecast.

Genetics is also contributing to making environmentally intrusive activities, such as mining and petroleum engineering, less damaging. In mining, engineered microorganisms assist in mineral leaching and metal concentration. In petroleum engineering, they ferment and emulsify the oil to help bring it to the surface.

Customizing materials at ever-smaller scales

The growing ability to manipulate materials at the molecular or atomic level is allowing manufacturers to customize materials for highly specific functions, such as environmental sensing and information processing. Genetics has made it possible to make more molecular structures and more complex materials using biological processes. For example, left-handed polymer molecules produced by genetics have properties that are different from their mirror opposites in nature, right-handed molecules. Their ability to attach to their opposites and prevent them from carrying out a harmful process—often referred to as jamming the lock—has made them a key tool in many pharmaceutical treatments.

Smart implants bridge the gap between the biological and electronic worlds. Implants with biomembranes control some of the body's biochemical functions, help target drug delivery, and are integral to synthetic organs and organoids (synthetic organs that induce the body to surround them or infiltrate them with tissues that produce hormones or perform other biological functions). Smart, inorganic materials are sometimes smarter than living biological ones today. Genetics produces polymers capable of bonding directly to living tissue for specialized applications, such as sensor implants to regulate blood pressure.

Japan has been particularly active in biosensors, which are devices that use immobilized biomolecules to interact with specific environmental chemicals. The activity can be detected and quantified through measuring changes in color, fluorescence, temperature, current, or voltage. Biosensors are routinely used in chemical processing and manufacturing for process monitoring and control, as well as further downstream in controlling industrial effluents.

Plants are bioengineered to produce raw materials for plastics, detergents, and food additives. For example, an engineered relative of rapeseed developed in 2017 produces a biodegradable plastic commonly used in food and beverage packaging.

Biologizing manufacturing

Manufacturing is becoming more like breeding. The expanding range of manufacturing applications includes molecular engineering for pharmaceuticals and other compounds, nanotechnology based on biological principles including self-assembly, rudimentary DNA chips, and biosensors. Biomanufacturing officially became a new subcategory within manufacturing in 2015—earning its own standard industrial classification (SIC). It now accounts for 1% of manufacturing under the revised classification scheme. Its share is expected to quadruple within a decade, especially as genetic engineering techniques are applied as part of custom production.

Enzymes, which were once too fragile for manufacturing applications, are now in widespread use. A Japanese approach pioneered in the 2000s is to subject enzymes to harsh conditions and cause them to mutate, so that they evolve an ability to withstand harsher conditions. A more recent innovation in 2021 was adapting microorganisms in the deep ocean that live on sulfur-based rather than oxygen-based systems for use in manufacturing processes involving sulfur.

A key consideration in biologizing is society's commitment to sustainability, which has driven a search for environmentally benign manufacturing strategies. Biological approaches, while typically slower than mechanistic ones, are more sustainable. The nondurable goods sector has been biologizing over the last two decades. All industrial enzymes are produced by genetic engineering today. Recombinant DNA has long been used in cheesemaking, wine-making, textiles, and paper production. Bioreactors, in which engineered living cells are used as biocatalysts, are now being used for new kinds of manufacturing, such as making new tree species, in addition to traditional applications in food processing, beverages, and chemicals. An especially attractive feature is that they can produce compounds that cannot easily be produced synthetically, due to high cost or environmental effects.

Nanotechnology is finally coming closer to delivery on some of its exotic promises last century. It is expected to be an increasingly useful tool for manufacturing in the next decade. Research has been focusing on methods for building materials and devices by manipulating atoms or molecules directly or through chemical or biological means. Self-assembly offers tremendous advantages in control and economy over conventional manufacturing. Self-assembled structures put themselves together based on attractive and repulsive forces between molecules, and could use a beaker on a table top rather than billion-dollar plants with clean rooms and vacuum chambers. Biological systems such as ribosomes remain the inspiration and models for assembler designs. Self-assembling nanomachines have been successfully programmed in the lab to build other nanomachines, but still too slowly to be commercially viable. The revolutionary potential is to change manufacturing from the top-down approach—using big materials and miniaturizing them by cutting, pounding, and winnowing—to a bottom-up approach—starting with atoms and biologically building them by self-assembly into molecules, and increasingly complex units.

Disappointing results in fueling alternative energy

Biomass energy has not lived up to the potential that many had forecast for it last century. In the United States and other World 1 nations, crop biomass has a small niche market. It has generally been unable to win out over competing uses for land—food and other commodity crops still typically earn more money than energy crops. In middle and destitute nations, the use of biomass has dropped almost by half, although this is explained by the fact that the unsustainable practice of using firewood and charcoal for fuel has been greatly reduced. The genetics revolution in biomass has bypassed many World 2 nations that lack the financial or technical resources to use it. Some middle nations, such as Thailand, however, are moving fast in these areas (see the middle and a destitute nation story later in this chapter.)

Energy from forests can be converted to natural gas with the aid of genetically engineered microorganisms. This gas can then be used in highly efficient gas-fired turbines. Genetics assists in energy conversion, including fermentation and other liquefaction techniques, as well as gasification tailored to feedstock species. An early technique was short-rotation wood-crop technology using genetic screening, physiological studies, and databases to improve yields. Genetics also plays a role in municipal solid waste conversion and processing, as well as producing energy from waste materials.

In the late 1990s, the discovery of microorganisms that thrive in petroleum provided a boost to enhanced oil recovery. The study of their structure and function led to the design of closely related engineered microorganisms that are pumped down into a well and sealed in. That leads to fermentation and, with the addition of water, promotes an emulsion that enhances production. Billions of barrels of oil in the United States have been recovered through enhanced recovery that otherwise would have remained in the ground.

Genetics, as pure information, links to information technology

Linkages between genetics and information technology are growing as researchers take advantage of the fact that genes are pure information. Information technology has long been a key enabling tool for genetics, handling the massive health-related genetics information in health smart cards, for example. A whole new discipline dubbed bioinformatics, or the science of biological computing, has sprung up to meet the enormous information needs of genome projects. The National Center for Biotechnology Information at the National Institutes of Health (NIH) and the European Bioinformatics Institute have worked together to integrate bioinformatic databases across the globe. Genome maps of any species are readily available to download from the Net practically anywhere. Other important applications over the last 35 years include:

- Artificial intelligence to determine changes in gene patterns

- Robots for doing genome sequencing and mapping and DNA replications

- Computer modeling and virtual reality for predicting molecular structures, which has been especially useful in pharmaceutical and in food processing for choosing enzymes, additives, and micronutrients

---

Evolutionary software engineering

Genetic algorithms are now commonly employed in computing. The principles of evolution are applied to computer coding. Pieces of computer code are viewed as analogous to chromosomes.

This approach has been used to program miniature robots by randomly generating the so-called chromosome codes and then selecting out the appropriate sequence when the robot behaved as desired. The chromosome pools of programs most closely matching the behavior are in turn combined in a process similar to reproduction.

Although the approach works, many trials are necessary to get the desired behavior. It is very slow and works only under limited conditions.

Genetics and information technology are now physically working together in advanced experimental computers. Biophotonic computers using biomolecules and photonic processors have set speed records for the fastest switching available. They are still in the laboratory, however, due to their high cost. Biomolecules appear capable of meeting the central hardware challenge of miniaturization, which has busied R&D labs since 2015 when Moore's Law (that computer capacity doubles every 18 months) finally collapsed as conventional semiconductor technology reached its physical limits. Bacteriorhodopsin's engineered descendants have the attractive features of serving as switches in logic gates, working in tandem with traditional silicon or gallium arsenide semiconductors. They are also being used in liquid crystal displays. Research continues into possible applications neural networks, although many scientists feel this particular path is an ill-fated detour. Analog biochips, which have been investigated since the early 1980s, have finally been yielding practical applications for tactile pattern recognition in computing environments using virtual reality over the last decade. They are synthetic organic organizations of molecules that in some cases can perform a chip function faster with greater storage density and smaller size.

Regulation supports genetics worldwide

Genetics continues to receive a great deal of regulatory attention. Fortunately the attention has been focused more on promoting better genetic science, technologies, businesses, and products, rather than churning old fears of runaway plagues or other genetic disasters.

Plant genetics and the associated risks are nothing new

Risk analyses of concerns about the release of genetically altered species concluded that humanity has long been altering plant genetics through traditional breeding practices. They added that no unique threat was posed with genetic manipulation. What is different today is the dramatic leap in capabilities to speed and enhance the process. Instead of waiting for generations for alterations to take effect, genes can be altered on the spot.

There have been localized problems with new breeds or new pest management strategies. Who can forget the Idaho potato blight of 2013, when an engineered microorganism targeting a potato pest turned out to rot the potatoes themselves. Fortunately the damage was contained to a couple of counties, and within three years the microorganism was wiped out.

The regulation of genetics has been a driver of the global management of global issues. This is a case where business has been out in front in calling for regulation. The regulatory situation 10 to 15 years ago was a mix of conflicting and confusing stipulations across hundreds of jurisdictions. What businesses were doing in one part of the world was illegal or prohibited in another. An early leader in international regulation was the United Nation's Codex Alimentarius Commission (CAC) that in 2015 adapted its charter of setting international standards for food products to include genetics or biotechnology products. Their voluntary standards were often adopted on a large enough scale to make de facto international standards. A promising development just 10 years ago was the renaming of HUGO, the Human Genome Organization, the international genome mapping agency, to GRO (Genetics Regulation Organization) —and the adoption of a new mandate, to harmonize conflicting genetics regulations. It is teaming with the sister regulatory bodies and is scheduled to promulgate the International Standardization Organization's latest work, ISO 46000 for genetics, along the revised previous lines of standards for manufacturing quality (ISO 9000), the environment (ISO 14000), telecommunications (ISO 22000), and energy (ISO 35000). It is a step forward, but the journey will be a long one.

DOTE leads the way in risk management

The Department of the Environment (DOTE) is the furthest ahead of any government agency in risk management. Over the last three decades, regulators have often responded to public fears of genetics inflamed by sensational media reports, rather than the available scientific evidence. Public fears generated political pressure that was in turn directed at regulatory agencies, sometimes forcing them to act counter to their best thinking.

Rather than wait for a crisis to develop, however, DOTE took advantage of research that clearly indicated that effective risk management should incorporate people's perceptions of a risk, even if they appear irrational or alarmist. They have been pioneering a shift in risk management from New Deal style paternalism of a 100 years ago, in which government knows best and takes care of everything, to an enabling model, in which regulators work closely with affected people and provide the most reliable information. People are involved in making the decisions that affect them. DOTE risk managers have been leading the way with this approach. Rather than dismissing people's fears, they are acknowledged. Often the fears are based on mistrust of the agency. It has taken years of providing credible information to build that trust, and maintaining it is an ongoing challenge.

Even the FDA, once one of the stodgiest paternalistic agencies, became so open and experimental to deal with the AIDS crisis last century, that it sometimes surged ahead of the activists demanding its reform. The same principle has applied to businesses that deal with DOTE and a local community together. Addressing constituent concerns and building a relationship based on trust also applies to businesses beyond risk management situations to being a key part of dealing with any customers or potential customers.

Because its effects are so far-reaching, jurisdiction for regulating genetics in the United States, unfortunately, remains dispersed across 10 agencies and institutions, although this was an improvement over the dozens of agencies a decade ago. Interagency cooperation is routine, leading to the gradual integration of state and local regulation at the federal level. In many states, it became politically popular to ban anything related to genetics. The first indicator of international resistance to gene-based change was banning dairy or beef products derived from cows given growth hormones. Later, legal restrictions on gene research, such as not being able to intervene in inheritance, were overturned. The antigenetics movement at the state and local levels gradually lost steam as genetics became demystified. People paid greater attention to it in schools, the workplace, and in public education campaigns.

The Departments of the Environment and Agriculture have overseen tens of thousands of field tests of engineered organisms since they began in the 1990s. A precedent-setting step in interagency cooperation also came in the 1990s when NIH's Recombinant DNA Advisory Committee voted its gene therapy subcommittee out of existence and forwarded new protocols to the Federal Drug Administration's (FDA's) Recombinant Advisory Committee (RAC).

Proposals to create a Department of Genetics or Biotechnology continue to spring up. So far, proponents have been unable to provide a compelling reason for consolidation. The strongest driver for consolidation at the national level turns out to be international pressure.

Cooperation spurs national and international genome programs

The human genome project was the big driver of today's advances in genetics. It was an international effort, although the bulk of the research was carried out in the United States. The genomes of most important species have been mapped, but work continues on mapping more of the planet's commercial species. The corn and cotton genomes were completed in 2000, rice (by Japan) in 2002, wheat (by Russia and the United States) in 2004, chicken and cattle in 2012, and the pig in 2014. One can think of the genome map's contribution to the understanding of species biology as comparable to that of the periodic table of elements for chemistry.

Gene hackers: the new breed

Despite tight controls over genes and the technologies for manipulating them, a new breed of hackers—gene hackers—has been raiding databanks and laboratories to create novel organisms. Incidents have been rare. The media attention they get makes them seem more frequent than they actually are.

There have been some gruesome results, some right out of science fiction—three headed frogs, stinky plants giving off incredibly foul odors, and of course, microorganisms such as the bacteria that poisoned Winchester, Massachusetts' water supply in 2019, killing a dozen people and making thousands ill. There are also heroic accounts. A gene hacker in 2007 secretly worked on a therapy for her fatally ill spouse, only to die herself when the effort went awry. This incident inspired the international blockbuster movie, *Altering Destiny*.

Automation greatly speeded the task of identifying and marking the three billion units of DNA that make up the almost 100,000 genes arranged over 23 pairs of chromosomes. If human genome were compiled in paper books, it would take 1,000 volumes the size of an old King James Bible to hold it all. Each human genome differs slightly from every other—roughly 0.1 of 1% or about three million nucleotides. Today's catalog of the human genome is a mosaic of a hypothetical average person. The human genome project was undertaken in the 1990s and was completed right on schedule in 2005. Whenever it looked like the goal would be missed, new advances, notably genomic mismatch scanning (GMS) and representational difference analysis (RDA) came along to speed the process.

Patent law evolves to keep up with genetics

U.S. law once gave the broadest protection to inventors, but the gradual harmonization of international patent law has lessened the differences among nations. Patent enforcement remains an issue as some countries, notably China and Korea, still resist. International agreement was necessary. For instance, the three basic tests are still (1) novelty, (2) not obvious, and (3) utility. The difference today is a stricter criterion for demonstrating utility. A precedent-setting case in 2001 was the invalidation of many cotton patents claimed by W.R. Grace, once a large multinational, because they were deemed to be too broad in seeking rights to any cotton created or modified by genetic engineering.

Patent law once prohibited the patenting of anything natural. The argument that eventually prevailed is that genes are not natural, even though the organism from which they are taken may be. NIH had sought to patent the human genome to prevent private entrepreneurs, and especially foreign capital, from controlling what was being created with U.S. public money. But when the U.S. Patent and Trademark Office rejected NIH's patent applications for 315 segments of gene codes in 1994, NIH decided not to appeal and to drop its other patent claims, believing that successful patents would impede research instead of stimulating it.

The graying royalty system

From the middle to the end of the last century, the most successful genetically manipulated food was hybrid corn, hybridized and grown by traditional, old-fashioned methods. The success of the product depended on two characteristics. The first was that the product itself was superior in all regards, and the second was that the farmer had to go back to the hybridizer each year for seed, thus tying it to an economic reward structure (the price of the seed). Virtually no genetic improvements were made in self-propagating crops, with the exception of several, notably rice, directed at poor nations' economic well-being. With the development of rapid, high-speed, low-cost DNA analyses, it became practical to identify the makeup of a carload or shipload of grain very easily, and the current national as well as international law provides a 25-year royalty fee for the developers of successful-in-commerce, self-propagated hybrid grains.

It is now estimated that 60% of all U.S.-grown and marketed grains and 34% of worldwide grains in international commerce are part of that royalty system. Royalties are in an unusual pattern; they start off low in the first five years of the introduction of the grain to propagate its use and they peak in the 12th to 17th year of use, and then gradually decline and fall to zero in the 26th year. In rank order of importance using the international royalty system are wheat, oats, rice, and rye.

The next part of the story involved a key researcher in patenting at NIH leaving to head a private, nonprofit venture, The Institute for Genomic Research (TIGR), and its profit-making ally, Human Genome Sciences (HGS), with ties to SmithKline as well. TIGR/HGS amassed a gargantuan amount of human expressed sequence tags (ESTs), which are gene fragments that can be used to rapidly track down genes of potential pharmaceutical value. The competition heated up as other big pharmaceutical firms, such as Merck, threatened to band together and bankroll a public database of ESTs. HGS eventually agreed to share the information with researchers if they gave HGS the right to be involved in any subsequent commercialization. A researcher looking for a defective gene similar to a certain known bacterial gene, for example, simply called HGS to see whether any of the genes in its files were similar, and, for a fee they would receive an answer and the necessary data. If the researcher subsequently used that information to develop a treatment for cancer or diabetes, HGS would seek a share in any subsequent marketing.

The impressive genetics tool kit

The basic tools of genetics have enabled advances in observing, locating, mapping, identifying, labeling, isolating, separating, describing, sequencing, altering, mutating, changing, moving, transferring, replicating, synthesizing, and designing. At the same time, homage should be paid to the many bacteria, the fruit fly Drosophila, and the mouse for being the testbeds to developing these capabilities.

THE WORLD 2 AND WORLD 3 NATIONS—A MIXED STORY FOR GENETICS

There have been some tremendous success stories as well as disappointments in the middle-income and destitute societies. Technology transfer remains a critical issue, especially for food production and distribution in the destitute nations. The hoped-for development of a "miracle" nutritional food has failed to materialize, but there have been sporadic success stories. Successful applications of genetics in the middle and destitute nations include:

- Transgenic livestock adapted to new, often harsh, environments

- Embryonic biomass industries based on engineered feedstock crops

- Near eradication of some infectious diseases, notably malaria

- A genetics-fueled second Green Revolution

2025

- Unarable lands brought back into production

- A tourism boom spurred in part by genetic tools for enhancing biodiversity

- Reforestation and ecological restoration preserve valuable sources of pharmaceuticals for export

- A growing presence in international seed and plant markets, including a breakthrough for the papermaking industry

Despite genetics advances, destitute nations such as Haiti and Egypt still experience occasional episodes of widespread starvation. Many World 2 and 3 nations' leaders maintain that agricultural genetics has actually hurt their economies by developing new crops that lessen the need for their indigenous ones. Farmers there have been hurt by substitutes for their unique export crops, brewed up in affluent nation bioreactors.

At the same time, transgenics have been exploited throughout Worlds 1 and 2. Engineered goats, for example, have prospered in several climates and terrains. In addition to being readily adaptable to new environments, they are leaner, their milk is less fatty, and they use less space than cows. Breeding out their predisposition to rapacity has made them an ideal livestock.

Many middle nations have developed niche industries in genetics, but they generally have not posed a threat to affluent nation dominance. Thailand's amazing recent breakthrough in biomass energy, however, may change this situation within the next decade. All in all, genetics has been neither a panacea nor a poison for the middle and destitute nations. It is likely that its effects will be felt more strongly over the next decade, as products and services from the affluent nations continue to spill over into Worlds 2 and 3.

Genetics has not played as large a role in health in these nations as in the affluent ones. Middle nations have made a transition to the mass medicine once practiced in the affluent nations. The transition to custom medicine is likely to be made over the next few decades as people become increasingly well-to-do and able to afford the latest technologies.

Genetics has been most useful at the public health level in combination with education and other measures. Most nations have instituted at least a rudimentary form of diagnostic genetics testing and counseling. The benefits include not only healthier populations, but smaller ones as well. More families are accepting the idea that they need not have as many children as their parents. Large families were often a hedge against the fact that many children would not survive. As the survival rates for each child increases, the need for a large family declines. Another public health-based driver of genetics medicine has been the problems of megacities. The close living and working quarters promotes the rapid spread of infectious diseases. Engineered immune system boosters have been effective in sporadic cases, but have not yet been distributed widely enough to achieve their potential.

An ironic side effect of genetics-led advances in medicine and public health in some World 2 and 3 nations is that genetic disorders have been accounting for an increased proportion of disease. These regions have a far greater percentage of consanguineous marriages (between blood relatives), roughly 20% to 50% of marriages in some parts of Asia and Africa, which lead to greater genetic complications. This greater share of genetic disease has increased the importance of developing cost-effective genetic therapies for these regions.

India's second Green Revolution

Genetics has been a tool for igniting a second Green Revolution in agriculture. India's prospects at the turn of the century were marred by a burgeoning population drawing on increasingly tired and overworked cropland. Disaster would most certainly have struck if not for the introduction of synthetic soil supplements, crop strains that could accommodate the land's existing conditions, and widespread application of integrated pest management techniques. Another important tool was bringing underutilized varieties of plants with high yields and high nutritional value into people's diets. Genetic engineering for protection against pests attacking food in storage—historically a critical shortcoming in World 3 food systems—improved food availability by 20%.

Genetics plays an important role in developing synthetic soils that combat erosion and in some cases have made the unarable arable. In combination with new strains of plants with high yield and fast growth, the yield from marginal lands shoots up.

India has also been making a foray into commercial genetic engineering of flowers. It is now the world's leading producer of petunias. These and other flowers have inbred pest and heat resistance critical to success in the Indian biosphere, thanks to genetic engineering.

Kenya capitalizes on its biodiversity riches

Kenya today is the biggest tourist attraction outside World 1. It has been turning around its desperate status as a destitute nation from the last century. It is moving into the middle cut of nations—based mostly on political and economic reform following the death of President Daniel Arap Moi—but in part due to its use of genetics as a tool for the conservation of its rich biodiversity. It has worked with environmental organizations in World 1 to pioneer the applications of genetics to help indigenous species adapt to their human-modified environments, as well as modifying the environments to better suit the animals. They have also pioneered the storage of gene samples of species. The tourism associated with wildlife has long been the nation's number one industry.

Lions and elephants were on the road to extinction in the wild at the turn of the century. Dealing with poaching was certainly a critical factor, but genetics had a saving role as well. In 2003, the hundreds of lions remaining in the wild were attacked with a virus that threatened to wipe them out. DNA typing identified the responsible microorganism, and as many lions as could be approached were given a genetically developed vaccine that enabled them to fight the virus off. With elephants, the breakthrough was using genetics to boost their food supply that had been dwindling for years.

The nation discovered that adopting a systemic approach to maintaining and enhancing the biosphere worked much better than piecemeal or bandaid solutions that tended to favor one species at the expense of another.

Brazil's pharmaceutical and chemical cornucopia

Like Kenya, Brazil has found economic opportunity in protecting and enhancing its biodiversity. Brazil's niche has been in pharmaceuticals and other chemicals. It is tapping its lush tropical forests, which are storehouses of over half the worlds' plant and animal species, where in the past they were recklessly cutting them down. Brazil has also diversified into genetic engineering of crops as sources of new pharmaceuticals to supplement its natural stores.

Genetics-based forest reclamation strategies have been central to the nation's success in exploiting its biodiversity. Genes for rapid growth have been engineered into native rainforest tree species. In tandem with environmental engineering and ecological restoration projects, 53% of the rainforests once thought to be lost forever, have been restored.

Thailand's strong commitment to becoming the genetics tiger

While other Asian tigers such as Singapore and Taiwan have pursued their fortunes as global leaders in information technology and financial services, Thailand has been pursing the genetics path. There has long been relatively strong support for biotechnology among the population. Thailand first built a presence in the international trade in seeds and plants, which was once largely in the hands of the United States, Netherlands, and Italy. A key development for this industry was the development of genetic treatments adapting seeds and plants to Thailand's relatively hot climate. Thailand, and to a smaller extent India, is rebalancing this trade from affluent nation dominance to a more equitable distribution.

Thailand has risen to be among the world's leading producers of biomass energy based on genetically engineering feedstock crops from agricultural wastes. This ideas has long been investigated, but only became a commercial winner in the last decade with the development of an engineered microorganism that converted ordinary waste into feedstock for fuel cells—an increasingly important energy source in remote areas throughout the world. In addition, just last year a group of Thai researchers working in tandem with a U.S. team, pioneered the biomass-fed "silkworm" microbes for producing uniform, high-quality fiber for papermaking.

Critical Developments, 2000-2025

Year	Development	Effect
2000	Widespread use of antisense technology using compounds to block genetic instructions.	Genetics-based medicine breaks into the main-. stream
2001	The hectare of farmland dedicated to raising transgenic animals goes into production.	Huge costs savings in raising farm animals as biofactories compared with building multi-million-dollar plants.
2004	Genetic Recording Act of 2004 passed.	Substantial safeguards for people's genetic information reduces social resistance to genetic testing.
2005	The human genome completely mapped.	The knowledge base for medical intervention is established.
2005	Routine genetic testing and	Entrepreneurs respond to, and saturate, market counseling available. demand.
2006	Corn and cotton genomes mapped.	Genomes of other important species continue to be completed.
2010	U.N.'s CAC standard for several bioprocessed legumes adopted as de facto international standard.	Demonstrates that international cooperation is possible and can benefit all.
2011	High Risk Insurance Act passed.	Government-backed insurance pool to cover people diagnosed with genetic disease or disorder.
2012	Deadly fungal rice blast eliminated.	A billion-dollar problem disappears.
2014	Restoration of the mastodon.	Entertainment value of genetics established.
2014	Malaria virtually wiped out in the destitute world.	One of the toughest killers is finally stopped.
2015	U.N.'s CAC adapts its charter to include standards-setting for genetically altered foods.	Begins the institutionalization of international regulation of genetics.

2015	Identifying biochemical transmission pathways of genetic information.	Building on the genome map to further progress toward treatments and remedies.
2017	GRO formed on the skeleton of HUGO to oversee international regulation of genetics.	Institutionalizing cooperation on genetics matters.
2019	Map for all genetically-based brain disorders completed.	Integration with brain science forms the basis for a new enabling technology.
2020	Widespread therapies for single-gene disorders.	Genetics knowledge is increasingly translated into effective actions.
2020	Identification of genetic predispositions nearly complete.	Substantial progress in the thorny task of separating nature from nurture.
2023	Chemical composition for all human genes fully identified.	The basis for therapies and treatments is fully in place.
2024	ISO 46000 standards for genetics promulgated.	ISO international standards have long tradition of aiding global commerce.
2024	First successful mass-market gene pill.	Overcomes the delivery problem that has hampered gene therapies.
2025	Genetics the second largest industry in the United States behind information technology.	Indicator of U.S. evolution into knowledge industries.

Unrealized Hopes and Fears

Event	Potential Effects
Runaway genetically engineered microorganism causing death and destruction.	Wipeout of large segments of people, plants, or other species.
Many human diseases wiped out through genetics.	Savings in health spending, but potential for a population explosion at the same time.
Pollution-eating microbes eliminate hazardous and nuclear waste dumps.	Resolution of key environmental issues affect the vitality of the planet.
Biomass energy crop breakthrough. environment.	Provide energy without harming the
Discovery and manipulation of an aging gene that dramatically boosts life expectancy.	Profound reshaping of societies accustomed to 85-year lifespans.

5

POWERING THREE WORLDS

Energy powers the world's industries, residences, businesses, and transportation systems. It is one of the four great enabling technologies—along with information technology, genetics, and materials—that have been fundamental drivers of science and technology developments over the last 30 years. Its central importance was often overshadowed by its casting as a villain in environmental problems. Reducing energy use absolutely and *per capita* has emerged as a global value. Most countries now accept the notion that economic growth and reduced energy use are compatible.

In the past, nations ignored energy efficiency at their own peril. Industries that were energy inefficient were at a disadvantage in the global marketplace, aside from the cost of environmental damage. Energy efficiency has become an important element of competitive economies. It is no coincidence that the world's most energy-efficient economies—Japan, the EC, and, more recently, the United States—have been the world's most successful economies as well.

Most people's daily lives have not been noticeably affected by the changes in energy over the last 30 years. The gradual reductions in energy use since the 1990s were subtle enough to go almost unnoticed. Only the crisis of 1999 jolted daily life. In the United States, gasoline lines returned, thermostats and energy-management-system (EMS) settings were programmed for maximum efficiency, and sales for old-fashioned woodstoves boomed. The crisis was less severe than those of the 1970s, but worse than in other nations that had taken stronger steps in the 1980s away from dependence on oil.

Energy efficiency was seen as an integral part of sustainability, as people became aware of the consequences of unchecked growth in energy use. Politically divisive views once labeled "doom and gloom" moved into the mainstream, as scientific evidence of their accuracy accumulated. This was a gradual process, which was occasionally jump-started by events such as the fall of the Saudi Kingdom and the scientific confirmation of global warming. Energy prices for the most part rose slowly and steadily. It was not until clear economic incentives for efficiency could be demonstrated that changes were made, and even then change was gradual.

The widespread concern for the environment's future led great numbers of people, primarily in the advanced nations, to both personal and legally mandated commitments to conservation.

The international energy picture—a world divided

World energy use is 464 quads today, up from 349 quads in 1990, a 33% increase, in contrast to a world population increase of 57%. Consumption would have been higher, given the increases in population and economic growth, if not for gains in energy efficiency. Although it was becoming apparent even in the 1990s that growing energy use would have unacceptable environmental—and economic—consequences, many were pessimistic about the prospects for constraining energy growth.

A note on numbers

Energy figures available in the 1990s were unreliable. Organizations used different methods and measures and often came up with different numbers for the same situation. Some estimates were based on surveys of individual countries, whereas others were derived by international organizations from unreliable national reports. Follow-up work typically revealed greater use than reported. Biomass energy use was not even included in most global accounting until after 2000. It turned out that biomass use steadily declined in middle and destitute nations as they shifted to more reliable sources.

The sharp contrast between the situations of the World 1 affluent and the World 2 and 3, middle and destitute nations dominated the last 30 years. In 1990, the affluent nations used 70% of the world's energy, more than twice as much energy as the middle and destitute nations. Today they account for only 40% of world energy use.

Most analysts expected World 1 nations to slowly reduce the growth rate of energy use and then go on to reduce overall energy use. World 2 and 3 nations, however, were on a much different course. Their soaring populations and rising economic growth rates led to the big increases in their energy use. It was inevitable.

It turned out that affluent nations reduced their energy use faster than expected, and middle and destitute nations' use grew more slowly than expected in the 1990s, largely due to conservation measures. Hopes that middle and destitute nations could leapfrog to advanced, efficient energy technologies were realized in only a few cases, notably China. More and more of these radical transitions have been taking place in the past decade.

The affluent nations' emerging transition to zero energy growth, and even net reduction, is similar to the demographic transition toward zero natural population growth that World 1 nations, other than the United States, have passed through. World 2 and 3 nations, holding 85% of the world's population, have been moving toward the demographic transition, but are at least a generation behind the affluent nations. Similarly, the middle and destitute nations are likely to be a generation behind in the energy transition.

How we got here

The key forces and events that shaped the past generation of energy circumstances are:

- Population and economic growth pressing on global energy consumption

- A prolonged shift away from fossil fuels to nuclear and sustainables

- Slow, steady progress in constraining energy's contribution to global warming

- Sustainability goals institutionalized in affluent nations; conflict between industrialization and the environment in middle and destitute nations

- Energy efficiency proved itself in the marketplace

- The Organization of Petroleum Exporting Countries (OPEC) collapsed amidst political turmoil

Population and economic growth pressing on global energy consumption

Faster population and economic growth rates combined with rising energy intensity to drive up energy use in the middle and destitute countries faster than the affluent countries. The population figures, the *per capita* energy use rates, and the quad numbers for the world are listed in the table below. China and the United States are included because they are the subjects of case studies included in this chapter.

2025

	Pop. 1992 (in billions)	Pop. 2025 (in billions)	*Per capita energy use: 2025 as % 1990*	quads 1990	quads 2025
World	5.6	8.4	91	349(+35)	464
World 1	1	1.3	66	238(+7)	184
World 2 and World 3	4.5	7.1	160	111(+28)	280
United States	.255	.328	66	85	69
China	1.2	1.6	160	29(+3)	64

The figure below compares energy consumption in the affluent and middle and destitute nations today and in 1990. ( ) = biomass quads

World 1, 2, and 3 Nation's Energy Consumption: 1990, 2025 In Quads

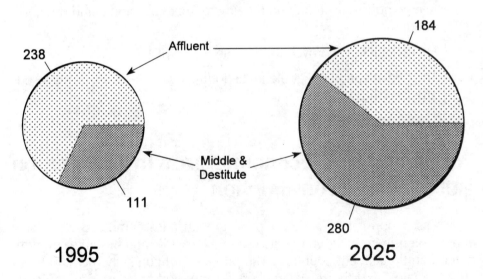

Today's consumption of 464 quads of energy would not have been a significant surprise to last century's mainstream energy forecasters; what would have been a surprise is the fact that World 1 nations now use far less energy than the middle and destitute nations. The 464 quads would have disappointed conservationists. Many projected that overall energy use might decline, as it did in the advanced nations. Forecasters on each extreme undervalued each other's primary assumptions. The mainstream forecasters underestimated the success of conservation and efficiency in the affluent nations, while conservationists underestimated the forces of population and economic growth in Worlds 2 and 3.

A prolonged shift away from fossil fuels to nuclear and sustainables

It has only been in the last 20 years that the long-term trend away from fossil fuels and toward nuclear energy and sustainables has intensified. Taking a longer 30-year perspective, coal and natural gas use increased 30% and 38% respectively, whereas oil use decreased 21%. The increases in nuclear power and sustainables have been far higher, at 235% and 162% respectively, but they started from a low base and have been expanding only recently. The changes in primary energy sources over the last 30 years are shown in the figure below.

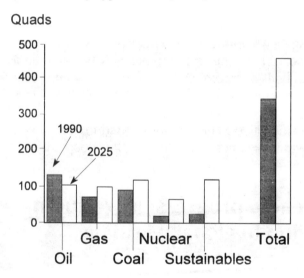

**World Energy use: 1990, 2025
By Type of Fuel, in Quads**

Technological advances boosted the prospects of nuclear power and sustainables. The nuclear industry's future appeared bleak in the 1980s and early 1990s. A sabotage incident at a Pakistani nuclear plant in 1999 initially seemed to seal the industry's fate. But extensive surveys revealed that the striking success of safety and containment measures actually raised people's confidence in nuclear power. The tradeoff between nuclear power's safety and waste disposal problems and fossil fuel's contribution to global warming began to be decided in favor of nuclear power—as global warming moved from the hypothetical to reality. People chose the known risks of nuclear power over the relatively unknown effects of warming.

The current generation of reactors addressed many of the issues that had held back the technology and the industry. Light-water reactor technology was improved and new inherently safe reactor designs, in which passive safety systems relying on natural forces such as gravity substituted for human or mechanical activation, were adopted. Significant steps have been taken to standardizing reactor designs. Greater automation and use of artificial intelligence and robotics reduced the number of workers and potential for human error and improved safety. Approval procedures were streamlined as a result of these encouraging developments.

Japan has recently introduced the world's first commercial fast-breeder reactor. Japan also remains the pioneer in reprocessing spent fuel. An international research and development (R&D) consortium for radioactive waste disposal recovered from a slow start to produce significant advances in disposal technology as well as developing institutional mechanisms to deal with siting problems. Member countries use teams of international experts to choose and inspect the best potential disposal sites across the globe. The findings are used as inputs by national governments in siting decisions.

Some economically depressed areas deemed suitable by the International Energy Agency readily accepted nuclear waste sites and even sought the facilities as a means to stimulate economic activity. Other areas went through intense political battles. A combination of incentives and penalties have been used to get these facilities operating. Tax credits, full reimbursement for relocation, guarantees for home values, and community development projects have been offered as incentives. Funding for needed improvements in recalcitrant communities was sometimes withheld until they cooperated.

Slow, steady progress in reducing energy's contribution to global warming

Fossil fuels were branded the primary culprits in the greenhouse effect in the 1980s. Global warming was scientifically confirmed by the International Global Warming Federation in 2010. A 30 centimeter sea-level rise since 2000 has led to local flooding problems as well as widespread destruction of wetlands and silting of rivers. These problems will be magnified and accompanied by new difficulties, such as the shifting of agricultural zones, if the anticipated additional 1 meter rise in sea level by 2050 occurs.

An increasing share of greenhouse gases has been coming from middle and destitute nations. Their greenhouse emissions increased as their energy use climbed. World 2 and 3 also burned more fossil fuels—mostly coal, some oil, and very little natural gas, the cleanest of the three.

In the affluent nations, Germany, and eventually the entire EC, took the lead in committing to reducing carbon dioxide emissions. Of course, as the EC began integrating the high-polluting Eastern European nations to different degrees, it was easy to demonstrate reduced emissions. The bulk of the German reductions since 1995, for example, have come from the former East Germany.

The formerly communist Eastern European countries had antiquated energy infrastructures that were built with little regard for efficiency or pollution control. Many EC nations have resisted and rejected membership for these countries because of the appalling state of their energy sectors. Bulgaria's application for full EC membership, for example, was rejected last year primarily for this reason.

The carbon tax that passed in the 1990s was effective in reducing greenhouse emissions. The industrial nations of Asia quickly followed the EC's lead in reducing greenhouse emissions, and the United States reluctantly did so years later. The United States stalled, waiting for more scientific evidence. When that came in 2010, U.S. leaders were able to take action. On the other hand, many U.S. corporations foresaw the coming confirmation and the likely regulations and antiwarming measures. They switched to cleaner technologies and gradually weaned themselves from oil and coal and toward natural gas and eventually to sustainables to avoid the anticipated regulations.

A by-product of efforts to combat warming was the formation of the International Global Warming Federation (IGWF) in Munich in 2000. It was created as an independent agency of the United Nations to coordinate international efforts against global warming. The IGWF formed a program to install nuclear reactors in some of the rapidly industrializing nations through a combination of grants and loans. Sanctions have been under consideration but have so far been avoided by negotiations and voluntary compliance.

Despite actions such as Sweden phasing out nuclear power in 2010, the use of nuclear energy grew. Nuclear technology posed an interesting dilemma to environmentalists. Although they almost universally opposed it early in its existence, perceptions began to change as global warming concerns grew. Because nuclear power generation does not emit greenhouse gases, it became attractive again in countries that previously turned away from it—such as the United States—after the turn of the century.

While perceptions changed, striking design advances vastly improved safety, efficiency, and reduced costs. Natural gas use picked up in the decades before and after the turn of the century, but it also contributes to warming, though less than other fossil fuels. Slow progress in developing markets and institutional mechanisms for sustainables left nuclear power as the most acceptable alternative.

Sustainability institutionalized in World 1 nations; conflict between industrialization and the environment in World 2 and 3 nations

Surveys of attitudes toward energy use have been tracked over the last 50 years or so by pollsters and social indicator groups. In affluent nations, they reveal a trend toward accepting the need for energy conservation as part of a larger social movement toward sustainability. The percentages of people agreeing with the statement "the government should have the authority to implement measures to enforce reductions in energy use" have steadily risen over the last 35 years to 76% today.

The clincher for sustainability and conservation was the economic gain from conservation. As evidence accumulated that conservation made good economic sense in the mid- and long-range, roughly 4 to 20 years, support for it grew rapidly. A key factor was the reformulation of GDP figures, which in turn affected business accounting practices, to include what were previously labeled externalities or social costs. When harm from global warming, acid rain, or smog were included in calculating the bottom line, conservation measures became much more credible. More important was the use of tax incentives, rebates, and credits by government to encourage the adoption of sustainables.

In middle and destitute nations, the story has been less positive. Sporadic energy shortages stalled many middle and destitute countries' efforts to industrialize over the last 35 years. Smog from vehicle fuel emissions periodically shut down metropolises such as Mexico City, Seoul, and Warsaw. Acid rain poisoned water supplies in Shanghai and Calcutta.

Through 2000, the terms of the debate were often framed as energy for jobs and economic growth versus environmental extremists who wanted to save plants and animals at the expense of people. In most cases, increasing energy use won. The typical energy options were coal, oil, hydroelectric, and biomass. They were deployed in ways that maximized their potential to damage the environment, at least around the turn of the century. Over the last decade or so, many polluters have shaped up as the costs of harming their environments increased and the benefits to economic growth diminished.

The tragic case of Poland

Poland was held up as a shining example of a country making the transition from a command to a free-market economy in the late 1990s. It appeared that the Polish economy was on sound footing and had turned the corner until the Panic of '99. The Polish stock market crashed that year as companies floundered with the final removal of government subsidies two years earlier. Pollution problems had reached extraordinary levels, and industries were hamstrung by a poor energy infrastructure and practically nonexistent efficiency measures. Reformist political leaders were voted out and replaced by xenophobic politicians pledging to rid the country of interference from "foreign saboteurs masquerading as aid and developmental personnel." Their nearly 10 years in power brought advances in energy technology and policy to a halt. As a result, Polish industry remains well behind its competitors in energy efficiency today.

Coal and oil combustion has led to levels of air pollution in cities of middle and destitute countries that were and still are unprecedented. Sustainables were not without side effects either. Vast stretches of countryside were disrupted by big hydroelectric projects. The adverse effects of these projects, felt chiefly during construction, included flooded communities, changed water tables, higher mercury levels in the water, disrupted fishing, and new diseases flourishing in still water. Biomass energy, especially fuelwood, contributed to deforestation.

The conflict was acute in Eastern Europe. These nations' energy industries were the worst *per capita* polluters in history, and their economies were in shambles. Leaders could not be persuaded that investments in pollution control and energy efficiency would make more economic sense in the longer term. Many political leaders exploited anti-environmentalism through demagoguery to boost their political fortunes. National leaders or international organizations that urged energy reforms were portrayed as interfering in internal matters and harming living standards. This sold well politically but kept these nations in poor shape far longer than necessary. Their energy-inefficient industries simply lost out in the global marketplace.

Energy efficiency proved itself in the marketplace

Conservation is gradually becoming a global value. Conservation measures have been successful for brief periods of time historically, typically in reaction to crisis. For instance, energy use in the United States remained constant from 1970 to 1986, largely in response to Arab oil embargoes. When the crisis passed, so did the rationale for conservation. What is different today is that energy efficiency makes environmental and economic sense, regardless of whether there is or may be a crisis. Reducing power demand has been shown to be cheaper than building new generating capacity. Also, Japan has forged a highly successful market niche by selling conservation and energy-efficient technologies and expertise. The United States missed the boat in marketing conservation technologies.

Some energy efficiency technologies that have been adopted

Structural shifts in the global economy, especially among the affluent nations, began to reduce energy use as well. The gradual transition from a predominantly industrial- to a knowledge-based global economy has correlated with a drop in energy intensity in global energy use. The drop has been more dramatic in affluent nations, and is only just beginning in middle and destitute nations, which have been going through a long period of growing energy intensity.

Longer-lived products

An important factor in reducing energy consumption was the advent of longer-lived products. Growing public acceptance and an increasing amount of legislation supporting recycling, reclamation, and remanufacturing increased the payoff from more durable goods. Automobiles, for example, used to have effective lifetimes of about five to seven years. By 2015 the lifetime of automobiles was an average of 15 to 20 years, as parts were reclaimed or the entire vehicle remanufactured. These processes are much less energy intensive than building a new auto.

Information and service industries are less energy intensive than their industrial predecessors. The industrial activities that remain have been automated with intelligent computer-integrated manufacturing, which has improved energy efficiency. Robotized factories can be scheduled to operate during off-peak hours, when power can be purchased at lower rates. Biotechnology and information technology have reduced the energy intensity of agriculture by almost 40% since 1990.

Information technology has been a key enabler of energy efficiency. Information continues to substitute for physical work. Demand-side management has become firmly institutionalized. Utilities and independents have leveraged information technology capabilities into dramatic drops in energy intensity.

Cogeneration was and still is an important element in rising rates of energy efficiency, especially in the industrial sector. The industrial sector long led the transportation and commercial and residential sectors in efficiency. The gap has been closing since the turn of the century.

OPEC collapsed amidst political turmoil

Desire to break free from the whims of OPEC and the volatility of Middle East politics led importing nations to search for alternative energy sources. The overthrow of the Saudi Kingdom in 1999 led to a worldwide rise in energy prices. Oil rose briefly to $40 per barrel. The price came down to $35 per barrel as a semblance of order was restored, but $20 per barrel oil has been history for 25 years. Small wars, terrorism, and the growing spread of radical Islamic fundamentalism undermined OPEC. It still exists, but with no real power.

The Commonwealth of Independent States, once the world's largest producer of oil, was unable to step into the void. By 2004, its oil production had fallen below 1990 levels. As the political situation has begun to stabilize over the last decade, production levels have been inching up, but are still well below peak year production of the 1980s.

Trade deficits based heavily on imported oil were another important factor in the decline of OPEC. Oil was often the single largest factor in trade deficits. The United States and other oil-importing nations decided that weaning from their dependence on oil was a necessary step to getting their economic house in order. In the interim, from roughly 1998-2010, the United States relied on the North American Free Trade Agreement with Mexico for supplying much of the oil.

WORLD 1 NATIONS: PROGRESS TO AN ENERGY TRANSITION

The global transition to a mix of clean fossil fuels, sustainables, nuclear power, and conservation technologies has been underway for 5 decades but has gained new momentum in the last 10 years. The transitions have not taken place uniformly. They have been influenced primarily in three ways:

- through crisis and shock
- by selecting a policy-driven path to sustainability
- in response to global warming

These forces affected the affluent nations to varying degrees. The United States was hit harder by the 1999 overthrow of the Saudi Kingdom than nations such as Japan, which invested heavily in alternatives to Middle East oil. The strong influence of environmental parties and interest groups in Germany meant that the policy-driven path applied to a greater extent there than elsewhere. Of the three influences, global warming came closest to having a uniform effect. The differences were in degrees of preparedness, as some nations better anticipated its effects.

The affluent societies' *per capita* energy consumption was at 66% of consumption in 1990. *Per capita* energy use in the United States is higher than in Germany and Japan, who were already more heavily invested in energy efficiency in the 1980s. There have been continual efficiency improvements in the industrial sectors, although at a slower rate. The biggest gains have come in the commercial and residential sectors, as information technology applications have significantly raised efficiency. Progress has been slowest in transportation. R&D and experimentation in this sector is booming, however, and payoffs are likely to come in the next decades.

A CASE STUDY— The United States catches up in efficiency

The United States remains the world's largest energy user. Government inaction, part complacency and part malfunction, prolonged a business-as-usual approach that ignored the need for an energy transition. It was not until the landmark Energy Transition Act of 2002 was passed that the U.S. government took a leadership role in directing the country towards reducing energy use. Up to that point, the United States lagged behind Japan in implementing efficiency measures and moving to alternatives. Today the United States has caught up and is only slightly less energy efficient than Japan and the EC. Measures adopted by the Japanese were often bellwethers for the United States. Additionally, these innovations as a rule first appeared in California, which has been heavily influenced by Japan due to its geographic proximity and has continually experimented with proactive regulation.

California: the bellwether state

The economic clout of California meant that companies could not ignore it. It was particularly innovative in transportation energy during the past 30 years. It was the first to effectively mandate the use of electric vehicles back in the 1990s. Automakers, although unhappy with the legislation, simply could not afford to ignore the large California market. California also was the first to implement a large-scale intelligent vehicle highway system as well as achieving the first embedded-power electric vehicle lane on a major highway. It continues to be a bellwether, innovating in energy policy and technology.

The fossil fuel scorecard: gas prospers, coal survives, oil fades

Natural gas grew the fastest of the fossil fuels due to its less harmful contribution to global warming, improvements in safety procedures, and its greater thermal efficiency. Gas has become relatively easy to recover and remains in abundant supply. The combined cycle gas turbine (CCGT) is an important break-through. It takes the exhaust gases from a gas turbine through a waste-heat recovery boiler to generate steam, which powers a turbine. Its conversion rate reached as high as 55%, compared with 40% for the best coal-fired generation and 35% and 43% for simple-cycle gas turbine and steam turbine plants, respectively. Clean coal technologies kept coal from losing ground, despite its reputation of being bad for the environment. Oil was the loser on two counts: its carbon dioxide emissions and the politics involved in obtaining it. The mix of fuels in the United States today and in 1990 is shown below.

U.S. Energy Consumption: 1990, 2025
By Type of Fuel, in Quads

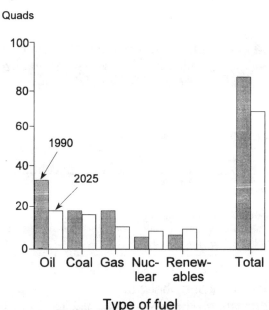

Nuclear power makes a comeback

Fear of global warming and its eventual scientific confirmation in 2010 paved the way for the wider social and political acceptance of nuclear power. Advanced, modular, and inherently safe designs radically improved safety compared with earlier plants. Plant life extension kept the industry alive until this new generation of reactor technology came on-line. The fine-tuning of waste-reprocessing techniques and improved means to dispose of unusable waste were critical. Continued worldwide use in the 1990s, with success, also eased the way for a renewed commitment to nuclear power by nations like the United States, which had practically abandoned it.

Standardizing reactors made economic sense but raised the issue of vulnerability to a mass shutdown. For example, a problem in one reactor could make it necessary to shut down all reactors while the problem is searched out. Such a situation has been avoided to date, but the issue remains important.

Fusion is still not commercially viable. Progress has been made. A pilot plant remains at least 15 years away.

Sustainables overcoming market and institutional obstacles

The costs of sustainables were often forecast to come down to levels competitive with fossil fuels and nuclear power, but usually did not due to an inability to implement them on suitable economies of scale. The failure to create a system for large-scale introduction resulted in costly haphazard applications. Success typically came in niche applications. Wind power, for example, succeeded in suitable regions of the United States and Europe. Yet the significant potential of photovoltaics was largely unrealized. It also succeeded in niche applications, but will be a much larger player. Photovoltaics started in applications difficult to wire to the electric grid, such as irrigation pumps, roadside signs, and remote facilities. A key to the future of photovoltaics is that significant cost reductions will come with manufacturing on large scales.

Solar technologies forge ahead

In Phoenix, an electric-utility-led consortium and city officials have recently arranged a large-scale infrastructure project to support solar technologies. They hope that solar power will supply over half the metropolitan area's power needs in 2030. Today it supplies about 5%.

Although progress has been disappointing, prospects are improving. The market for solar thermal and photovoltaics looks most promising, as they have advanced further than other solar technologies. Solar thermal plants, which produce steam for electricity or heat, have been used primarily as supplemental sources. They work only during sunlight and use storage technologies that have not been more cost-effective to date than competitors. State and local governments are beginning to be sold on the payoffs from photovoltaics and solar thermal.

Big hydropower projects declined the most of the sustainables, due to their harmful environmental effects described earlier and their high initial capital costs. Dam safety issues also led to the decommissioning of many older plants. Ocean energy and geothermal energy have yet to make a substantial contribution to power production, succeeding only in suitable geographic areas. Important advances in biotechnology have improved the attractiveness of crop biomass, in which crops are grown and harvested exclusively for energy use. So far, however, crop biomass has been unable to compete successfully with other uses for the land.

The figure below shows the shifts in end uses of energy in the United States.

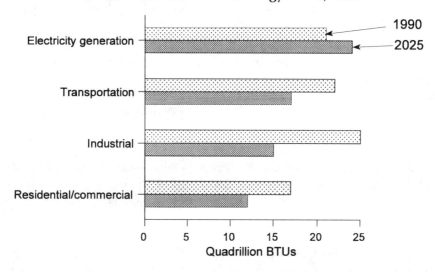

United States End Uses of Energy: 1990, 2025

Electrification continues to grow as alternatives falter

Electricity remains the primary means for delivering energy. It is approaching half of U.S. energy use. National commitments to sustainables and nuclear power ensure that electrification will continue to grow, since energy from these sources is most efficiently delivered by the electric power grid.

The addition of Mexico to the U.S.-Canadian electric power grid 10 years ago created the North American Power Network (NAPNET), the largest and most complex power network in the world. Negotiations are underway to connect Central and South America as well. Superconductive transmission wires are becoming cost-effective. The first 300-kilometer line laid in 2019 connected three rural counties of Georgia to Atlanta.

The hydrogen economy

Hydrogen has continually been touted as a possible successor to electricity by its growing number of proponents. When its technical prospects flagged in the 1990s and 2000s, it was kept alive as an alternative to electricity by the persistent belief by many that electromagnetic fields led to serious health effects—despite substantial scientific evidence to the contrary. These fears kept the possibility of hydrogen in public debate until some of the technical obstacles were overcome. Until more substantial benefits can be demonstrated, however, it is unlikely to overcome the inertia of sunk investment in electricity.

2025

Electricity has had few practical competitors. Hydrogen (see box) and fuel cells were brought up most often. Second generation fuel cells became serious components of the energy mix in the marketplace at the turn of the century. This generation made notable steps forward from the first. And today's third generation is similarly advanced from the second. Their primary application so far, however, has been for portable power needs, such as construction sites. Their costs will have to come down further if they are to compete successfully with the electric power grid.

Battery and energy storage has produced steady progress, but no breakthroughs. Ranges for batteries for electric vehicles have quadrupled from 120 kilometers before recharging in 1990 to about 500 kilometers miles today. In energy storage, several compressed air energy storage plants have been built since the success of the Alabama plant. Pumped storage is still used as backup by many utilities.

The deregulation of electric utilities is largely completed. Competition had important effects on the entire electric system. Unbundling, for example, enabled the split of generation from transmission and distribution. This vastly increased the number of players involved. The increased competition improved the quality and reliability of the service. Utility transmission lines were made effective common carriers when the National Energy Act of 1992 was passed, loosening the 1935 Public Utility Holding Company Act. Consumers now choose their electricity supplier in a way similar to how they pick their long-distance telephone carrier.

Wheeling became an important means for making up power shortfalls and avoiding the need to build new capacity. Power-rich regions such as eastern Canada supplied not only the U.S. Northeast, but could have their power wheeled across the continent in emergency situations. Critics have long argued that increasing the complexity and connectivity of the network increased its vulnerability to breakdown, sabotage, or terrorism. No major incidents have occurred to date.

The consolidation of utilities predicted by many analysts has not happened on a large scale in either the United States or the EC. The numbers of independent power producers, continue to grow. Utilities have been using independents to supply expanded capacity. This was especially critical at the turn of the century in the United States, when independents stepped in to meet the capacity shortfalls that would have occurred due to the utilities' reluctance to build new capacity. Independent sales to utilities accounted for 15% of all electricity by 2010.

The means for delivering electricity has been radically reshaped by the use of information technology. Power distribution uses a computerized auction system based on electronic data interchange technology. Utility computers offer power at variable quality and rates to business, residential, industrial, and now transportation customer's computers, which make purchases based on programming.

Power sales based on the quality of the product have been an important innovation for 30 years. Users now purchase different grades according to the task at hand—buying high quality to run sensitive high-tech equipment and lower quality for other tasks. This greatly increases the overall efficiency of energy conversion.

Residential and commercial energy efficiency got a boost from infotech

Efficiency measures in this sector have enjoyed the greatest success over the last 35 years, largely the result of information technology applications. Smart homes are now standard in the United States. All new homes are built with a computer control center that monitors or regulates energy use among other functions. Severe efficiency measures have been passed into law. The advent of manufactured housing facilitated this transition, as it is much easier to standardize manufactured housing features than site-built ones. The cost of new homes rose roughly 10% as a result of efficiency regulations, but the resultant energy savings averaged 40% greater than homes built before the regulations. This is a massive lifetime savings.

Sorting out the building-regulation mess

Until the 2002 overhaul of housing codes and regulations, builders were sometimes subject to multiple jurisdictions with conflicting standards. The 2002 Building Energy Code Rationalization Act harmonized the energy-related codes and regulations into a single national policy.

Energy management systems improve efficiency

Saving energy is not something that takes a conscious effort these days. Textbooks contain charming historical anecdotes about people turning down their thermostats, wearing sweaters around the home, and being careful to turn off lights. These were seen as significant sacrifices. These practices relaxed, however, when energy prices came down again. Today, of course, these kinds of "sacrifices" are unnecessary. EMSs regulate all facets of home energy use. Temperature standards are mandated by law. Lighting, appliances and all other power uses are carefully monitored. When someone leaves a room, for example, the lighting automatically ceases. Variations can be programmed in, so long as they comply with legal standards. For example, the coffeemaker can be set for specific times, and lighting needs can be customized to the individual—more lighting for Johnnie and less for Amy. The savings over homes without EMSs average 32% to 72%.

There are still wide disparities in how homes are heated in different regions of the United States, but more are moving away from oil and toward natural gas and electric. The retrofit market has proven especially lucrative. The 60% to 70% savings in energy use that became common in retrofitting home and office buildings around 2000 created a large demand for retrofitting.

Gains in the commercial sector were even more impressive than in residential. Total building energy use was cut by a third, despite the fact that U.S. building use had been increasing through the early 1990s. Businesses got the message—with help from government tax incentives—that investments in efficiency led to significant long-term savings. They sought to sharpen this edge as the struggle to stay competitive intensified.

Industrial energy-saving innovations continue

Energy intensity has been declining in the industrial sector since the 1970s. The rate of savings has decreased, as the easy savings have already been made. There has been a structural economic switch to less-energy-intense products as well. Consumer appetites for energy-intensive durable goods have lessened as they have been satiated to a considerable extent. The growing application of the 3Rs—recycling, reclamation, and remanufacturing—has further decreased energy intensity. Each of the three takes less energy than starting a new product from scratch.

Information technology, particularly artificial intelligence in the 1990s and 2000s, led to sophisticated ways of saving energy. Energy management systems maximized space conditioning to unprecedented levels. Complex simulation models enabled further optimization of industrial processes once thought to have reached the limits of efficiency.

Industrial systems have become fully mechatronic. Stand-alone operations are nearly extinct. Microprocessors have been embedded in all products to enable communications and better coordination and save energy. Machines in intelligent computer-integrated manufacturing systems, for example, signal control systems when they are about to fail. Operations are then adjusted until the needed maintenance can be performed or replacements made. Today, industrial energy use per dollar of industrial output is 40% of the 1980 amounts in constant dollars.

Transportation has been the laggard in energy efficiency

Transportation continues to be the most inefficient and environmentally harmful end use of energy. It has been the most reliant on oil and has therefore been down the rockiest path of the three sectors. On the other hand, it has probably been the most innovative of the three. The search to reduce harmful engine emissions, as well as for alternatives to oil-based fuels, has spawned a great deal of innovative R&D, experimentation and trials.

People have an almost bewildering array of transportation options from which to choose. No single winner has emerged from experiments with alternative vehicle fuels. Each has its advantages and drawbacks. The most significant change is the widespread adoption of hybrid vehicles. Vehicles powered solely by gasoline were outlawed 14 years ago, in 2011. Gasoline is still used, but only in hybrids. There have been, and still are, many types of hybrids. They range from the predominant gasoline-electric to ethanol/methanol-electric and reformulated gasoline-electric, to compressed natural gas-gasoline, hydrogen-gasoline, and hydrogen-electric, to the occasional solar-electric.

Experimentation with gas hybrid engines are ongoing as well. Two-stroke engines and direct-injection diesels have become conventional. Stirling engines are becoming popular for larger vehicles, and ceramic low-heat rejection engines are catching on as their prices come down.

The debate about which fuel or power configuration should be standardized will likely continue for at least the next decade. It is inefficient to make so many kinds of vehicles, and the fuel supply infrastructure problems are troublesome. Some form of electric hybrid is likely to emerge as the vehicle of choice.

The Big Three's Advanced Battery Consortium formed in the late 1980s achieved some technical advances. More importantly, it set the precedent of cooperation among automakers. Today, two of the three are left, but they have nine autonomous divisions. Global alliances have virtually erased distinctions between U.S.- or foreign-made vehicles anyway.

Alternatives to the automobile have yet to become significant in overall, day-to-day travel. The lack of coherent government policy has led to a hodgepodge assortment of transportation minisystems. High-speed rail and maglev systems have been successful only under limited circumstances. The Texas Triangle system, linking Houston, Dallas, and San Antonio, for example, easily outpaces car travel. But there are few similar geographic situations that enable rail to compete with vehicle travel on a cost and convenience basis. Intelligent vehicle highway systems have proved an effective means of personal vehicle travel, and helped it to remain the predominant mode of local and medium-distance travel. Personal rapid transit and older mass-transit systems serve only niche markets, typically in older, densely populated cities.

WORLD 2 AND 3 NATIONS—ARE THEIR ENERGY PATTERNS SUSTAINABLE?

Population growth and urbanization, along with economic growth and industrialization, drove up *per capita* energy consumption to 160% of 1990 levels. This compares with affluent nations reducing their *per capita* consumption to 66% of 1990 levels. Recent indications are that the middle and destitute nations' *per capita* consumption has peaked as a result of the combined effects of slowing rates of population growth and the growing implementation of energy efficiency measures.

The World 2 and 3 nations have had to rely on fossil fuels, particularly coal and oil, to a greater degree than World 1 nations, as they have been unable to make the capital investments necessary for alternatives. Proximity to resources is more important in middle and destitute nations that often lacked the infrastructures needed for transportation and distribution.

Their use of biomass declined roughly by 50%. Up to the turn of the century, in particular, most of the biomass use was fuelwood. This unsustainable practice gradually declined as more sophisticated forms of power became available. The biotech revolution in biomass, for example, has so far generally bypassed middle and destitute nations that are simply unable to afford the technology or lack the market or institutional mechanisms to distribute it on a scale that would make a substantial difference.

Affluent nations that have taken advantage of genetics to speed the growth of biomass feedstocks and genetically create microorganisms that convert waste materials into useful fuels—as part of their 3Rs programs—are taking advantage of business opportunities for technology transfer in the middle and destitute nations. This technology transfer is beginning to have a positive effect today and should grow in importance over the next decade. The few people still relying on traditional biomass as their principal energy source are typically rural villagers.

The primary side effect of increased energy use in middle and destitute nations is their contribution to global warming. On a positive note, the International Global Warming Federation has helped middle and destitute nations to shift from reliance on coal and oil. The loan and grant programs to finance construction of small nuclear plants or sustainables begun in the IGWF's early days have been a success story, although many would argue that its efforts have been on too small a scale.

A CASE STUDY—China's expanding energy use threatens its future

China has been the middle and destitute world's largest energy consumer for decades. It uses three times as much energy as India, the second largest energy user in the middle and destitute world. It is now neck-and-neck with the United States. Its *per capita* consumption, happily, is less than that of seven other large World 2 and 3 countries.

China's economy grew at 5% annually, almost double the average for middle and destitute countries for the period 1990 to 2025. Yet, its energy use grew at an average annual rate of 2.2%. Efficiency measures accounted for the difference. Without the efficiency savings, China would have used an additional 38 quads of energy in 2025 on top of the 64 quads that it now consumes.

The World 1 nations, through the IGWF, transferred efficiency technologies, including clean coal and nuclear power technologies, to China in the interest of reducing its greenhouse emissions. China became the world's number one contributor to global warming just after the turn of the century. China's leaders refused to take substantial voluntary actions to reduce greenhouse emissions, pointing out that the affluent nations developed through hydrocarbon energy. But they did respond favorably to the IGWF incentives.

Many of China's early efficiency gains were easy, because it was extremely inefficient to begin with. Tremendous savings have been realized with computerization in factories, improved building integrity, better use of industrial waste heat, and more efficient boilers, machinery, and electronics, as well as consumer goods.

Although energy efficiency has improved, it remains below the affluent nations' levels. For example, its energy intensity is still two-and-a-half times greater than Japan's, the world's most energy-efficient economy. The Chinese government enacted plans to cut its energy intensity first by a third and then by half from 1990 levels. It made substantial investments in its energy infrastructure, some years as high as 40% of annual public investment funds. But government subsidies of coal and oil, continued until 2008, kept their use inefficient. A new generation of leaders more willing to rely on market incentives came into power at that time. Although the transition to nonsubsidized power was often bumpy, it has played out well for today.

The evolving mix of energy sources

The most significant difference in China's energy situation today and 30 years ago is that a nuclear industry has developed. Although begun primarily with the aid and assistance of the IGWF, a native Chinese industry is now established and has projected impressive growth over the next 25 years. The mix of energy sources today is compared with the 1990 mix in the figure:

China's Energy Consumption: 1990, 2025 By Type of Fuel, in Quads

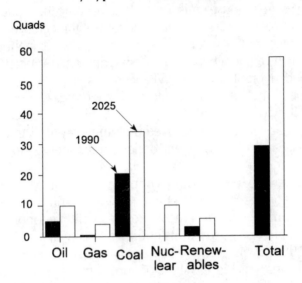

Fossil fuels: mostly coal, some oil, and little gas

China has been the world's largest coal producer and consumer for decades. It derives 75% of its energy from coal. This makes China atypical of middle and destitute countries, where coal production and consumption have been concentrated in a relatively small number of countries. China, India, South Africa, and Korea combined account for 96% of middle and destitute nations' coal production. Despite continually increasing coal production, China had almost continuous energy shortfalls. Over the last 30 years, many factories worked short weeks because of a lack of electricity.

China's neighbors provided some clean coal technology in the interest of avoiding cross-border pollution. Simple coal cleaning technologies removed gross amounts of impurities. Coal is now washed 100% of the time, a significant improvement from less than one-fifth of the time in the 1990s. Not performing this relatively simple task leads to high sulfur emissions. Many more advanced technologies were beyond most of Chinese industry's means. Even small savings, however, went a long way. Ironically, 40% of China's transportation energy in 1990 was coal burned to transport coal. The electrification of China's railways lessened this paradox.

The underground mining that accounted for more than three-fourths of China's coal production caused land subsidence and acid drainage, which contaminated local water supplies and damaged aquifers. Surface mining led to the removal of large amounts of topsoil, leading to erosion, siltation, and water contamination, and destroyed much of the surrounding ecosphere. On a positive note, these problems have created a growing number of successful land restoration businesses.

In the late 1990s, China began substantial oil imports. The government responded slowly to its inability to meet its domestic needs. New fields were developed with foreign assistance and created several regional booms.

Natural gas use has doubled in China since 1990, but it started from an extremely low base. It has not been well exploited due to the lack of supporting infrastructure. This waste was common in middle and destitute countries despite the fact that they often had rich natural gas resources. For example, in 1990 Nigeria flared off 21 billion cubic meters of gas, which would have been enough energy to meet all the country's commercial energy needs.

A nuclear industry was born

The Chinese nuclear industry got off the ground in the 1990s with three reactors coming on-line. IGWF's support of nuclear power spurred its advancement in selected areas of the middle and destitute world. Japan in particular aided the Chinese nuclear industry, as it was in its interest to help wean China from coal, as much of the pollution from the coal use reached Japan. It also provided business for the Japanese nuclear industry.

Sustainables are underused

Biomass is no longer the predominant sustainable fuel of middle and destitute countries. China's situation differed in that it always used less biomass and relied more on crop waste and animal dung, rather than the fuelwood predominant in the average middle and destitute country. Unfortunately, this reliance on fuelwood biomass in the past has led to serious environmental difficulties in rural areas, such as reduction of soil fertility, soil erosion, and serious deforestation. The resultant breakdown of ecological balance has lowered the ecosphere's resistance to damage caused by floods or droughts.

The rest of the sustainables picture remains one of underused potential. China's large geothermal potential remains underdeveloped. Small, inefficient hydroelectric generators supply roughly half of the electricity used in rural areas. Some wind turbines have been deployed, mainly in Inner Mongolia. Proposals to use wind and solar power plants to produce hydrogen in the generally sunny and windy, vast central and northwestern desert regions and pipe it east, remain on the drawing board. Photovoltaics have been used on a small scale.

Demand for energy services exploded in China with the fall of the old power structure after the turn of the century. Only significant steps in improving energy efficiency kept energy growth rates at manageable levels for the available energy infrastructure and saved the environment from a worse fate than the current unacceptable levels of acid rain, smog, water pollution, and the threat of sea-level rise in the thriving coastal areas.

Electricity use spirals upward

China's electricity use has grown more rapidly than that in most other middle and destitute nations. This faster growth can be directly attributed to the government's push to develop its manufacturing base. The thriving coastal manufacturing industries have been primary factors in increasing electricity demand. Over three-fourths of China's electricity has been going to its industrial sector. Over 90% of homes now have electricity, whereas 35 years ago only 65% did. Some remote rural areas still do not have electricity and rely on burning coal for heat and cooking. Coal has by far been the largest fuel input to the electric system.

Residential and commercial demand outstrips supply

This sector accounts for roughly 40% of China's fuel use. Residential and commercial energy requirements still have not been satisfied. China's race to raise its living standards has relied on energy-intensive activities. Mechanized agriculture, manufacturing, and building up the infrastructure are highly energy intensive. Growing urbanization has also created pressure on the electric system. Satisfying people's growing demands for consumer goods and services drove up energy demand as well.

Differences in energy use between worlds 1 and 2

Energy use is reasonably similar between middle and affluent countries in terms of quantity used for electric lighting and appliances, industrial goods, and automobiles. The differences between the countries are due largely to the relative share of traditional villagers and the urban dwellers, and in the forms and quantities of energy used by those who are making the transition between these two extremes.

The typical middle country family's energy use path has been from biomass to liquid fuels, such as kerosene, to electricity. Energy use in traditional villages throughout the middle and destitute world has been fairly similar in terms of quantity used, sources (biomass and muscle), and services provided (cooking and subsistence agriculture). The Chinese path has been different. People relied primarily on coal rather than biomass for heating and cooking and skipped the liquid fuels stage, leapfrogging directly to electricity in most cases.

The dawn of electricity

Many Chinese homes have only recently been hooked to an electric power grid. Up to this point, coal was often burned right in the home for warmth or cooking. The labor process of starting and maintaining the coal fire is not likely to be missed. The fascination with things electric continues to mount. Pent-up consumer demand is exploding. Store shelves are often cleaned out, and power grids fail regularly in areas newly brought on-line. It has taken an average of five years to establish reliable power and saturate markets with electric appliances.

Industrial power remains unreliable

China is the third largest commercial energy user in the world, accounting for 10% of the world total. Yet almost half of its workers are still in agriculture, even though the industrial base has been expanding rapidly over the last 35 years. Unreliability still characterizes the industrial power system in China. Losses sustained by industry due to unreliable electric power supplies are estimated to be 2% of annual GDP. The quality and reliability of the electricity remains poor, which has hurt China's attempts to become a world leader in manufacturing.

China differed from most middle and destitute nations in its greater share of energy for manufacturing. The typical middle and destitute nation had industrial-process heat and cooking as the largest energy services—each accounting for about one-third of all energy consumed. This pattern also contrasts with the United States, where transportation and heating and cooling are the biggest energy users.

Transportation getting a bigger, but still small, share of energy

Transportation accounts for only a small fraction of energy use in China today, although its share is continuing to grow. Heavy investments in the transportation infrastructure, including natural gas pipelines as well, will enable much greater mobility for the population and boost transportation's share of energy use. The big gap between fuel supply and demand is one of the important factors that hinder the development of an automotive industry, and transportation in general.

The density of railways and highways per person is still only 1/20th of that in the United States. The number of automobiles has increased tenfold from one per thousand people to one per hundred. China does not use electric vehicles. It relies on inefficient diesel and gasoline vehicles. The last coal-fired train was retired just last year. This is a remarkable achievement, given that only 10% of China's railways were electrified in 1990.

Critical Developments, 1994-2025

Year	Development	Effect
2000	Sabotage leads to meltdown of nuclear plant in Pakistan.	Containment procedures are successful and improve nuclear prospects.
2000	Islamic fundamentalists overthrow Saudi ruling family.	OPEC unity disintegrates.
2000	International Global Warming Federation forms.	Transfers technologies to alleviate greenhouse emissions.
2001	15km/liter vehicles mandated by U.S. Congress.	Increased efficiency of current fleet and prepared the way for alternative fuels.
2002	U.S. Energy Transition Act passed; massive conservation campaign.	The energy transition takes off.
2003	Two percent of California's new vehicles are electric.	Other states begin to encourage electric vehicles.
2004	The Commonwealth of Independent States' oil production falls below 1990 level.	Encourages shifts away from oil.
2009	Nuclear nations agree to submit to Internal Energy Agency standards, after grandfather clause added.	Further boost to improving prospects for nuclear power.
2010	Global warming confirmed as significant.	Laggards in reducing greenhouse gases forced to begin to act.
2010	Global population surges past 7 billion.	The race between sustainability and population is highlighted.
2011	Single-fuel gasoline vehicles outlawed in the United States.	Alternatives get badly needed support.
2023	International Global Warming Federation able to impose sanctions.	Not been used to date, but are a threat.

Unrealized Hopes and Fears

Event	Potential Effects
Solar power satellites for power on Earth.	Would have collected energy from sun and transmitted it to Earth via microwaves or laser beams.
Commercial fusion power.	Has been touted as the answer to energy problems, but has proved elusive.
Superconductivity.	Small-scale deployments for power transmission only; revolutionary potential not realized.
Gold's hypothesis.	The large methane deposits posited by Gold have yet to be found.
Hydrogen.	This potential successor to electricity has yet to make significant inroads.

6

THE WORLD OF THINGS

Today science is close to fulfilling a dream: to put molecules together with great precision and reliability to create new materials with nearly any desired capabilities or characteristics. Like the typesetter of a century ago, materials scientists can assemble materials molecule by molecule to meet a need or create a capability. Within five years, they will do so atom by atom when materials of such precision are called for.

Materials scientists and engineers already can assemble materials for thousands of different applications and to wide-ranging and complex specifications. Their expertise and artistry are in using knowledge and computer power to find ways to make materials quickly. Imagine a single machine that could do the creative work for them and also be the engine for producing any material so concerned.

Today's polymer composition machines and desktop manufacturing anticipate such development. The ancestors of desktop manufacturing were the automated eyeglass lens grinding equipment and key-making machines of the last century. Today, machines for custom making small parts of metal, ceramic, or polymers are in widespread use. Such machines become even more practical with miniature and nanomachines.

What has brought materials science this far is a combination of advanced technologies and growing body of knowledge in molecular technology. Some forces of change shaping materials over the past three decades are:

- Photonics for manipulating, modeling, and analyzing information, allows materials design, prototyping, and testing on computers.

- Advances in molecular science through combined advances in chemistry, physics, and biochemistry. Genetics played an unexpected role in the 1990s and 2000s in advancing molecular science, as researchers studied "how nature does it" and mimicked those processes in the laboratory and factory.

- <u>Instruments</u> that make atomic and molecular structures and phenomena and their behaviors visible and measurable. Especially important was the arrival of in situ, nondestructive, real-time testing and monitoring.

- <u>Processing</u> capabilities making unprecedented purity and compositional specificity possible.

Materials are integral to the changes that have reshaped the world, from medicine to transportation to clothing to space travel.

Materials changes have brought a permanent net energy savings across affluent World 1 societies. Lighter-weight, safer vehicles, better insulation, faster conductors, improved materials processing, and other changes have helped the affluent countries reduce their energy use *per capita* to 66% of 1990 levels. The table below shows some fundamental shifts in materials science and technology over the past generation.

Fundamental Changes that Shaped Materials and Materials Use

1995	2025
Wasteful use of materials; 3Rs (recycling, reclamation, and remanufacturing) the exception	Closed-loop materials uses; 3Rs the rule
Short product lifetimes	Extended product lifetimes
Trial uses of composites	Compositing is the basis of many materials in commerce
Experimental uses of ceramics; use in small parts and coatings	Ceramics a dominant material in engines and other devices, structures, and surface applications
Study of materials in nature	Biomimetics, imitating natural materials, finds wide spread uses
Surfaces generally inert and decorative	Surfaces put to work with new functions
Most products are large, and all are bigger than 2 millimeters	Micromachines and nanodevices in wide application
Most materials dumb and unresponsive	Smart materials and dynamic structures commonplace
Explorations of materials in unique combinations	Constant innovation in combining capabilities
Bulk processes, imprecise, difficult to customize	Batch processes for high precision, customization

A list of the Stookey Awards for innovative materials gives a useful timeline of materials development since 2005. The Stookey, named for Donald Stookey, the inventor of photochromatic glass, is a Steuben-glass figure of

Daedalus. The box below lists the materials technologies that have won Stookeys over the past 20 years.

The Stookey Awards: 2005 - 2025

2005	Injectable osteophilic polymers for bone regeneration
2006	Perfection of repulpable adhesives for paper cartons, etc.
2007	Fullerene-based lattice molecules for complex catalytic reactions
2008	The Halloran process for producing halide glasses for optical fiber—achieved greater purity to parts per hundred trillion
2009	Superplasticization of wood, making wood fibers that can be formed by conventional plastics methods; e.g., molding, casting, extrusion, pultrusion, and compositing
2010	Machinable advanced structural ceramics tolerant of extreme conditions
2011	Noncytotoxic medical adhesives replacing cyanoacrylates, not rejected by the body
2012	Thermotropic (moldable) liquid-crystal polymers with greater transverse strength, widespread uses in electrooptical and optical systems
2013	Synthetic gemstone coatings, including sapphire, emerald, and ruby; primarily used in durable decorative coatings
2014	[Jury finds no development worthy of the award. The Stookey panel was subsequently sacked by the executive committee and replaced]
2015	A process for reusing office paper–"the reverse Xerox"–a polymer paper is passed through a machine that removes ink and toner, produces blank sheets for reuse
2016	Synthetic rubber-based aerogels used in ultralightweight padding and insulation where brittle materials function poorly. May be bent, formed, etc. Also used in toys.
2017	Piezofabrics–the basis for virtual reality tactile sensation and response suits
2018	Superconducting metals operating at 175 kelvin, making greater industrial use possible
2019	The Mercer Compositor for selective deposition of molecules and atoms in prescribed forms and combinations. The precursor to anticipated atomic composition machines.
2020	Permeselective membranes that isolate toxins in ground and surface water. Captures and binds up a broad class of toxins for safe disposal. Can make water potable with one pass
2021	A process to factory-grow synthetic cotton
2022	Polymer muscle replacements
2023	Biomass-fed "silkworm" microbes that produce uniform, high-quality fiber for paper-making
2024	The Chalmers-Woo class of quasi-composite metapolymers–expected to offer hardness, machinability, control of conduction, alternative finishes, and injection moldability
2025	Multi-directional epitaxy–allows near-net shape and net shape production of synthetic molecules, coatings, and micromachinery

2025

A critical shift seen by the 2010s was a full-fledged commitment to sustainable materials use in the advanced societies. People realized that waste was harming them and their communities through the cost of disposal and destructive environmental effects. This realization meshed with the growing recognition that human life on Earth could not go on consuming materials and energy at 20th century rates and maintain a healthy planet. As the public in the advanced nations became more aware, governments adopted national accounting practices that included more accurate assessments of energy and materials resources used.

Looking back 30 or 40 years, it is clear how the material makeup of goods in commerce has changed. Designers once merely substituted new materials for old in the same products. Then they learned to redesign products, taking advantage of new capabilities in the materials. Subsequently, those materials made entirely new products possible.

Most materials changes are not visible to the casual user or the consumer because:

- Designers substitute one material for another in the same application with hidden design changes. The aesthetics and tactile nature of the new material may match the old. Thus, there are synthetic wood, polymer paper, and Permacotton™.

- Materials are embedded, composited, or present in microscopic quantities, doing their jobs behind the scenes, i.e., beyond sight or feel.

- The capabilities and characteristics brought by surface and internal structure may be essential to function, yet invisible to the user. For example, there are the honeycombed crystal lattice of polyceramics and the whiskers or wafers in matrix composites.

Materials, more than most manufactured goods, cross borders and economic strata. Even the most advanced materials may find primary or secondary uses in the poorest economies. This may happen in their second, third, or fourth uses, as goods or materials are reused. Since the turn of the century, the 3Rs—recycling, reclamation, and remanufacturing—have become integral to materials from product design through use and disposal. The 3Rs is a concept of affluence. Increasingly deep-seated traditions of extended use and reuse of goods in the middle-income and destitute countries are merging with a global recognition of the need to reduce waste. There are stories of carbon-epoxy jet fuselages turned into people's houses or into paca (rodent) pens in Ghana and Burundi.

WORLD 1 SOCIETIES HARNESS MATERIALS SCIENCE AND TECHNOLOGY

The world's affluent societies experienced revolutionary change in how materials are used across society over the past four decades. New materials technology reshaped everyday objects, altered economies, made and ruined fortunes, prolonged human lives, saved energy, and allowed extended space travel, for example. The United States, along with Japan, Germany, and the United Kingdom, spearheaded the scientific and technological change. Dozens of other countries joined with them in putting materials to work in new ways and with new combinations of capabilities. Environmental issues and attitudes were a central shaper of change, along with new technological capabilities in materials and materials processing.

A CASE STUDY: U.S. society revolutionized by materials technology

In only the most affluent countries can advanced science and technology flower. The United States exemplifies the cross-cutting transformations made possible by the arrival of new materials. It illustrates the fundamental shifts that have reshaped materials science and technology since the 1990s, and how modern materials are used across society.

Imagine, or remember if you are old enough, 1.5 metric ton, fuel-hogging automobiles, drafty houses, stain-covered clothing, peeling paint, frozen hands or fingers, changing tires every 65,000 kilometers, rusting bridges, blood donations, etc. All those things were commonplace 35 years ago.

FUNDAMENTAL SHIFTS IN MATERIALS TECHNOLOGY

Fundamental shifts in the material world over the past 30 years are summarized in the box and described below.

Materials use for sustainable life

Sustainable materials use arose as a public issue late in the last century. The rule of thumb today is "use it at least twice." There is considerably more pressure, however, for extended lifetimes and repeated reuse.

Fundamental shifts in materials, 1995-2025

Materials use for sustainable life
- Extended product lifetimes
- Materials reuse
- Safety
- Environmental friendliness

New and newly important materials categories
- The composites revolution
- The ceramics revolution
- Mimicking nature
- Putting surfaces to work
- Making devices smaller
- New forms and uses of traditional materials

Making materials smarter
- All-new combinations of capabilities
- Processes maximize precision and customization

Fundamental to sustainable materials use is preserving value added. Designers, joining the cause, are making manufactured goods that have practical lifetimes of five times as long as their 1990s forebears. More of the public today, missionized by the sustainability movement, takes pride in recounting stories of how long they have owned their cars, refrigerators, watches, clothes, computers, etc. Those anxious for new goods can always rest assured their used goods will find second and third owners. Modularity in design and construction, from small devices to buildings, makes materials and component reuse easier. When a module wears out or fails, you just swap in a new or remanufactured replacement. That was rare 30 years ago.

Corporations have had to adjust their philosophy and finances to accommodate extended lifetimes and reuse. Goods designed to last longer can be sold at higher prices, but sales volume is down. On the other hand, the waste recovery, reclamation, and resale industries are flourishing. Reusing materials not only saves on the materials themselves, but it also saves energy and human labor.

The schematic of the materials cycle shows the now well-established flow-back loop, which began to emerge in the 1980s.

A Schematic Materials Cycle

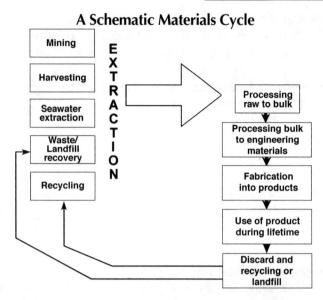

Extended product lifetimes

Driven by economics and regulation, goods today last longer and are designed for iterative reuse. Adapting to style and technological change is sometimes difficult. Part of the genius of designers is allowing for stylistic changes and anticipated technological changes in materials choice and construction.

Extended lifetimes reduce the lifetime costs of materials. Greater quality is designed and built-in. Initial costs, on the other hand, are higher.

A #16F Automotive Rib Module

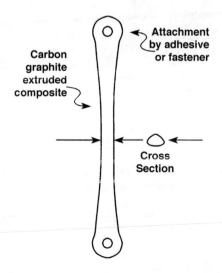

2025

The outer panels of vehicles, the surfaces of walls, the colors of carpets and clothing, for example, may all be designed for restyling, whereas interior frameworks, substrates, and other basic materials are built for longer-use lifetimes.

Most ingenious are the designers who find new uses for composite material forms and frameworks. These expensive materials are difficult to recycle and lose 90% of their value if recycled. It helps that many today are designed from basic modules that have wide use in commerce. A #16F rib module from your vehicle, for example, could find a later use in a refrigerator, utility tunnel, or roof support. See the illustration. This makes ease of disassembly a necessity for successful products. This practice gained momentum with the International Agreement on Standards in Construction, signed in 2011.

U.S. legislation to promote sustainable materials uses

2000 — The Allen Act—Outlawed the export of hazardous waste materials, without a demonstration that the materials would be safely handled in the receiving country.

2000 — The Commodity Plastics Standardization Act—Restricted packaging and other disposable and bulk applications polymers to 11 types and also required them to be separable from plastics and other materials.

2000 — The Manufacturing Energy Conservation Act—Offered incentives and technology programs to reduce energy consumption in manufacturing, including in recycling processes.

2003 — The Morrison Act (The Antileaching Act)—Required approval of all new materials as safe from leaching of toxics in solid waste, ash, or if used on or in the ground. Alternatively, makers could demonstrate that they would recover the goods for safe disposal prior to their entering the solid waste stream.

2006 — The Durable Goods Reclamation Act—Made the makers of motor vehicles, large appliances and certain other manufactures responsible for the cost of disposal of their goods.

2010 — The Solid Waste Abatement Act—required all municipalities to cut solid wastes by 70% of volume by 2020.

2011 — The Semayle Act—Adopted the International Materials Characterization Standards and the associated materials handling and processing standards.

2015 — The Materials Conservation Act (The Closed Loop Act)—Gave tax incentives to manufacturers that reduced waste to less than five percent of raw materials by weight.

2018 — The Consumer's Act—included provisions to require all manufactured hard goods to have a label with mean time to failure and average costs of five most frequent repairs.

Taggants enable automated materials reclamation

The now standard inclusion of taggants in polymers, ceramics, glasses, metals, and composites is the key to cost effective reclamation and recycling. The taggants not only identify the material, they identify parts and modules for proper remanufacturing and reuse. Scarcely practiced 20 years ago, chemical tagging is an expectation across industry and a legal requirement in dozens of countries. Although sensors on materials handling equipment can identify and grade most materials, taggants make high-speed reprocessing more effective and accurate.

Materials reuse

Waste grading and categorization, despite its centrality to manufacturing and municipal reclamation programs, is still a new concept. It was not until 2013 that the United States established standards for wastes to simplify their resale and reuse.

The waste grading and categorization industry that was spawned from this practice was already a billion-dollar industry by the late 1990s. Today, it has reached over a hundred-billion dollars.

Automated handling and transportation equipment makes it possible to collect, segregate, and redistribute solids and liquid wastes. Most U.S. municipalities achieve 80% reclaimed or recycled wastes, including biomass. The record is 96%, set in Guilford, Connecticut. Those figures are prior to final waste burning, which creates a marginally useful slag once toxics are segregated.

The sustainable use of materials has risen steadily in importance in World 1 societies for several reasons. First, raw materials may be costly and scarce, making their durability and their ongoing reuse essential. Second, materials manufacture and the manufacture of goods from materials is expensive and energy intensive. Third, people recognized in the affluent societies in the 1980s and 1990s that waste is a physical burden on society, and minimizing waste is an effective strategy for preserving the environment for ecological and aesthetic reasons. This recognition came first because of the local aesthetic and cost impacts of excessive waste. Later, vigorous public-interest-group campaigns, such as Sustain Our Society (SOS) and Project Help Eliminate Waste! (PHEW!), built public support for structural change in society and industry.

Reusing materials, and especially remanufacturing them, is energy-efficient and cost-effective. They preserve value added and save energy. The past three decades have seen growth in these areas, with a growing emphasis on design for reuse. Interestingly, materials reuse is a deep-seated practice in traditional societies. World 1 countries have continued to learn lessons from poor countries on how to scavenge and reuse materials. Industry also now has greater capabilities in tertiary recovery, in which materials are converted back to the base feedstock chemicals so pristine materials may be made from them.

Recycling and remanufacturing now affects 88% of U.S. vehicles. Up to 60% of vehicle parts by weight are taken for remanufacture. Another 37% of the weight can be recycled. With this, the expected lifespan for a new vehicle is 19 years, whereas today the average age is 13, up from about 7 years in 1990.

Safety

A material's safety extends from its creation by environmentally and occupationally safe methods to its use by people young and old, smart and ignorant, to its safe disposal or reuse. It is not enough to count on people handling and disposing of materials properly. An affliction of earlier advanced materials, such as polymer composites, was that they were not always safe when burned. Most advanced materials today either have no toxic effects when burned or are treated, designed, or packaged to make them safe in fires.

Among innovations for safety are polymers that soften on sharp impact. First used in vehicle dashboards in the 2010s, advanced forms are found in athletic shoes, flooring, furniture, protective equipment, and elsewhere.

Critical to materials safety is the prevention of chemical emissions (outgassing) from materials, particularly from polymers. Originally an issue with vinyls and formaldehyde last century, it seemed that every new material used to come under suspicion. Today most materials' designs limit or eliminate outgassing.

Environmental friendliness

In the United States and other World 1 countries, it is usually not enough for materials to be conserved and reused. The public, out of combined environmental and safety concerns, rejects materials that harm the air, water, animals, plant, and human life.

In 32 signatory countries as of May 2024, materials by law must not leach toxics or otherwise harm soils, air, and water supplies. The Convention on Environmental Balance permits such materials only with the maker's agreement to recover 98% of them. Storage batteries, for example, may employ toxic acids and metals only if the maker has an acceptable, documentable plan to take back 99.6% of used units for safe dismantlement and/or reuse.

New and newly important materials categories

Materials rise and fall like movie stars as they enter commerce with new capabilities, new uses, and revolutionary designs and still must compete with yet newer materials. Among the superstars of the past three decades are several composites, ceramics, and biomimetic materials.

Strength and weight

The remarkable gains made in strength to weight ratios in materials can be illustrated by the following example:

Cross section

Suspending a 25 ton weight vertically from cast iron rods (a common structural material two centuries ago) would require a rods at least 24mm by 24mm in cross section. Using a 1980s vintage high-strength polymer fiber would require fibers with about a 9 mm by 9 mm cross sections. Using today's strongest carbon or aramid fiber would require fibers only 2mm by 2mm in cross section. The thinner fibers are, of course, only a fraction of the weight of the 1980s or earlier fiber.

The composites revolution

Composites are almost any combination of materials and forms. For example, they include carbon whiskers, dendrites, or fibers combined in carbon matrices, or fibers in a metal matrix. Compositing is a way to bring the capabilities of two or more materials together in one.

The workhorses of the composites revolution, developed in the 1980s, are carbon fiber reinforced composites. By the turn of the century, 30% to 40% by weight of most civilian aircraft consisted of these composites in place of metals. Pleasure boats saw a similar changeover to composites from wood and fiberglass over the past 30 years. Other aircraft composites are aluminum-containing silicon carbide ceramic fibers and carbon-fiber-reinforced polyetheretherketone. The costs of these composites have dropped about 50% since 2010, making them attractive competitors for use in aircraft and other vehicles.

Composites developed in the 1980s and 1990s revolutionized materials in manufacturing and aerospace. But because those predominantly fiber-reinforced composites had lots of internal interfaces that weaken a material, scientists sought further ways to make the materials stronger and more stable. A first step was altering the morphology of fibers, e.g., with the use of dendrites and wafers. A recent solution was the liquid-crystal-based molecular composite. In a polymer-based molecular composite, uniformly dispersed rigid macromolecules become backbones for flexible-coil polymers that attach to them. The materials have high-impact resistance, fracture toughness, and compressive strength. They can be configured to have electronic and optical properties.

A historic difficulty with composites was the need to orient reinforcing fibers in complex forms. Aircraft wings and fuselages, for example, would be wound with fibers to create the needed strength across their forms. Several composites technologies made this process easier and cheaper, and sometimes unnecessary. Techniques for winding fibers for composites have become highly refined and automated. Also, inducing extremely long reinforcing molecules, dendrites, wafers, and whiskers to grow in matrices is now possible using closely tuned magnetic fields. Other reinforcing molecules are grown in a melt or mix, and form the strong spines of a material. Fiber orientation can be controlled in three dimensions.

The ceramics development story

Ceramics, favored for being lighter and running at higher temperatures than metals, found increasing use in engines in the 1990s and 2000s. They also wear less and thus impart less friction to engines.

Ceramic coatings were available earliest, in the mid-1990s.

Monolithic ceramics have been available since around 2000.

Ceramic composites were commercialized in the mid-2000s. These were in use in 46% of vehicle engines by 2011.

Compositing ceramics solved problems of brittleness and added further strength, heat resistance, and other properties.

Ceramics take over for metals

Advanced ceramics' heat tolerance, toughness, and dimensional stability won them early stardom in such applications as catalytic converters and computer data disks in the 1980s and 1990s. Then materials scientists turned to giving ceramics tensile fracture toughness and to finding better ways to make them. Near net shape and net shape processing made ceramics economical for wider use by 2009 in such places as automobile engines. Estimates of energy savings from the use of ceramic engines in diesel vehicles run between 37% and 60%.

Ceramics have it all:

Dielectric	Optical conducting
Ferroelectric	Optoelectronic
Piezoelectric	Wear resistant
Radiation resistant	Low friction
Semiconducting	Corrosion resistant
Superconducting	Chemically adsorbing
Ionic conducting	Biologically compatible
Magnetic	Thermally conductive
Translucent	

Ceramic matrix composites for jet engines came into use in the first decade of this century, replacing metals. In use, they ran far hotter than metal engines and led to energy savings. Advanced versions of those same materials are used today.

Top-grade ceramic materials must be void free, contain no agglomerations of unmixed material, and have no chemical impurities. Chemical vapor deposition, melt atomization, and other processes create the pure, tiny particles needed. Sintering and hot isostatic pressing of the particles create the final ceramic forms. Sometimes designed-in voids are used to strengthen ceramic forms. Processors achieve additional capabilities by devising ceramic composites, such as silicon carbide in alumina, with their own inherent strengths.

Ceramic-ceramic reinforced composites are one of the hardest, most heat-resistant, and most thermal-shock-resistant materials available. The fibers reinforcing these materials lend them creep resistance, preventing shape changes under high heat combined with stress. They continue to be favored for applications such as high- pressure/high-heat nozzles, dies, and turbine wheels. Some common ceramics applications are:

- Engine blocks and associated valves, ducts, pistons, bearings

- Cutting tools, blades, and dies, including household scissors that never need resharpening

- Filters for exhausts, catalyzing, or precipitating materials

- Medical implants including synthetic bone and teeth

- Coatings for any application requiring abrasion and heat resistance

- Substrates for materials such as optoelectronics requiring high dimensional stability

Mimicking nature

The greatest lesson learned in materials science over the past four decades may be that nature knows more about materials than people do. Sturdy seashells, adhesives used by barnacles, human bone, feathers, and the self-assembling and reproducing molecules, proteins, and enzymes of cells, for example, show the cleverness of nature in making efficient, perfectly suited materials for every need.

Manufactured materials mimic nature

- Plant stalk structures inspire the design of honeycombed sandwich panels, bridges, and pipelines

- Strong, resilient cellular materials mimic wood and plant internal structures for structural supports and device housings

- Ceramics microstructure may be based on the tough shells of sea organisms, for stronger tools, devices, cladding

- Synthetic enzyme catalysts for targeted medical treatment

- Artificial spleen and pancreas tissues

- Featherette™, a featherlike material that is lightweight, breathes, and repels water for rainwear, fabric roofs

---

Materials scientists have learned how to do many things that nature does, sometimes better and usually faster. Today's storehouse of materials available to humankind includes dozens that directly mimic nature.

Putting surfaces to work

Since the mid- or late 1990s, materials surfaces have become integral to the functions of products. Making the surfaces of materials active and even smart adds a capability where none existed before. Thirty years ago, the average device was smart on the inside, if at all. Today we have bioactive medical implants; membranes for water, food, or medical filtering; catalysts; and sensors—all relying on active, functional surfaces.

Within composites, surfaces play important roles at the interfaces of matrices and reinforcing particles or fibers. Mechanical strength may come from designed-in surface morphologies and chemical bonding at interfaces.

Making devices smaller

Materials made possible the progressive miniaturization of devices through the 1980s and 1990s. By 2000, silicon-based micromechanics became practical, and devices were miniaturized further, some to the thickness of a human hair. This miniaturization trend continues today as scientists develop practical applications for nanodevices, carrying science down to molecular levels.

Micromechanics has a long history. In the 1980s, scientists worked at microscopic scales to produce machines with practical purposes, such as switches, pumps, valves, and levers. With these micromachines, used in optical, electronic, and medical applications, scientists anticipated the rise of practical nanotechnology and much smaller devices in the 2010s. Some applications of today's nanotechnology are:

- Molecular magnets for switches in equipment

- Molecular electronic devices including electronic chips

- "Chip-board" memory molecules

- Self-assembling ribosome analogs, integrated with transistors for biosensors

- Designer molecules introduced into the body to promote hormone production or scavenge toxins

New forms and uses of traditional materials

A theme running through developments of the past 40 years is new forms and uses of traditional materials. We first saw wood as a chemical feedstock in the last century. That use became increasingly sophisticated in combination with rising understanding of biochemistry and genetics. Trees and other plants are still efficient producers of useful feedstock chemicals. Molded, extruded, cast, and other forms of plastic wood and other lignin/cellulose-based concoctions have put wood to hundreds of new uses.

Traditional materials include plastics, rubber, leather, wood, stone, glass, and ceramics. Today these materials are synthesized, composited, coated, or impregnated with other materials. Some are made active or smart, and some have found new uses because they can be produced with greater refinement than ever before.

Periodic fads and fashions revering "natural" materials reinforce demand for these traditional materials. But consumers are equally demanding that the materials be superlative; they do not want yesterday's materials to have yesterday's problems. Wood that rots and glass that shatters, for example, would find few markets today.

Making materials smarter

Materials scientists continue to look for ways to make materials smarter. Everyone is familiar with smart materials such as shape-memory polymers, composites, and metals. Others are:

- Dynamic materials that adjust their properties to conditions and needs, such as opacity changes based on light

- Stress-alert paints and coatings that sense and announce structural change or damage to themselves

- Selective membranes for sequestering toxins in air, water, and food

- Synthetic molecules with chemical missions such as detoxifying waste, monitoring blood chemistry

2025

- Biosensors for implanted sensors, etc.

- Neurally stimulated artificial muscle

Also important to commerce are things made smart through mechatronics—the embedment of sensors and actuators in materials and structures. Today's dynamic structures are based on this technology. Responsive vehicle frames, dynamic buildings, smart implants and prostheses in medicine are examples of these technologies.

An international materials characterization scheme

The variety and complexity of materials in commerce, plus the regular practice of customized preparation, led the ISO in 2009 to establish a clear standard for materials characterization. The scheme includes the material's properties and capabilities, surface geometry and chemistry, crystallography, fiber orientation, composition, etc.

All-new combinations of capabilities

Putting capabilities together in any desired combination is increasingly possible. There can be high strength with light weight, transparency with magnetism, bioreactivity on an inside surface with an inert outer surface, high tensile strength along one axis and friable along others.

There is increasingly the sense that one can dare materials engineers to make a new material. A designer or customer may approach an engineer and say, "I want a machinable glass with microtubules running horizontally through it, I want it red, and I want it to absorb nitrogen across its vertical surfaces." The designer or customer, choosing to have a lower-grade material, might add: "make it for 0.39 ECU a kilo."

Processes maximize precision and customization

Today's materials and materials-engineering strategies let people have products closely tailored to their needs. Materials production processes are precise to parts per billion in chemical makeup and to dimensional tolerances in the micron range, if needed.

Electrochemical processing makes wide compositional choice possible

Electrochemistry is particularly useful for creating materials of highly specific morphology and composition, using subtly tuned electrical current to influence what elements are laid down in what order and configuration. Varying temperatures from high heat to cryogenic open up wide ranges of kinetic processes for the electrochemist.

Last century, the materials business was characterized by large-scale commodity production. Today, the tide has turned. A large share of production is for a single customer or batch-sized.

MATERIALS ACROSS U.S. SOCIETY

This section illustrates the state of materials in the United States today. It examines some social, economic, technological, and institutional forces involved with materials use. It also looks at the basic functions of human life— food, clothing, shelter, etc.— and examines materials use in each.

Society constantly rethinks materials choices as technology changes, but also as the needs and values of society change. Choice is shaped especially by cost, environmental concerns, and regulations. Conservation measures are concentrated on bulk materials, those that create the biggest energy and resource waste.

Today, for example, there are 37,000 different plastics available for use in commerce, up from about 15,000 in 1990. The bulk of the 37,000 are small-volume, special-application materials, or materials that engineers developed for which there is no current use. At the same time, social and regulatory pressure has resulted in limiting the choice of plastics, metals, and glasses used in high volumes, to make recycling economical. Of the tens of thousands of plastics available, only 11 are allowed in containers, food wrappers, packaging, and other materials for disposable use or bulk applications such as insulation.

Smarter, safer, more efficient shelter

It is often in housing that new materials first find $50-billion-plus markets. Central to materials for housing today are safety, reliability, and environmental soundness. Safety from fire has long been a tenet of home construction, and researchers have transformed traditional and new materials from highly flammable to flame resistant and even flame proof. Similarly, materials are evaluated for the safety of their smoke. Polymer and polymer composite building and finishing materials are now routinely self-extinguishing. Death from fire or smoke in home fires today is unusual compared with 30 years ago. Only 137 deaths from home fires occurred in the United States in 2024. Death usually occurs when an old house burns.

Upper-income homes are likely to have dozens of smart features built into them. Smart and responsive structures and finishes, enabled by materials innovations and mechatronics, make homes more comfortable, economical, and safe. The house diagrams below show examples of these technologies.

Materials Used in a Typical Modern House

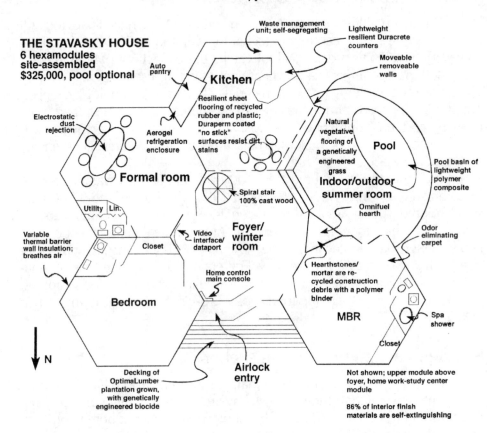

THE STAVASKY HOUSE
6 hexamodules
site-assembled
$325,000, pool optional

Waste management unit; self-segregating

Lightweight resilient Duracrete counters

Moveable removeable walls

Auto pantry

Kitchen

Electrostatic dust rejection

Aerogel refrigeration enclosure

Resilient sheet flooring of recycled rubber and plastic; Duraperm coated "no stick" surfaces resist dirt, stains

Natural vegetative flooring of a genetically engineered grass

Pool

Pool basin of lightweight polymer composite

Formal room

Indoor/outdoor summer room

Spiral stair 100% cast wood

Omnifuel hearth

Odor eliminating carpet

Variable thermal barrier wall insulation; breathes air

Utility Lin.

Closet

Video interface/ dataport

Foyer/ winter room

Hearthstones/ mortar are re-cycled construction debris with a polymer binder

Home control main console

Bedroom

MBR

Spa shower

Closet

N

Decking of OptimaLumber plantation grown, with genetically engineered biocide

Airlock entry

Not shown; upper module above foyer, home work-study center module

86% of interior finish materials are self-extinguishing

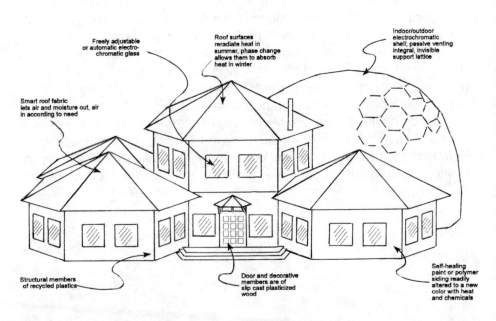

Roof surfaces reradiate heat in summer, phase change allows them to absorb heat in winter

Indoor/outdoor electrochromatic shell; passive venting integral, invisible support lattice

Freely adjustable or automatic electro-chromatic glass

Smart roof fabric lets air and moisture out, air in according to need

Structural members of recycled plastics

Door and decorative members are of slip cast plasticized wood

Self-healing paint or polymer siding readily altered to a new color with heat and chemicals

---

Materials in use in commercial and industrial structures

- Fluoropolymer paints that bond strongly with materials for a permanent coating. Used on concretes, polymers, polymer composites, metals, ceramics.

- Composite structural members, such as trusses and beams save space and weight in large-building construction. The members are reused.

- Polymer concretes, the resin-impregnated portland cements, and other cements make greater strength and durability available in construction and allow wider aesthetic choice, such as the inclusion of color and textures.

- Optically transparent conducting materials used for signs and smart windows.

- Electrically conductive polymers used in security coatings for windows and wall covering composite fibers, where office electronics eavesdropping could be intercepted, these coatings trap electromagnetic signals.

Materials improve business and industrial structures

Like houses, commercial buildings are smarter, safer, and more efficient than ever before. Information technology, coupled with new materials and design principles, has reshaped these structures. Smartness comes with flexibility (and vice versa), and more and more organizations find it useful and necessary to reconfigure their space frequently for changing needs. Movable partitions and floors, reroutable utilities, and options for changing finishes are important to these users.

Steel and concrete were the traditional materials of construction in the last century. They still are important components of buildings, industrial structures, and other large works. Composites, however, have steadily replaced steel and concrete in structures by outcompeting them. Steel made a steady march to higher strength over the past three decades, keeping it competitive in some applications with composites. The dominant steels are fine-grained, low-alloy steels.

Concretes gained from combined-in polymer additives, etc. Today's polymer concretes are 50% stronger and far more wear resistant than those in use 30 years ago. At the same time, engineers manipulate the morphology of concrete, giving it microscopically porous structure that also has the capability to form shapes, arches, etc.

Efficiency and sustainability shape transportation technology

The world of transportation relies on materials as diverse as superconductors and polymer concretes in mass applications such as railbeds or in small but critical applications such as valves and optical switches in vehicles.

The traditional material of transportation used to be steel. Steel structures, rails, and vehicle bodies dominated transportation in the last century.

Steel thrived in vehicles, competing with alternative materials for most of the last century, until lighter- weight aluminums and then composites and polymers displaced it. Steel continues to be favored for nonmaglev rails and some civil structure applications. Weldability and toughness make steel the material of choice where there are extreme conditions and frequent need for repairs. It is still favored for large-scale construction, such as in oil rigs, ship hulls, pipelines, and buildings, as well as in concrete reinforcement. The scarcity of chromium and the availability of alternatives such as ceramic-coated materials have reduced stainless steel use by 57% since 2000. The history of change in jet engine materials, depicted in the graph below, shows how new materials can supplant old ones.

Major Trends in Key Materials Used in Jet Aircraft Engines 1960 - 2025

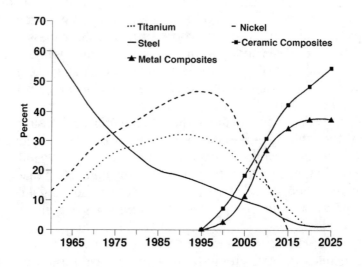

Extensive change came in vehicles starting in about 1990, with the first experiments in disassembling cars for parts reclamation and materials recycling. By 1990, about 75% of a vehicle could readily be taken for parts reclamation and materials recycling. Unfortunately, not many vehicles were reused. Vehicle redesign, greater durability, and materials choice brought the potential share of reusable vehicles to 97% by last year. About 88% of vehicles are taken apart for recycling and parts reuse. A share of the remaining 12% are sold whole to overseas markets.

Some innovative materials for transportation developed in the past three decades are:

- Solid lubricants that need be applied only once. These are particularly important in high-heat engines where liquid lubricants would be volatile.

- Lightweight composites for aircraft and motor vehicles. Carbon-epoxy composites are the stars in this field. Lighter aircraft and vehicle frames have produced dramatic fuel savings—up to 36% for aircraft and as much as 50% savings for motor vehicles.

- Fluoropolymer-based car paints that are extremely durable and tough, protecting polymer and composite panels, as well as metals.

- Conductive polymers used for lightweight batteries for vehicles.

- Piezoelectric film coatings on vehicles (also machinery) which allow continuous inspection of structural integrity through monitoring vibrations and sounds throughout the structure.

- Ceramic engine materials including the sialons (silicon, aluminum, oxygen, nitrogen), which are toughened and protected by ion bombardment or laser glazing. These engines tolerate extremes of heat and burn fuel more efficiently than their predecessors.

- Superconducting magnets installed in highly efficient motors, generators, and turbines. Their use in power transmission is still limited by cost and temperature requirements. Ambient temperature superconductivity on mass scales is perhaps five years away. Transmission within machinery and plants is near.

- Hydrogen occlusive alloys for hydrogen-powered automobiles. The alloys make it possible to store hydrogen fuel. Hydrogen vehicles are 5% of vehicles on the road today in Europe, Japan, and North America.

Food and agriculture: packaging materials shape the business

Convenience drives the use of materials in the food business. With safety as a given, it is the ease of handling, preparation, and discard that shape how materials are used. Today's self-cooking packages, for example, save steps, time, energy, and aggravation for the user. Even greater quality can be had with robot chefs—automated kitchen appliances that prepare food and are linked to automated pantries and food-ordering networks.

Convenience also requires ease of recycling. Since recycling laws came into effect in the 1990s, consumers have stridently demanded recyclable packaging from food vendors and punished the companies that did not comply. In North America, the standard today is that any disposable plastic must be made of 1 of 11 standard recyclable plastics. Limiting packaging to a handful of materials, while maintaining convenience in use and making packaging smarter, has been difficult. Smart packaging must use either bar codes or smart labels that can be recycled with plastic packaging, recovered in the recycling plant, or easily removed and disposed of separately.

The food industry uses packaging creatively to enhance the image of its products and to wow the consumer. Some food packaging innovations of the past 10 years are:

- Barrier films for food packaging, which, for example, limit the intrusion of oxygen, light, or moisture, to protect foods and prevent decay. Some are biocidal, preserving foods without added chemicals.

- Smart, self-announcing packaging labels, using liquid crystals and piezomaterials.

- Freezer-to-cooker recyclable packaging.

- Freshness-monitoring packaging materials and freshness "chips."

- Microchip instructions and other information on package labels that burn off safely when recycled with the material.

Agriculture relies on advanced and commodity materials as well. Some recent developments are:

- Polymer fabrics for containment agriculture structures. The materials offer passive heating and cooling capabilities. Also used is the featherlike synthetic Featherette™.

- Lightweight composites used in agricultural machinery to reduce the weight of machines, protecting soil and crop roots. They also significantly reduce energy consumption and thus costs.

- Water desalination that is now cost competitive with other water sources. Membrane technology and greater energy efficiency made this possible, with the added benefit of extracting useful minerals with little extra energy input.

- Vermin-proof and moisture-proof reusable unitized storage containers, with size grades for produce, low-volume products, and commodity products. Designed for easy cleaning and to work with automated logistics systems.

Clothing and fabric go high tech

Today's fabrics are comparatively lighter weight, stronger, more stain-resistant, fade-free, and breathable compared with those used in the 1990s. Some are smart. Drawing on composite layup technology, weaving machines now make seamless garments where 20 years ago and earlier, materials would be hand-pieced and pattern-matched. Now tailoring has converged with weaving, and cutting is minimized. Today's automated tailoring systems rely on the ability to custom weave a garment to precise dimensions and preferences.

Clothing materials innovation must never get in the way of design and aesthetics. Thus the most successful of today's fabrics and accessory materials can be varied in form, color, texture, and finish almost infinitely while maintaining certain desired characteristics such as comfort, thermal properties, washability, and feel. Other developments are:

- Customized weaving and patterning of cloth, designed, approved, and ordered via computer

- Leathers and simulated leathers with integral color that look good even when scuffed or gouged

- Shape-memory polymers seamlessly woven into fabrics for strength, stays, and other inclusions. These can learn the wearer's shape, then be fixed, and returned to that shape with each cleaning.

- Fabrics for quick change interior design applications, such as ColoRight™, which can have its color changed electronically

- Conductive polymers find use today as antistatic coatings for clothing and carpets

Materials choice is closely linked to energy consumption

Materials are essential to the generation and use of energy. Superconductors and photovoltaics dominate today's changing energy picture. At the same time, more efficient filters, catalysts, and scrubbers for hydrocarbon energy plants benefit from the materials revolution.

Practical superconductor applications were slow to arrive, but fast to find widespread commercial uses in the United States The first practical superconductors in 1998 to 2010 were in electronics and later in industrial electrical machinery, uses where they could be maintained at the required temperatures with liquid nitrogen. Their early-announced promise for use in power transmission and maglev trains was slow to be realized. The materials required too much cooling in applications over hundreds of kilometers. Superconductive coils for use in power storage came into use in the United States in 2005, and in Europe in 2007.

Materials choices in energy are driven by cost savings and efficiency. Key developments are lighter-weight materials for transportation, and materials that let engines burn fuel more efficiently, usually at higher temperatures. All-new energy sources have come from materials as well. Piezoelectric, wind, hydro generation, and photovoltaics are notable examples. Some energy materials are:

- Cost-competitive solar cells, available since the mid-1990s, with much greater generating capability. Photovoltaic silicon crystal sheets can now be grown for use with little variation in quality. Despite their great promise, solar cells still provide only 4% percent of U.S. power, though they are nearly ubiquitous in mobile situations and in remote locations more than 9 kilometers from main power lines.

- Turbines and engines relying on extremely high temperature-tolerant metals, ceramics, and composites, with high strengths and dimensional stabilities. These engines can operate at 2500°C.

- Wind and hydropower generation using structures incorporating piezoelectric sheets and surfaces delivers 1.7% of U.S. electricity and is expected to reach 12% by 2035.

- Lightweight batteries first perfected by the Advanced Battery Consortium in the United States in 1997 are universal in electric motor vehicles.

- Coating/structural materials for storage and transmission of hydrogen.

- Aerogels for insulation formed to net shape and sandwiched in continuous processes.

- Superconducting films, wires, and other materials.

- Superconductive materials in magnets used in turbines make for billions in cost savings and gains in energy efficiency in generation. At the use end, these materials make for greater efficiency in motors.

- Glass ceramics are the preferred medium for storing radioactive wastes. The materials are relatively insoluble and do not allow radioactive materials to leach out. Scientists in the next five years will test the feasibility of disposing of nuclear waste in the geological plate interface, the subduction zones, injecting marble-sized glass ceramics to 1,200 meters.

Information technology: long silicon-dominated, sees new contenders

Advances in computing and other information technology are convergent with materials developments. Cycles of change in information technology accompany the development of new materials. Computing and telecommunications used copper in the 1970s and 1980s, silica in the 1980s to 2010s, and silica and gallium arsenide in the past 20 years. Silica may yet win against the assault by the more expensive gallium arsenide with its recently perfected optical uses.

Materials transform telecommunications—A 150-year record

1878 to 1970	Electromechanical and human switching using copper
1963 to 2005	Electronic switching using copper and other conductors and silicon-based semiconductors
1995 to 2015	Optoelectronics switching, combining semiconductors, e.g., lithium niobate
2005 to 2025	All-photonic switching using optical silica, metal fluorides for fiber, gallium arsenide, and other semiconductors in switches, repeaters, etc.

MBE dominates semiconductor manufacture

Since molecular beam epitaxy (MBE) came into widespread industrial use in 1999, it has seen steady improvements. Today, semiconductors can be grown to final desired shape through MBE processes that are effective in three dimensions to high tolerances. Complex heterostructures are possible.

The greatest shifts in the past generation have come in the changeover from electronic to optical communications. Speed, efficiency, and energy savings from going optical paid for the research and capital necessary for the transition.

Researchers quickly achieved 98.4% of the theoretical maximum transmission rates for silica-based optical fiber. Then they turned to other materials, such as metal fluorides, to enhance optical purity and transmissibility.

At the functional ends of optical communications lines, emitters, repeaters, and detectors require other semiconductors. Materials scientists have steadily refined complex materials of indium, gallium, arsenic, and phosphorous, laid down through epitaxy, chemical vapor deposition, or atomic composition, for generating radiation, laser light, and other purposes.

Advances in optical fibers and computer chips are rooted squarely in advances in materials capabilities and materials processing. Integrated circuits today are orders of magnitude more complex than those of the 1970s and 1980s. Electron beam lithography can now create devices on chips with sizes of 0.1 millionths of a meter.

Magnetism

Exploring methods for better magnetic data storage in the 1990s and 2000s, scientists discovered new properties of magnetism and its effects on materials. Properties such as anisotropy, galvanomagnetic effects, magnetooptic effects, and what goes on at magnet and antimagnet interfaces found useful application across science. Molecular manipulation and engineering, photonic switching and signal manipulation, and tunable microwave ovens, are examples.

Processing technology has created all-new materials. In the information technology area, most important in the past three decades have been the products of epitaxy, including semiconductor films that behave as multiquantum wells in which phenomena not found in nature occur. These brought nearly unlimited flexibility to the design of integrated optical circuitry. Today there are experimental efforts to use vapor deposition in three dimensions to create quantum wires and quantum dots, in addition to quantum wells. Most attractive is creating a highly pure material using processes such as epitaxy and vapor deposition, and then using ion beams to alter characteristics in controlled regions of the blank to change their function. Thus a monolithic blank could have seamlessly integrated superconducting, semiconducting, insulating, etc., regions.

Critical developments include:

- Flexible computer and photonic chips and microcircuitry materials that restore themselves when stressed, so that they can resume function. Used in mechatronics applications.

- Diamond-based electronic and photonic components that can run at faster frequencies and higher temperatures than silicon, gallium arsenide, and other optical materials.

- Smaller, faster chips based on nanomaterials such as molecular magnets and switches.

Innovation in materials for environmental preservation and remediation

As humankind learns to manage the environment, more specialized materials come into play for use in the environment and to intervene between humans and the environment. Thus materials are used in nature, in factories, in the home, etc., to preserve or improve the environment.

In most societies, regulatory and social pressures have made it imperative that materials be chosen carefully for every use to which they are put. Their entire life cycles bear scrutiny with respect to the environment.

Waste-free or low-waste manufacturing processes are enabled by today's materials processing strategies. Computers ensure optimal use from extraction through assembly. By-products and wastes are fed back into feedstocks, used elsewhere, or sold.

Health care depends on biocompatibles

Advanced health-care technologies rely on materials for implants and prostheses, surgical repair materials, drug delivery, and synthetic biochemical materials. A key to the design and choice of such materials is their biocompatibility, along with their compatibility with pharmaceuticals and other materials in use. Often the materials must be extraordinarily pure and safe. They may be in the body for a lifetime. Biomimicking materials such as artificial hormones, proteins, and bone, are critical to modern medicine. Among today's biocompatible materials technologies are:

- Nonthrombogenic (non-clot-inducing) polymers, glass ceramics, and other materials for implants. Materials prepared for small-diameter vascular grafts of less than 1 mm diameter have been particularly important.

- Polymer backbones for inducing nerve regeneration.

- Bioreabsorbable materials for temporary implants.

- Biocompatible materials, chiefly ceramics and polymer-ceramics are nonvolatile, nonreactive, and accommodated by the body, which can alter the materials, for example, "growing" bone on a synthetic substrate. Bone repair and regeneration may then be electrically induced.

- Cage compounds, such as fullerenes and dendrites, that can hold atoms and molecules for targeted delivery inside cells and tissues.

- Polymer molecules that substitute for hemoglobin, which were practical by 2002.

- Shape-memory polymers for artificial muscles and joints, which were first demonstrated in 2003 and are now in widespread use.

- Corrosion- and fatigue-resistant glass ceramics and other composites for use inside the body that will not need to be replaced. They do not react with the body (unless designed to do so) and are as durable as bone. Soon some of these implants will be osteophilic, self-repairing, or relying on the body to make repairs to them.

- Cyanoacrylates and synthetic protein adhesives, which are the mainstays of the surgical adhesives business. These allow more refined wound closing and may be injected or introduced by catheter, allowing faster healing than traditional suturing.

- Organoids—synthetic organs that induce the body to surround them or infiltrate them with tissues and that produce hormones or perform other biological functions. The synthetic spleen and pancreas are organoids.

- Biomembranes, including implantable ones, that make it possible to control some human biochemical functions, targeted drug delivery, and synthetic organs, such as the artificial kidney.

Materials change the shape of infrastructure and construction

Durability, flexibility, and efficiency in construction affect materials choices in infrastructure and other civil engineering works. Architects, designers, and engineers have unprecedented variety in materials choice. Over the past decades, they have built more structures for combined longer lifetimes and flexibility for changing functions and ancillary technologies. Some key infrastructure and construction materials technologies are:

- Practical diamond coatings for optical fibers, which lend strength and corrosion resistance.

- Pavements that are higher strength, extruded or cast in place, and fiber reinforced are gradually replacing highways, roadways, sidewalks, and runways around the world.

- Flexible polymer concretes and other materials for flexible pavements that "ride" on ground surfaces without losing their structural integrity.

- Macrodefect-free construction concretes which provide greater flexibility for use in paving, pipes, structures, window and door frames, and table tops.

- In situ recycling of concrete, which is made possible by the new concretes coupled with advanced reprocessing techniques. This practice has become routine in the United States, Europe, and Japan.

Managing the globe

Humankind's attempts to manage the globe are evident in the myriad programs and technologies to preserve and enhance the global environment. Materials have enabled much of this work, including the management of water supplies, air quality, and remediation of ecologies. Examples include:

- In-ground polymer membranes that protect aquifers and soil by preventing contact with tainted soils, water, or wastes.

- Soil stabilizers and erosion abatement materials, including geotextiles and soil adjuvants.

- Soil adjuvants based on a catalytic polymer backbone that reverse soil laterization in part through chemical action. These materials are highly effective but have been slow to be adopted in the destitute tropical countries where they are most needed.

- Landscape reclamation technologies, using soil boosters, erosion barriers, and soil stabilizers.

- Macroengineering structures, such as the Kanawara Dam in Shikoku, Japan, the New Lower Mississippi Control System, and the tidal enhancement programs to reclaim the Adriatic.

- Dynamic structures, including dams, bridges, seawalls, pipelines, and channels, designed to respond to and harmonize better with nature.

- Climate modification strategies, such as the rain guides of the central Rockies and the thermal updraft zones of the western Sahara.

Leisure and entertainment: new materials create new possibilities

Nearly any interesting or new material can become a plaything or the center of a new game or leisure activity. The smart materials have been popular in the past 20 years as the basis for some sports. Mechatronics, bringing the ability to animate just about anything, results in constant development of new toys.

Materials at play in the world of toys

New and old materials are constantly being put to new uses in toys. Some are the basis of all-new toys, and others improve or change well-known toys. Some recent toys and games based on materials are:

- Piezoelectric rubber balls—they make noises when they bounce.

- Photonic play jewelry—Laser light dances and glitters from the facets of the jewels.

- Crazy cushions—Supersponges (aka rubber aerogels) are see-through. The Superbounce™ suit for kids is a rubber aerogel suit for the torso and legs that lets kids bounce as if on a trampoline.

- Photon racetracks—2-centimeter-thick optical cords are the racetracks. You squeeze the pulse laser trigger and see whose light pulse, visible as it travels, hits the target first. Light in the cords moves at about a meter per second.

- Slamamals™—shape-memory plastics in these animal figures allow kids to squash and reshape the figures at will and watch as they return to their original shapes.

- Electrotag shirts—Piezoelectric fabrics are used to make shirts that are worn for games of tag. When a person is tagged, everyone hears the shirt make a squeak.

- Virtual reality suits—Piezoelectric fabrics are used to make these expensive suits that not only sense motion and pressure from the wearer, but also respond to the virtual environment with the wearer feeling realistic sensations on all body surfaces.

- Stickysand™—electrostatically attracting play sand sticks lightly together and holds a shape temporarily.

- Pliantex™ materials make games like shadow lacrosse safer for players. Their sticks soften on impact while maintaining their ability to catch and hurl smart balls during games.

Materials are also integral to art. New materials are at play in nearly every modern art form, from sculpture with advanced polymers and liquid crystals, to photonic art drawing on the optical materials and liquid crystals. Among the most intriguing developments of the past 10 years is the "Bonsai" sculpture—a sculptural form in which the material grows, and the artist reshapes and reforms the piece every few days or weeks. Artists are famous for discovering aesthetic uses for new materials developed for industrial purposes.

Materials give manufacturing new flexibility and latitude

Materials changed what things are manufactured over the past three decades, but also changed how things are manufactured. New materials came into use in industry in the equipment used. Simultaneously, materials processing technology reshaped factories. Some developments were:

- Ceramic and diamond coatings on bearings, shafts, and other mechanical parts that limit wear, make equipment tolerant of high temperatures, give it dimensional precision, and that allow higher speeds of operation, finer tolerances, and greater strength.

- Inert materials for handling other materials without contaminating the mix.

- Dimensionally stable materials for dies, blanks, molds, forms, etc.

- Ultraflexible continuous adjustment mechanisms for materials feedstocks.

- Continuous, real-time, nondestructive sensing and monitoring through sensors that tolerate harsh conditions in process lines.

- Software designs for compositing approaches. Artificial intelligence optimizes configuration so that maximum strength and other qualities are achieved efficiently.

- Manufacture of ceramics, metals, and composites with custom fibers and whiskers orientation.

- Solid materials, including metal oxides, used to lubricate ceramic engine parts for operating at temperatures of 500° C plus.

- Electrically conductive polymers used in EMI (electromagnetic interference) shielding, among other uses.

MATERIALS IN WORLD 2 COUNTRIES

The world's middle-income countries enjoy some benefits of today's most advanced materials and miss out on others. Often governed by cost, they may use old-fashioned metals, polymers, glass, and concretes, for example, in construction. On the other hand, high-speed ground and air transportation, major infrastructure elements such as pipelines, and hard goods such as vehicles and appliances are likely to be made on advanced designs, using advanced materials.

In dozens of middle-income countries, people are embracing photovoltaics with an enthusiasm not seen since the arrival 40 years ago of audio-tape cassette players. There is a runaway market for portable home panels that can power a small houseful of lights, a television, and a radio. There is the potential that some World 2 regions will leapfrog World 1 countries in going to all renewable energy sources for home and small-enterprise energy consumption. Two hundred thirty-seven million of the low-cost units have been sold since they came on the market.

Despite the potential for renewable energy, environmental regulation in the middle-income countries is spotty. Some countries are signatories to international agreements to limit pollution and materials waste. Others refuse to sign, believing they have the right to reap the fruits of development before they clean up their lifeways.

Middle-income economics favors materials reuse. Recycling and reclamation make goods cheaper in the home country. Reclaimed materials also have ready markets in other countries, as do cheaper goods made from recycled or reconditioned materials. The table below summarizes materials use across World 2 societies by briefly examining the same categories of materials across society used earlier for the United States. There are, of course, a mix of richer and poorer people in every middle-income country, so what the table describes is necessarily an average.

Materials Use in World 2 Countries

Shelter	Traditional wood, metal, and polymer concretes are dominant. Extruded recycled plastics in construction members are replacing wood timbers.
Business and industrial structures	Advanced steels and glasses in use; high tech fabric structures are popular because of relatively low cost; small enterprises, which are dominant, favor traditional, familiar materials but also make heavy use of reclaimed materials of all categories.

Transportation	Carbon fiber reinforced composites for aircraft and high-speed rail structures; ceramic engines in trucks and other motor vehicles are generally imported.
Food and agriculture	Artificial soils and soil boosters and adjuvants are used in agriculture; membrane technologies are used for processing foods.
Clothing	The better-off individuals typically buy foreign-made clothing. There is some production of cheaper, mid-tech fabrics of polymer fibers, e.g., Polyflex[tm].
Energy	Lightweight composites for frames of aircraft and rail vehicles bring energy savings in transportation. There is some experimentation with lightweight electric vehicles, and photovoltaic panels are increasingly used, particularly in rural areas.
Information technology	Silica-based computing and optical technologies dominate equipment produced in these countries. About 46% of the middle-income countries are networked with optical fiber to homes and businesses; most main trunk connections are either fiber or wireless.
Environment	Remediation technologies are appearing in dozens of middle-income countries intent on reclaiming spoiled lands and restoring resources such as forests.
Healthcare	High-tech implants are rare; surgical adhesives, synthetic hemoglobin, artificial skin, and artificial arteries are more common.
Infrastructure and construction	Polymer concretes dominate civil works; steel is still more common than composites in new construction. There is frequent use of recycled materials such as glass, sludge, rubber, and slag, for paving.
Leisure and entertainment	Sporting goods, typically produced for export, often employ advanced polymers. Most leisure involves traditional activities.
Manufacturing	Global firms typically use advanced practices in their middle-tier country installations; nationals often build plants with imported used equipment that is up to 25 years old.

MATERIALS IN THE WORLD 3 COUNTRIES

In these poorest of the world's countries, many materials practices would not have looked out of place in the 1960s in the affluent countries. The arrival of modern materials is uneven, so that while some industries and infrastructure uses the most up-to-date materials, others use traditional materials.

It is not uncommon in World 3 to see carbon epoxy composites in high-speed rail lines passing thatched roof, concrete or mud huts. Cast-off goods such as clothing, vehicles, and appliances from the middle-income and affluent world are imported for reuse. Of modern materials, polymer fibers like Polyflex^tm have the greatest success because they are inexpensive. Where most advanced technologies appear is decided by large businesses, government, or international organizations.

In World 3, advanced materials have little chance for a market unless they are extremely inexpensive and offer some new capability. In South Asia and Africa, the Chinese have found a nearly insatiable demand for cookware made from shatterproof, cheap, heat-tolerant polymers. This cookware can be used on electric, gas, or wood stoves and ovens, and lasts two to five years before embrittling. People have bought over a billion of the plastic pots and pans so far.

The poor countries are also frequent buyers of reclaimed and recycled materials from World 2 and World 1 countries. They likewise have thriving activities in the scavenging of wastes from old landfills, industrial scrap heaps, and municipal waste-processing centers. By some estimates, 37% to 54% of urban solid waste in the destitute countries is collected for sale or reuse completely outside the formal economy.

The table below illustrates the state of materials for this part of the world's population by briefly examining each of the categories of materials across society used earlier for the United States.

Materials Use in the Poor Countries

Shelter

Traditional materials dominate, but advanced polymers and reinforced concretes are commonly used if they were introduced through international aid programs. Fabric roofs and structures have become popular.

Business and industrial structures

Small enterprises are likely to use traditional materials; large businesses may be as advanced as their in affluent country counterparts, depending on profitability and ownership. International corporations operate advanced facilities in many of these countries.

Transportation

Pedal and foot transportation dominate. Roads are made of a mix of materials, some using advanced paving technologies and materials and others traditional materials or dirt. Vehicles and aircraft are imported, and, consequently, high-tech materials have been introduced.

Food and agriculture

Emphasis is on the transportation and storage of food to minimize spoilage and loss to vermin. Innovations in silos and cargo containers include material coatings that ward off pests. Modularized cargo containers also help in storage, transport, and delivery logistics; at the user end, plastic bags are ubiquitous along with recycled food storage containers. Home food storage includes airtight and insulated polymer containers that were first introduced by the U.S. AID program.

Clothing

The majority of the poor wear cheap clothing, usually of polyester or Polyflex™, mass-produced in Asia. Some plants use recycled polymers for fiber production. In some countries, fabric from locally grown cotton is common.

Energy

As in the middle-income countries, photovoltaics have seen rapid growth in the poor countries. People who can afford the systems are likely to turn to them out of frustration with intermittent power grid service. The systems by necessity are portable, so they can be locked up or carried away to avoid theft. Small and large businesses are major users. Communities may have a shared panel for energy in a pump house or community center.

Information technology

Most countries have yet to install optical fiber networks. Computing technology comes from the advanced nations.

Environment

Particularly relying on international environmental programs and technology transfer, these countries often have access to advanced materials for use in environmental protection and remediation. Costs still limit their adoption of more technologies.

Health care

Advanced health-care technology is available only to the wealthy. Materials for implants and prostheses, where used, are of last-century vintage.

Infrastructure and construction

Paving materials consist of traditional materials with some use of advanced concretes. Modular and factory-built structures are largely absent.

Leisure and entertainment

Sports equipment, typically imported, brings some advanced materials. Most activities involve traditional materials and common, low-technology materials.

Manufacturing   Mid-technology materials are used in large-scale operations; materials processing technology is well behind that of World 1.

Critical Developments 1990-2025

Year	Development	Effect
1990	BMW and Volkswagen establish experimental plants for automobile disassembly.	First large-scale industry efforts to create closed-loop materials systems in manufactured goods.
1998	The Allen Act.	Outlawed the export of hazardous wastes without a demonstration that the materials would be safely handled in the recipient country; drove waste-handling organizations to search for new processes for making hazardous waste safe or reusable.
2005-7	Superconducting power coils come into use in the United States and Europe.	First routine applications of superconductors in energy.
2009	ISO establishes materials characterization standards covering composites and other advanced materials.	Greater capabilities for recycling and reclamation internationally; ease of finding new applications of materials, managing materials choice in product design.
2011	International Agreement on Standards in Construction.	Established guidelines for the use of standardized materials and parts in construction, appliances, vehicles, and other durable goods.
2011	The Semayle Act.	Adopted the International Materials Characterization Standards and the associated materials handling and processing standards; expected to lead to greater international exchange and sale of wastes and reclaimed materials, raising the efficiency of resource use worldwide.
2013	Recycling and Reclamation Act.	Established industry and community guidelines for designing reuse into materials systems, from manufacturing to waste removal and reclamation.
2010s	First large-scale applications taking full advantage of nanotechnology.	New capabilities in medicine, materials processing, electronics, and photonics.
2024	Convention on Environmental Balance gets its 30th through 32nd signatories.	Convention begins to achieve critical mass in efforts to make materials and other resource use more sustainable, and to limit environmental damage from materials in use.

Unrealized Hopes and Fears

Event	Potential Effects
Catastrophic shortages of raw materials; petroleum crisis shoots up the prices of petrochemical-based materials.	Return to traditional materials; natural fibers, wood, ligninocellulose as a polymer feedstock; innovation in plant and animal fiber production; continued preeminence of glass, ceramics, and metals.
Complete closed-loop materials use in the affluent nations; society is virtually waste-free.	An order of magnitude drop in energy costs in manufacturing, and in raw materials savings, cost savings in preserving value added, saving costs of waste disposal.
Ambient temperature superconductors.	Superconductors outcompete optical fiber, proving faster for communications; an ambient temperature superconductor would reverse the fortunes of photonic communication and restore the dominance of electronic communication.
Discovery of safety problems with composites or composites manufacturing.	Return to preeminence of more traditional materials such as plastics, concrete, wood, and metals.
A U.S. ban on once-through plastics, unilaterally handed down by the government.	Industry decries the law, threatens to raise prices on nearly all consumer goods; taps public support against the ban.
A machine is developed to make nearly any material, atom by atom, from a stock of elements at low cost.	Materials engineering is turned on its head, mostprocessing is replaced by the technology; subsequent drive to develop processes that recycle any material back into its constituent elements.

7

WORKING TOWARD A SUSTAINABLE WORLD

Environmentalism is the underpinning of global society that has emerged over the last 40 years to influence business, government, and daily life. It reflects a worldwide orientation to a systems approach that connects nature and people.

Sustainability is the organizing concept that rallied support for environmentalism. Sustainability redefines the relationship of humanity to the planet from one of open-ended exploitation to one of dynamic balance between short- and long-term human needs. Humankind's actions in this balance continue to shift from controlling to managing.

The transition to sustainability is creating more regulated societies worldwide today compared to 1990. As with any significant social change, hardship and austerity were anticipated. Those fears were unfounded. Although the progress along the learning curve to sustainability has been bumpy, it has not been harsh. Some World 1 affluent nations, such as the United States, Japan, and those in the EC, are past the transition. Sustainable practices are firmly embedded in the daily lives of people in these nations. Surveys consistently show that people in these countries feel they are better off today than they were a generation ago. They are also more optimistic about the future.

The worldwide market for environmental goods and services has grown from $200 billion in 1990 to over $2 trillion today. In the United States, for example, 30% of GDP growth over the last 30 years has been devoted to environmental cleanup or preservation. The change in national accounting systems to include environmental costs—once seen as minor externalities—lowered GDP growth for a while. Previously, the use of resources such as forests, fisheries, and farmland, was counted solely as consumption that contributed to GDP growth. Under the new system of national accounts, natural resources are assigned economic value as assets and their use counts as depreciation.

The progress made by World 1 nations in environmental matters, such as stopping chlorofluorocarbon (CFC) production, cleaning water supplies, and reducing solid waste, is often overshadowed by the negatives in less well-off nations. Problems that are well in hand in affluent nations still plague many middle-income and poor nations. Deforestation, erosion, and toxins in the air, water, and waste dumps are part of people's lives in these nations. Wars, such as the Bangladesh-Assam, India conflicts of 2001 and 2009, the famine in Sudan in 1999, the cholera epidemic in Guatemala in 2009, and the millions of refugees throughout the last 30 years have often been triggered or worsened by environmental degradation.

The focus on the environment has evolved

Over the last 35 years, the global approach to the environment has evolved from cataloging the problems to actions to deal with them. A strong anticipatory capability is in place worldwide. Environmental interest groups have been instrumental in pushing environmental issues onto policymakers' agendas.

Today global society is moving toward preventing problems. As restoration becomes more firmly in hand, it is expected that enhancement will be the next step. Enhancement includes managing the planet, such as replenishing aquifers, overseeing species evolution, and avoiding natural disasters.

The Evolution of Environmental Strategies

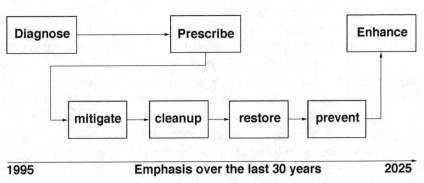

The key environmental themes of the last 35 years have been:

- sustainability as a fundamental global value

- new issues and concerns

- new strategies and tools

- science improves capabilities for dealing with the environment, with ecology as the core environmental science

- increasing management of the global commons

Sustainability as a fundamental global value

Sustainability is loosely defined as meeting the needs of the present generation without compromising the ability to meet the needs of future generations. A sustainable society has informational, social, and institutional feedback mechanisms to keep population growth and resource use within the constraints of carrying capacity. Sustainability is not a new idea, but bringing the world from acceptance to implementation is new. Developing and refining what these sustainable living criteria are continues. Limits to our ability to use the planet's resources are now acknowledged. The most commonly used number for the Earth's ultimate carrying capacity is about 14 billion people. Compared with today's 8.4 billion, we are already close to the limit.

Lessons from overshooting carrying capacity

Fisheries were an early example of how exponential growth in demand could exceed capacity. Aquaculture has had to substitute for and help bring back areas that were overfished and depleted. Capture fishing became much more targeted, largely in response to pressure from environmental interest groups such as Greenpeace. Switching from a yield-maximizing to a conservation orientation developed better stock assessment, better understanding of relationships between fish populations and exploitation, and between habitat degradation, natural environmental cycles and conditions, and populations.

A key distinction is made today between growth and development. To grow is to increase in size by the assimilation or accretion of materials; to develop is to expand or realize the potential of existing materials. For example, U.S. energy use has not grown from 1990 to 2025. Available energy is put to more efficient use and has led to increased economic development.

Birth control technologies may be turning the corner in combating overpopulation

Exploding population growth has long been implicated as the primary culprit of environmental degradation. Too many people pressure too few resources. Many countries approached the limits of their carrying capacity in the 2000s, while others surpassed them. Although population control measures failed in the past, there is hope that future calamities from surpassing carrying capacities will be avoided.

Demographic models had been predicting disaster. They projected growth accurately, due to the robustness of population patterns. They also targeted urbanization and led to efforts to prepare for the inevitable influx of people. Businesses continually used microdemographics to target markets but also included environmental and political risk analyses of the viability of areas to support their numbers. For example, Dow chose not to invest in Calcutta in 2004, despite the large attractive market, because its risk analyses revealed that the supporting infrastructure would be overwhelmed and likely lead to political turmoil, which indeed was the case just five years later.

Why sustainability?

The transition to sustainability was viewed as inevitable by the turn of the century. Governments accepted the choice between planning for an orderly transition or taking their chances with chaos and disorder. Population and economic growth were outstripping the carrying- and waste-absorbing capacities of biological and technological support systems.

Sustainability caught on because it has the three key elements of a durable, long-term, social movement: a description of the present situation, an account of the means of resolution, and an image of some desired future state. It links theoretical areas and appeals to a broad and diverse public. Although it takes on a quasi-religious fervor among some adherents, to most it is simply common sense.

Protecting the environment by adopting sustainable living practices is firmly ingrained as a widespread value in affluent and some middle-income nations. Poorer nations, dealing with pressing survival needs, feel that they must choose economic growth needs over those of the environment when the two are seen in conflict. Sustainability is not a fixed event, but a process. It began in World 1 nations as a fringe idea and has gradually moved into the mainstream. Sustainable practices became first established in affluent nations because their people were willing and able to spend money to support the concept.

Even though World 2 and 3 countries often act otherwise, sustainability has been taking root there as well. These countries have responded to World 1 countries' calls to protect the environment by saying they will do what they can afford. When the affluent countries began providing substantial funds after the turn of the century, the middle and destitute countries followed through in adopting environmentally friendly practices. When money is not available, however, the environment loses out to economic needs.

New issues and concerns

Population control tops the environmental agenda today, because overpopulation drives so many other environmental issues. Global warming and nuclear matters, primarily waste disposal, have been predominant concerns for decades. The International Global Warming Federation was formed in 2000 to address global warming, and nuclear matters were given higher priority by a strengthened International Atomic Energy Agency after the turn of the century.

The global scope of warming, nuclear-waste disposal, and other issues call many practices into question. For example, in addressing global warming, agriculture, forestry, and natural resources practices have been examined. A primary goal has been to restore the global carbon balance. The shift in environmental concerns between the 1990s and 2025 is shown in the table below.

The Shift in Environmental Concerns: 1990, 2025

Top Environmental Concerns: 1990	Top Environmental Concerns: 2025
1. global warming	1. population control*
2. nuclear matters	2. global warming and associated effects
3. ozone layer deterioration	3. nuclear matters
4. tropical forest shrinkage	4. enhancement and managment* • natural disasters • guiding evolution • replenishing aquifers • macroengineering projects
5. soil erosion (desertification)	5. sustainable agriculture
6. unsatisfied pressure for sustainable agriculture	6. effects of biotechnology*
7. nonrational energy production, distribution, and allocation	7. deforestation
8. acid rain	8. orbital debris
9. decline of quality of life in rural areas throughout the world	9. water rights*
10. decline in quality of life in urban areas, especially in World 3 countries	10. ocean pollution*
11. preserving Antarctica	11. immune system overload from toxics, mold, and fungi
12. badly managed middle and destitute industrialization with associated trans-border problems	12. ozone depletion
13. litter and debris in orbit in near space	13. hazardous-waste transportation*
14. mining wastes and mine excavations	14. exportation of polluting technologies*
15. worldwide attacks on coral reefs	15. battery disposal*
16. side effects of civil works such as dams and canals	16. soil erosion
17. coastal areas, affected by economic development and pollutants	17. noise pollution*

* newly important issues

The focus has shifted from problem identification to problem resolution. Issues that remain in roughly the same order of priority that they were in in 1990 are more likely to be resolved today. Similarly, more progress has been made in dealing with issues that fell in the rankings, such as ozone depletion, soil erosion, and deforestation.

The greatest gains have been made in energy matters, acid rain, improved quality of life in rural and urban areas, preserving Antarctica, managing industrialization and civil works, mining wastes and excavations, and preserving coral reefs and coastal areas. These issues are no longer primary concerns. Only one issue, orbital debris, rose in the rankings from 1990 to 2025. This is explained by the construction of the International Space Station and Moon Base, as well as the proliferation of satellite networks.

Eight new issues have emerged as top environmental concerns, in addition to population control. They were of concern in 1990, but not as prominently. Their importance today reflects intensifying effects and progress made on resolving other issues. They also raise new concerns as they create new effects. Biotechnology, for example, is being used on a much larger scale and with wider applications, leading to new concerns.

More sophisticated technologies can detect ever more subtle effects. In ocean pollution for example, improved measurements have identified problems that 1990s technology could not detect.

Other important environmental issues, although not among the top 17, are computer disposal, vehicle disposal, decommissioning or moving and reusing offshore platforms, overprotected animals (as a negative side effect of preservation), static electricity, biomagnetic effects, deteriorating suburbs, aesthetics and visibility, urban vermin, algae blooms, and odors.

Issue emergence: ocean pollution

The well-being of oceans was long taken for granted until sophisticated oceanography technologies developed in the 2000s sounded an alarm. Oceans that were once self-regulating and able to process the effects of human activities became overwhelmed. Sick oceans resulted from many small-scale problems coming together and overwhelming the ocean's defenses. Waste dumping, although illegal, continued clandestinely. Runoff from polluted rivers harmed ocean life. Red tides, blooms of microscopic algae, became far more common and killed off shellfish populations and harmed humans who ate contaminated shellfish. Overfishing continued to deplete fish stocks. Drift nets still used by some nations swept up ocean life indiscriminately. Coral reefs, key processors of the world's CO_2, were bleached, or killed off, by a combination of poor fishing and tourism practices and pollution. The microorganisms living on the thin surface layer of oceans that are key elements of the food chain were threatened as well. The combination of these factors put ocean pollution near the top of many lists of chief environmental concerns.

Ongoing issue: orbital debris

The space station and spacecraft were frequent targets of small orbital debris, which sometimes inflicted substantial damage due to the high speeds, about 35,000 kilometers per hour that objects travel in space. There were thousands of big objects, such as spent rocket stages and dead satellites, as well as screwdrivers and other objects in orbit, most about 300 to 400 miles above the Earth, along with trillions of tiny objects, such as paint flecks. When a tiny piece of an abandoned satellite crashed through a space station window during construction, extra shielding requirements became mandatory and the completion of the Earth-based Orbital Object Radar Tracking System (OORTS) was pushed forward to 1999. OORTS helped track the debris and avoided all but an occasional collision.

New strategies and tools

Progress in environmental matters has been considerable in affluent nations and smaller in middle and destitute nations. Money, expertise, and technology from World 1 are slowly being transferred to poorer nations in the interest of improving the planet's health. A four-step strategy, outlined in the box below, forms today's framework for considering environmental issues.

Plan to Address Environmental Issues 2025

Issues are typically dealt with by more than one strategy. The issues below are categorized according to where the primary emphasis is concentrated today.

1. *Mitigation* of current pollution. The affluent nations are moving away from mitigation to prevention. In the middle and destitute nations, however, mitigation still occurs much more than prevention.

- smog
- acid rain
- indoor air
- solid waste
- toxics

- noise
- macroengineering
- degraded infrastructure
- biomagnetic effects
- static electricity

- urban vermin
- new effects from biotechnology
- natural disasters
- odors

2. *Cleanup* of current polluting activities. The affluent nations have completed a great deal of their cleanup. Attention in cleanup has shifted to the middle and destitute nations.

- nuclear waste
- hazardous waste
- ocean dumping

- wastewater
- orbital debris
- computer disposal

- battery disposal
- decommissioning nuclear plants
- algae blooms

3. *Restoration* of degraded areas. Restoration considers interactions between communities and ecosystems. Large-scale restoration projects have been undertaken and completed in affluent nations, and on an experimental basis in middle and destitute nations.

- strip mining/wastes
- groundwater contamination

- de/reforestation
- tropical forests
- aesthetics

- erosion/desertification
- coral reefs
- coastal areas
- abandoned industrial sites

4. *Prevention* of future problems, typically at the source. At this point, almost exclusively an affluent country activity.

- global warming
- ozone depletion

- preserving biodiversity
- resource depletion

- oil spills
- nature reserves

Source: Adapted from the United Nations Environment Program, *State of the Environment 2020*, (NY: UNEP, 2020), p. 23.

2025

The previous method of dealing with air, land, and water issues as separate categories was recognized as flawed, because in many cases, cleaning up a problem in one area merely transferred the effects to another. Today's framework is an operational, technologically oriented framework to achieve sustainability system wide. The social and technological tools that were deployed in support of this strategy are:

Technological Tools

- information technology and remote sensing
- the 3Rs[1]
- cleaner and more efficient energy technologies
- environmentally friendly manufacturing
- biotechnology
- sustainable agriculture

Social Tools

- international agreements
- new regulations
- stricter regulatory enforcement
- new national accounts
- greening of business
- government plans
- market incentives
- trade policies
- environmental curricula
- setting acceptable, reasonable risk standards

[1] recycling breaks products down to primary materials; reclamation is the salvaging and reuse of parts and components of complex systems; remanufacturing is sending whole units back to the factory for refurbishing.

The tools are highlighted in the U.S. case study later in this chapter.

Ecology, the core environmental science

A new family of environmental sciences is in place today with ecology at its core. The scientific store of understanding about the environment has been growing exponentially. Today's more refined monitoring of environmental conditions has led to better understanding of naturally occurring cycles, such as drought and flooding as well as fire and regrowth. Sophisticated information systems and networks have brought together the large amount of data gathered in the 1990s, facilitated its conversion into information in the 2000s, and generated knowledge in the 2010s that in turn has led to today's growing wisdom in environmental applications.

Growing Sophistication about the Environment

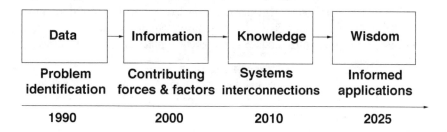

The contributions of the environmental sciences to the environment are summarized in the table below.

Some Contributions of Science to the Environment

Science	Contribution to the Environment
Ecology	Ecology matures into the core science of environmentalism, evolving from the micro to the macro. It produces solid scientific understanding of progressively larger biotas over longer periods of time.
Biochemistry	Biochemistry improves understanding of effects of hazardous materials on humans. It also maps the interactions of the biotic and abiotic environment.
Sensing	Local and remote sensors monitor the environment, often in large systems coordinated by information technology. The Earth Observation System as part of the Resource Tracking Model is a pioneering system.
Geology	Geology contributes to more informed mining and waste disposal, restores damaged ecosystems such as underground aquifers, and plays an especially important role in the disposal and storage of radioactive waste. Prediction of earthquakes and other natural disasters is accurate to hours, days, or weeks before the event.
Forestry	Reforestation has been dramatically sped up by advanced ecology and forestry techniques aided by biotechnology. Strategies include specialized fertilizers, soil treatments, harvesting technologies, and genetic enhancement.
Climatology	Close to full-scale regional climate management is made pactical by better models in three-dimensional (3-D) and real-time. Simulations now extend 200 years or more.
Oceanography	Ocean interactions with climate are better understood. The new discipline of coastal geography flourishes. Marine engineering was revisited and is a vital tool in preparations for the effects of sea-level rise from global warming. Fisheries management is restoring overfished areas.

Soil science	Soil science provides valuable input to restorative ecology, dealing with coal mines, former manufacturing facilities, nuclear sites, or degraded forest areas. Decontamination, remediation of depleted or acid soils, erosion prevention through methods of cultivation, drainage systems, groundcovers, and soil stabilization through additives or synthetic supplements are other areas aided by soil science.
Toxicology	Toxicologists are able to inventory the massive amounts of toxics from many sources introduced into the environment. Sophisticated instruments detect increasingly minute toxins and enable more accurate tracking to the source.

Increasing management of the global commons

It is axiomatic that global problems require global solutions. Global warming, nuclear-waste disposal, ozone depletion, deforestation, and ocean pollution have transborder effects that are beyond the means of individual nations to address. The monitoring and tracking capabilities of the Resource Tracking Model (RTM), an international network of sensors, computers, and databases, is the linchpin of international agreements. It assures signatories that cheating will not go undetected. For example, the international dumping of toxins is closely regulated, especially since the adoption of the 2012 Convention on Hazardous Waste Export, which outlawed the export of hazardous wastes if there is no demonstration that the materials would be safely handled in the recipient country.

A precedent for global environmental agreements was set by the Montreal Accords in 1987. The scale of the agreements has expanded from including mostly affluent nations to bringing World 2 and 3 nations aboard. The degree of their participation and cooperation is directly related to the affluent nations providing vital financial and technical assistance.

Global management and sustainabilty are complementary concepts. The international scientific community has been giving increased attention to global, geophysical, and biological issues for decades. Recently, 21,356 student scientists from universities across the globe signed a declaration to work toward totally managing the planet within the next 100 years.

Mechanisms similar to the U.S. environmental impact statement of the 1980s and 1990s, but far more streamlined and user-friendly, are used by most World 1 countries. Simpler and less comprehensive ones are used in World 2 and World 3, the middle and destitute nations. Negotiations to develop international environmental standards have been underway for five years now. It appears that at least another decade will be required to reach agreement.

Macroengineering has been a useful technological tool for global management. For example, earthquake prevention systems traversing hundreds of kilometers of terrain have been constructed along the San Andreas fault, and tidal enhancement programs to reclaim the Adriatic are underway.

In three years, the United States, Korea, and Japan will perform the first experiments to dispose of nuclear waste by embedding it in glass marbles and injecting it into the subduction zones between the geological plates, which would then carry the wastes below the crust of the Earth.

Progress toward sustainability varied around the world

Environmental problems come from different circumstances in different regions. They arise primarily from extravagent lifestyles in World 1 nations and from population growth in World 2 and Word 3 nations. Affluent nations are reducing their consumption and waste creation. Over the last 30 years, World 2 nations mirrored many of the harmful environmental practices of the affluent nations when they industrialized. For example, their energy and pesticide use increased. They are roughly a generation behind the affluent nations today in environmental responsibilities.

An alternative approach to population control: block the immigration outlet

Emigrants fleeing environment-based problems in their countries are no longer welcome across the world. World 1 and, increasingly, World 2 nations stopped further immigration of these environmental refugees into their countries. The reasoning behind the no-immigration policies is that taking in refugees from mismanaged countries merely enabled these countries to continue their errant ways. The affluent nations feel that they have served as a safety valve and prolonged misery in the long term by playing that role. Refusing to take refugees forces nations to face their exploding populations and enact reforms. Although there is no formal international policy, there are many national policies.

People with desirable job skills, or lots of money, however, are still welcomed in many nations.

In many cases, environmental problems deriving from poverty and population growth spiralled World 3 societies into collapse or near collapse. Short-term survival needs overwhelmed the carrying capacity of some countries.

On the road back from disaster

Haiti, profiled in a case study below, is a prominent example of countries in collapse. The famines, disease, refugees, and deaths indicative of collapse have occurred or are occurring across boundaries, in single nations, or parts of nations. For instance, significant portions of sub-Saharan Africa, Bangladesh, and the São Paolo metropolitan area of Brazil have been classified as in collapse.

Brazil demonstrates the interrelated effects of rural and urban devastation. São Paolo has problems typical of overpopulated urban areas with overwhelmed infrastructures, and heavy air pollution, while surrounding rural areas are deforested. Villagers fleeing to escape the declining crop productivity of the countryside further exacerbate Sâo Paolo's infrastructure problems.

Slowly recovering

Russia and Eastern Europe are slowly undoing the damage from their centrally planned past, in which the commitment to industrialization inflicted massive environmental stress. In the Commonwealth of Independent States (CIS), for example, there is extensive river contamination, the drying up of the Aral Sea, the shrinkage of the Caspian Sea, and serious pollution of Lake Baikal. The story is similar, and in some places even worse, in Eastern Europe.

Getting better all the time

China and Southeast Asia are positive examples of the early stages of a transition to sustainability. China's population growth is under better control than that of much of Worlds 2 and 3. Industrial development, once virtually unconstrained regarding the environment, is being brought in line with sustainability principles. Energy conservation measures and the transition away from fossil fuels are well underway. Improving living standards, urbanization, and increasing environmental awareness are leading to pressure to clean up the environment.

The success stories

The environmental situations of the EC, Japan, and the United States are the globe's bright spots. Their practices are in rough parity today. The EC countries are the leading practitioners of sustainability within their borders. Japan leads in exporting environmental services and products, and the United States has achieved the best balance between internal and external practices.

The EC has had the world's brightest environmental prospects over the last 35 years. A crucial factor was the influence of their environmentally-based political parties, e.g., the Greens. Although the Green Party has disappeared and its members have integrated into the dominant parties, their environmental principles are now mainstream. The most outstanding country in implementing sustainability is the Netherlands, which totally manages its country in line with sustainabilty. The stronger environmental countries of the EC have forced the weaker ones, such as Spain, Portugal, and Greece, to meet the environmental standards of the community as a condition of trade.

Japan has long been the leader in developing markets for environmental technology and expertise. The nation's businesses and government leaders were the first to recognize the enormous potential of the environmental market. Domestic businesses were also among the leaders in adopting sustainable practices. The Fanuc robotics factory, once famous for being the first totally automated factory, is now recognized for being the first waste-free factory. All its materials are either recycled, reclaimed, remanufactured, or used as inputs for its on-site waste-to-energy facility.

The United States made up lost time with its Selling Sustainability campaign. Businesses that had adopted a wait-and-see approach to sustainability became alarmed at the success of the EC and Japan in capturing new markets and improving the competitiveness of their domestic industries. Business leaders came to accept environmental regulations because they have been applied fairly and have not been a competitive disadvantage. Most businesses enthusiastically endorsed the Selling Sustainability campaign and mobilized their resources to improve their own practices. Today, U.S. businesses are among the world's most environmentally friendly and internationally competitive.

Where environmental issues stand

Progress in dealing with environmental issues varies widely across the globe. Restoration and prevention is underway in World 1. World 2 nations are primarily in the cleanup stage. Destitute nations are either not yet addressing their environmental issues or are mitigating their effects.

The assessment below analyzes the progress of the affluent, middle, and destitute nations on the top 17 environmental issues today. The table is not intended to suggest that all nations within each category are acting in unison. The middle-income nations in particular range widely in their scores. Some middle nations' performance resemble the affluent nations, and others are closer to World 3 conditions.

Environmental Issues Assessment Stages
of Environmental Activities

Issue	World 1	World 2	World 3
1. population control	4	1	1
2. global warming	4	2	1
3. nuclear matters	2	1	0
4. enhancement and managing	1	0	0
5. sustainable agriculture	3	2	1
6. effects of biotechnology	1	1	0
7. deforestation	3	2	1
8. orbital debris	2	0	0
9. water rights	3	2	1
10. ocean pollution	2	1	1
11. immune system overload from toxics, mold, and fungi	2	1	1
12. ozone depletion	4	2	1
13. hazardous-waste transportation	4	2	1
14. exportation of polluting technologies	4	0	0
15. battery disposal	2	2	1
16. soil erosion	3	2	1
17. noise pollution	4	1	0

Key

The five stages of environmental management are:

Stage 0	Little or no action underway
Stage 1	Primarily mitigation
Stage 2	Primarily cleanup
Stage 3	Primarily restoration
Stage 4	Primarily prevention

WORLD 1—PROGRESS ON ALL FRONTS

The story in these nations is one of considerable progress. Significant environmental problems either have been or are being mitigated, cleaned up, restored, or prevented. Thirty-five years ago, more problems were being identified than addressed. Those that were being addressed, were mostly at the mitigation and cleanup stages. There has been gradual progress up the scale from mitigation and cleanup to restoring degraded areas and preventing future difficulties.

Environmental awareness worldwide was initially promoted by environmental public interest groups. Their extensive lobbying and education programs eventually nudged environmental issues onto government agendas and into schools. Their success has reduced their numbers today, although they are still a potent political force.

The U.S. environmental movement was arguably the strongest worldwide at the end of the century. There were over 10,000 environmental groups in the United States in the 1990s. But the United States government lagged the EC and Japan in enacting sustainability policies. The EC took the lead in using tax policy by instituting the carbon tax. Similar initiatives in the United States initially failed. Japan was first in capitalizing on green business opportunities. Again, the U.S. government initially failed to provide incentives to help U.S. businesses get sustainable.

After the turn of the century, U.S. government leaders realized that they no longer could ignore the success of the EC and Japan. Where the United States once had the ability to force other countries to play by its rules, it now had to go along with its competitors or lose market share. The EC market is larger than the United States. The combination of EC and Japanese policies set a precedent that the United States could not ignore.

California had led the way in the United States, especially in moving to alternative energy sources. In 2002, the Selling Sustainability campaign officially signaled that the United States was committed to sustainability.

A CASE STUDY— The United States at the forefront of sustainability

A four-pronged strategy of mitigation, cleanup, restoration, and prevention characterizes the U.S. approach. The goal is to do less mitigation and cleanup and more restoration and prevention. In the past 35 years, the first two strategies have been predominant. Restoration and prevention efforts picked up over the last decade and are gradually becoming more prominent. An emerging strategy is enhancement. The cloud seeding project over agriculture zones in the Midwest is an example of enhancement.

2025

An experimental approach is required for dealing with new problems and issues. The multifaceted nature of environmental problems requires comprehensive, systemic, and holistic strategies. The box below illustrates the systemic nature of environmental problems.

Multiple approaches to global warming

Some global-scale issues require action on multiple fronts. Global warming, for example, has been addressed by each of the four primary strategies for dealing with environmental problems. Efforts are ongoing to change the source, modify the effects, as well as accommodate and adapt. There have been distinct effects by region, hence different strategies as well.

Global warming has affected ecologies, sea levels, coastlines, and polar ice. Ecological and climate zones, and therefore agricultural zones, are shifting, with warmer belts moving north, as global warming continues and rainfall patterns change. These changes have in turn led to broad ecological shifts and small-scale biota disruption. The 30cm sea-level rise due primarily to ocean warming and the melting of the Antarctic icecap since 2000 threatens to destroy wetlands, silt up river channels, and clog river deltas. Although the rise is significant in itself, perhaps the more important effects come from the effects of severe weather, which in turn are intensified as a result of the sea-level rise.

National and international agreements on global warming proliferated. For example there was the passage of the Clean Air Act Amendments in the United States in 1990, the formation of the International Global Warming Federation in 2000, and the 2011 International Agreement on the Troposphere, to name a few.

Strategy	Effects
Mitigation	Strategies range from cloud seeding, artificial wind barriers for creating thermal zones to increase rainfall, to building civil works in threatened coastal zones, or to neutralizing or purge the bad actors in situ. For example, one proposal is to use lasers to dissociate CFCs. Other strategies include placing solar mirrors in space to deflect sunlight. Another is deliberately polluting the upper atmosphere with soot or dust. A long-term, last-chance option of using biotech to adapt plants and animals to the changed conditions has been proposed but not implemented. Under research are mechanisms for controlling the annual formation and dissolution of polar ice.
Cleanup	The world's poor countries cannot effectively build the civil works necessary to hold back the sea if it rises further. Damage has been moderate so far, with the exception of large-scale flooding in Bangladesh in 2016 and the Maldives in 2019.
Restoration	Detection of phytoplankton depletion by the Earth Observation system has led to biotechnology-based remediation efforts. Projects have been underway to determine the optimum balance of crop, genetics, rainfall and soil treatments.
Prevention	Dealing with warming at the source included measures to ban all new CFCs worldwide in 2005. Remote and other sensing devices monitor compliance. Another strategy is to move people out of harm's way. Many people in coastal communities have moved to higher ground.

Tools used for achieving sustainability

Social and technological tools support the four-pronged strategy to achieve sustainable societies.

Ecoterrorism

Disrupting the environment is a new terrorist tactic used worldwide. Terrorists feel that damaging a region's water supply, for example, is more effective than taking hostages because protecting the environment has been a high priority of most governments. The poisoning of the Washington, D.C., water supply in 2001 was a public relations success from the terrorists' point of view, although no one was poisoned and the switch to other sources was fairly smooth. The reservoir was cleaned and back to normal within a month. Other strategies, actual or threatened, include burning old-growth forests, releasing predators into crop and grazing areas, and kidnapping protected species.

International agreements and summits for the global commons

The United States participates in scores of formal and informal international agreements regarding the environment. The environmental arena provides the strongest evidence of the trend toward global management. The agreements provide the institutional mechanisms first to negotiate accords among affluent nations and, more recently, to enable the affluent nations to transfer funds and technology to the middle and developing nations to encourage them to adopt sustainable practices. The clear trend is toward international supervision of global commons, shaping use, regulation, costing, and access.

The RTM database is the foundation for global environmental management. It is eagerly tapped by the public and private sectors, as well as individual citizens. The UN's Sustainable Development Commission of the 1990s was a focal point of international efforts to achieve sustainability.

National plans combine regulations, market incentives, and stricter enforcement

Today's Department of the Environment is firmly in control of the government's sustainability efforts. The old Environmental Protection Agency (EPA) was re-formed into a cabinet-level department in 2001. Programs to assess the U.S. stock of natural resources, such as the Geology Survey, Biological Survey, Forest Survey, etc., were coalesced under the DOTE. Although initially suffering from turf battles, a new leadership headed by Secretary Jennifer Goldman has successfully asserted itself and brought recalcitrant agencies in line.

Triage for rendering international environmental assistance

Criteria for rendering assistance	Criteria for not rendering assistance
• a little assistance can almost surely help	• the issue, although acute, is of no large global concern
• where help is rendered, it is of generally high importance within that country	• the in-country situation is of such a nature that attempts to act are likely to have little or no lasting effect
• help is likely to have some permanent or long-lasting effect, and be institutionalized for follow-through	• it is a case of continuing, enlarging crises
• the polity is prepared to act to protect the future	• the polity is de facto indifferent or in competent to act to protect the future

The tools for implementing a "do what you can" policy are:

• targeted education and indoctrination

• internal political requirements as conditions for aid and assistance

• an action agenda of political, economic, and demographic reform mutually agreed on

• strict accountability measures

• rapid feedback to/from the donor

National commissions to develop visions of sustainable societies proliferated over the last 20 years. They drew on extensive environmental databases and networks, as well as the successful example of Japan's New Earth 21 plan. Innovative regulations were needed to carry out their recommendations. Market incentives have been a useful supplement in many cases. The enforcement of regulations is much stronger today than in the past, when sufficient regulations often existed but were ignored with impunity due to lack of resources for enforcement.

World 1 countries still try to evade the issue of technology transfer to aid Worlds 2 and 3 by arguing that new, environmentally sound technologies are within the province of private industry. Evasions on this ground are becoming less effective, because the days of unfettered free trade are long gone. Trade policies now incorporate environmental values. A particular strength of the United States is the arrangement of government-funded precommercial research consortiums for developing environmental technology for middle and destitute nations.

New methods of national accounting internalize externalities

The once so-called externalities are now internalized in national accounts. The costs of pollution are part of the bottom line. Including these external costs has created incentives and led to actions to reduce them. Environmentally friendly projects that would have been too expensive under the old accounting system are now money-makers. Items that were once virtually free, like clean water and air, are no longer free under the new accounting system. Their use is more carefully managed as a result.

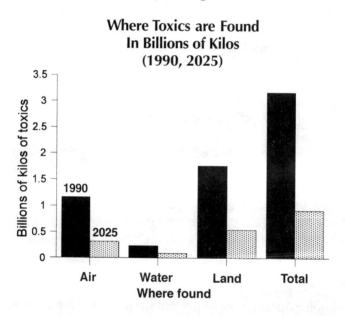

Where Toxics are Found In Billions of Kilos (1990, 2025)

Information technology and remote sensing provide sophisticated global databases and models

The global store of information about the environment has increased exponentially over the previous century. Much of this data used to be raw and unexamined as sensing technology was providing data faster than it could be usefully assimilated. Advances in information technology are at last catching up with the overload, and are leading to profound gains in knowledge. Global models are much more sophisticated than the relatively simplistic model used in the 1972 classic *Limits to Growth*.

Environmental crime detection

Sensing technologies backed by massive databases are key tools in the hands of environmental detectives. Multiple strategies make it difficult to evade responsibility for polluting. Vehicle emissions, for example, are tested at inspection, in parking garages, and by roadside detectors. Although some are able to beat the systems, most find the costs of compliance are lower than those for environmental violations.

Environmental police tangle with a criminal underworld that will, for a price, help companies get around environmental laws. Foiling toxic detectors and illegally dumping waste are examples of the large range of polluting crimes. Detectives employ forensic soil scientists to find out who is contaminating soil.

A vast array of remote sensing instruments are available: multispectral instrument arrays for surveying ecological damage from a specific event, infrared imaging for monitoring, high-resolution radiometers for crop assessment, imaging radars and scatterometers for high-resolution wind field mapping and ocean current mapping, and ozone sensors for the ground and stratosphere.

Sensing technologies play a key role in integrating environmental monitoring systems. Capabilities for detecting toxic materials, for example, are much greater today and helped reduce them. The box above demonstrates the success in reducing toxics. Making spy satellite data public in 1999 gave an additional boost to the store of environmental research knowledge. Sensing also enables in situ pollution detectors and earthquake detection.

Biotech as a tool

Biotechnology is the most important single new scientific capability for dealing with environmental problems. Its effects are often indirect in that it is an enabling technology for many sectors of the environment, such as agriculture and forestry.

Release of biotech products: successes and failures, but no disasters

Fears of a genetically altered microorganism running amok have proved unfounded. There has been no large-scale wipeout of a plant or animal population. This does not mean that all biotech trials have been a success. Complex interactions often lead to failed products. A first drought-resistant strain of wheat, for example, turned out to be susceptible to pests that traditional strains resist.

Biotechnology has a significant role in cleanup, for example, by detecting toxics or in engineering microorganisms that degrade specific toxics. Genetically altered or enhanced microorganisms are integral to processing and treating waste. Pollution control with biotechnology is now a multibillion-dollar industry.

Proposals to genetically manipulate plants and animals so that they are better suited to their local environments are gaining adherents.

Setting acceptable, reasonable risk standards has been an ongoing battle

It is difficult for people to accept the idea that not all problems can be solved and that priorities have to be set about which ones will be addressed, given the available resources. Setting priorities is a long arduous task. It is still going on, but it is becoming less critical as the primary problems are being addressed.

Lots of money had been wasted by overly strict standards or by bureaucratic procedures that locked millions of dollars into cleanup projects that would have been better spent elsewhere. Funds are now allocated on a weighted basis according to the four-step framework of cleanup, mitigation, restoration, and prevention. Although most funds are still for cleanup, prevention will soon become the top-funded item.

No definitive solution has been found for a principal risk issue: nuclear waste. It is still primarily in the cleanup stage. Research has been ongoing but unsuccessful in preventing waste at the source. The key technical issue in cleanup has been preventing radioactive material from contaminating groundwater. The cleanup itself has been slow. No long-term storage sites existed until after the turn of the century.

Today, the International Atomic Energy Agency monitors the international storage facilities that temporarily store the waste in specially constructed buildings. This interim solution is buying time for researchers to develop long-term alternatives.

The most dangerous radioactive waste continues to be commercial power-plant fuel. It accounts for only a tiny percentage of waste volume, but for over 95% of the radioactivity of commercial and military waste combined. Nuclear arms reductions have sharply reduced waste productions from weapons programs.

The most progress has been made in disposing of low-level wastes. Their volume has been significantly reduced by compacting, shredding, and incineration. Intermediate and high-level wastes have been dealt with by a combination of vitrification, transmutation, reprocessing, and synroc (sealing waste within synthetic rock).

The Banks-Hyde process developed in 2004 dealt with the difficult task of separating hazardous organic chemicals from radioactive waste.

On the prevention side, strict safety procedures for transporting the waste to storage sites have so far been effective in avoiding any incidents.

Strategies for Disposing of Radioactive Wastes

Strategy	Description	Status
Processing		
Reprocessing	The chemical dissolution of spent reactor fuel to recover fissile material.	Reusing irradiated fuel for power production also leaves behind potent radioisotopes. The overall volume of radioactive waste actually expands. Many countries, including France, the United Kingdom, the United States, and Japan, are now reprocessing.
Transmutation	Converts waste to shorter-lived isotopes through neutron bombardment.	All the countries doing reprocessing use transmutation to reduce radioactivity of the remaining waste.
Compaction, shredding, and incinerating	Used only for low-level waste.	Standard procedures worldwide.
Reracking	Replacing the original aluminum storage racks that hold fuel assemblies in storage pools with new racks with better materials such as borated steel to triple capacity.	This technique has been used longer than hoped for. Improved materials capabilities have kept it viable.
Dry casks	Dry storage has the advantage of not requiring an active cooling system.	Can be safe for interim storage for at least 100 years. The chief strategy at the moment.
Microwaves	Concentrates and solidifies highly radioactive liquid waste.	Used sparingly; glassification more popular.
Glassification	Highly radioactive materials are separated from other components of stored liquid waste and mixed with molten glass. The borosilicate glass cools and hardens inside steel canisters buried in the Earth.	France has been glassifying since 1978 and Belgium since 1985. The U.S. Yucca Mountain site uses this method as well. It has been the preferred method for long-term storage.
Synroc	Sealing waste within synthetic rock created from three titanate minerals and a little metal alloy.	Synroc so far appears more durable, more resistant to leaching and irradiation, and more suitable for burial in deep boreholes.

End disposal		
Geological burial	Bury waste deep in specially constructed repositories.	Many sites have begun storing waste in the last 15 years. Two sites are operating in the United States, at Yucca Mountain and the salt caverns near Carlsbad, NM. More sites are planned than have been completed worldwide.
Seabed burial in subduction zones	Injecting vitrified waste into sub-duction zones and let plate tectonics carry it to the Earth's mantle.	The first trial, involving the United States, Korea, and Japan is scheduled for 2028.
Space disposal	Rocket waste into solar orbit.	Proscribed by the International Space Agency in 2010.
Specially constructed building	Store waste above ground for up to 100 years.	The primary means today due to lack of suitable alternatives.

Environmental risk analyses are now a routine element of business decisions affecting the environment. Negotiations between businesses and local residents have evolved from an adversarial to a bargaining approach. It has taken a long time to build the trust that exists in most circumstances today.

Another significant change has been the government's role in some areas of risk analysis, which is shifting from a paternalistic "we'll take care of everything" approach to an enabling "here's the information, you decide" strategy. This has not yet taken firm root in the environmental area. The high priority attached to environmental cleanup, and the proven need for strict enforcement measures, have kept government in a paternalistic role in this area longer than many business leaders would like.

Environmental education, skills, and techniques significantly upgraded

Education begun in the 1990s bears fruit today. Schoolchildren are taught principles of sustainable development and how to care for the environment. More importantly, field experiments are an integral part of the curriculum. Students learn by doing, and as adults they are much more aggressive in acting against harmful environmental practices.

Local and international groups continue to research issues, educate the public, litigate when necessary, and organize citizens to press local and national governments to abandon environmentally destructive policies. The media has finally come to grips with the importance of environmental issues. Multimedia news formats include daily sections on the environment.

Environmental engineering is an area of booming job prospects. Hundreds of thousands of new jobs have been created for environmental professionals. International environmental science and engineering programs are in place under UN auspices. Education systems are adjusting worldwide to provide the needed skills.

The expertise of civil engineers today includes monitoring the infrastructure's effect on the environment. Environmental monitoring capabilities are built into the infrastructure, such as in roadside emissions testing. Closed-loop recycling of wastes has been successfully adopted in hundreds of communities across the country.

Business earns green from going green

The Japanese are the world leaders in the environmental technology market. Their business practices are environmentally friendly and they aggressively seize markets for environmental technologies and services. For example, they have long been leaders in exporting green manufacturing technology, and they are continually expanding into newer and broader areas. Japan benefits from its long-term focus on process rather than product technologies. Changes to benefit the environment take place mainly at the process level, which the Japanese are still more skilled at than Americans.

Booming market for environmental audits

Growing environmental regulations and a desire by some businesses and consumers to be proactive in going green created a burgeoning demand for environmental audits. In response, a wild proliferation of environmental auditing firms, often of shaky credibility, sprung up. The industry gradually consolidated into a number of solid, reputable firms. The passage of the 1999 Environmental Audit Certification Act by the U.S. government created performance guidelines.

In the United States, many industrial companies once targeted by environmentalists have become leaders in environmental technologies. New business opportunities are continually ripening overseas in areas such as perpetual-use containers, solar scooters, air purifiers, photovoltaics, thermally efficient materials, wind generators, maglev rail cars, solar heaters, and closed-loop water systems.

The CERES IV (Coalition for Environmentally Responsible Economies) principles, adapted by public-private commissions from the 1980s Valdez Principles, became popular with businesses in the 2000s that were using them to drive efforts to increase their efficiency. Businesses adopted them at first to enhance their public image, but businesses today are less reactive to environmental regulation and more proactive because it has been demonstrated again and again that it makes good business sense to be at the forefront of sustainability.

An important change was a shift away from overemphasis on obligations to stockholder returns. Taking their cue from Japanese and German business groups that were consistently winning market shares over the long term, U.S. public and private leaders supported the passage of the 2004 Bradshaw Act, which adjusted fiscal, securities, and tax policy to encourage companies to plan for the long term.

Labeling products as green was an effective tool late last century as sustainabilty was catching on. Environmentally sensitive consumers will alter their buying habits if given relevant information about products. Surveys continually showed that people are willing to pay a little more if it would benefit the environment.

The 3Rs reduce waste

Once-through societies are a relic. Programs to reduce waste have taken root in affluent societies. Simple curbside recycling of paper, cans, glass, and plastics mushroomed into measures mandating that manufacturers take responsibility for end use disposal of their products. Germany is the pioneer here. It has a comprehensive program designed to force companies to assume responsibility for disposing of the packaging used with their products.

Where Waste Ends Up

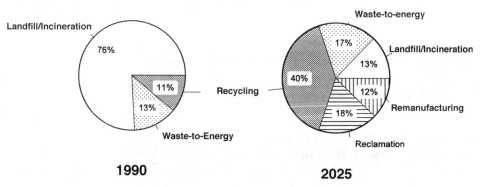

1990   **2025**

Figures for reclamation and remanufacturing not available for 1990

The key element in the success of 3Rs programs is stimulating demand for 3Rs products. Early recycling programs suffered from glutted markets, and they needed government subsidies to survive. Surpluses of recycled materials earned less on the market than it took to recycle them. Business alliances and the federal government's commitment to promote recycled materials (later to reclaimed and remanufactured materials) eventually stimulated a market large enough to make the 3Rs programs economically attractive.

The primary enabling capability for the 3Rs is the assembly of the Resource Tracking Model, an offshoot of the EOS. The RTM provides reliable estimates of a region's carrying capacity. The Melbourne Waste Minimization Conference is currently negotiating an international agreement on the 3Rs.

Taggants (chemcial identifiers) built into products greatly increased sorting efficiency for recycling. Once a labor-intensive task, it is now automated.

Incentives and disincentives are particularly effective stimulants in reducing waste. For example, requiring people to pay for trash disposal by the kilo reduced waste significantly. People are amazingly innovative when the opportunity to save money presents itself. Unfortunately, some people are using their innovative ability to avoid paying their fair share.

Manufacturing practices get the green seal of approval

The manufacturing sector is a great success story in the effort to clean up industrial practices in World 1. The transition to sustainability was made in this sector of the economy first. It became apparent during conversion that gains in efficiency from green practices paid for themselves quickly. Studies found that the net effect of environmental regulations on productivity was a small decline at first and a gradual gain over the longer term. The Zero Waste Proclamation, which was signed by 579 manufacturers in 2020 with an announced goal of zero waste by 2030, is representative of manufacturers' enthusiasm for waste reduction.

An effective tool provided by government is the DOTE's on-line system. It provides companies access to case studies of successful implementation of cleaner production practice across industry. It is eagerly tapped by thousands of businesses, who in turn provided funding to keep it up to date. Plans to make it available overseas met political obstacles, but it was done in 2011.

Energy cleans up its act

Reduced energy use provides the strongest indication of the state of a nation's environment. No other sector of the economy has such direct impact on the environment. The last 35 years have provided strong evidence for this connection. The remarkable reduction in energy use by the United States and other affluent nations correlates with their strong environmental performance. World 2 nations that increased their energy use typically added to their environmental problems.

Fears of global warming prompted the U.S. government to pass incentives to encourage a transition away from fossil fuels. The resurgence of the nuclear industry was boosted by government tax incentives as well. The Selling Sustainability campaign included a massive commitment to energy conservation.

Structural shifts in the economy toward information-based activities reduced energy use. Information technology, in terms of personal computer (PC) use, accounted for roughly 5% of electricity use in 1990. Electrical energy use in information applications has tripled along with a thousandfold increase in computing power per unit of energy.

The laggard in energy efficiency continues to be the transportation sector. The country is still heavily reliant on automobiles. Studies have continually proven that the U.S. automobile-based transportation system is less efficient than the rail-based maglev and mass-transit systems of Europe and Japan. Substantial gains have been made, however, in improving automobile efficiency. Single-fuel gasoline vehicles were banned in 2011, and their gas-electric, ethanol/methanol electric, gas, hydrogen gas/electric hybrid successors are far more efficient. Electric vehicles today can travel almost 500 kilometers per charge, up from 120 in the 1990s.

Sustainable agriculture and forestry practices not widespread enough

Biotechnology, automation, and integrated pest management are integral components of a revolution in agricultural efficiency. Today's farms conserve and recycle water, soil, and waste. Sensors and monitoring systems provide real-time updates of ecosphere interactions. Agroecology principles, once popular mainly in curricula, have made their way onto the farm.

Drought-, disease-, and pest-resistant crops are standout achievements of the biotechnicians. Creating crops that require less water is particularly important as battles over water rights heat up in California and other states. In the Midwest, macroengineers have been busy for the last dozen years replenishing the Ogallala Aquifer by diverting the Mississippi.

Integrated pest management is indispensable to the total ecosystem orientation. It has shifted from general and broad spectrum attacks to targeting specific species and their behavior patterns. The shift has been away from control and eradication and toward management and natural ecosystem balance.

Speeding forest regeneration

Forests that would take hundreds of years to mature naturally have been cut to tens of years. Japan, once a deforestation miscreant, has transferred technology over the last 35 years that has recreated thousands of forests. Even rain forests, once thought to be nonrecoverable, have been regenerated.

Forestry is also driven to balance resource needs with environmental needs. Government regulation, such as banning clear-cutting in 1999, enforced this balance. Forests became valued for their roles in converting CO_2, erosion prevention, aesthetics, and recreation. Of course, the switch away from wood products to new materials in many bulk applications was an important factor as well. Technological advances have enabled forestry to become a sophisticated, integrated applied science today. It safely and productively extracts wood and other resources from forests without destroying or depleting them.

2025

A relatively seamless transition to sustainability

There was no jarring incident or action that can be said to be the dividing line between unsustainable and sustainable living practices. The Selling Sustainability campaign is pointed to by historians as the transition point, but it did not jar our daily lives.

Selling sustainability

The multimillion-dollar national information and advertising campaign, Selling Sustainability, reached people in multimedia formats. This public/private project championed by the president and the leaders of many Fortune 100 companies sought to win people over to sustainability. The organizers accepted the social science argument that social and economic change always starts with individuals, even when it occurs within large organizations. If the public accepted the notion, the argument went, they would hold politicians accountable at the ballot box.

It was not a smooth process. Particularly effective were the electronic town hall meetings, in which people shared their experiences in converting to sustainability. People realized that their efforts were not alone and drew strength from numbers. These efforts were reinforced by a combination of incentives and disincentives. People made a virtue of things over which they had no choice.

Without structural reform, of course, individual efforts would have been useless. The most effective instrument was environmental taxes. The government tool kit for restructuring the economy for sustainability included tax policies, subsidies, regulations, R&D funding, and procurement policies.

Daily routines changed gradually. Many consumer products fell out of use: some brands or types of motor oils, paints, freon-based refrigerators and air conditioners, antiperspirants, colognes, nail-polish removers, windshield washer fluids, shampoos, and hair spray to name a few. Although many still gripe about the loss of a favorite product, replacements have satisfied all but the most die-hard consumers.

Some people are directly affected by warming-induced sea-level rise. Many summer beach homes and resorts are now nearly worthless. Fishermen and farmers are in danger of losing their sources of income or having to relocate to find new ones. Some coastal businesses have already had to relocate. Many insurance companies were bankrupted or pulled out of high-risk coastal areas. Those that remain are still in court haggling over warming-related claims that continue to pour in.

Home and business environmental audits are standard practice. Although some older buildings were grandfathered, new ones are built and maintained to high efficiency and durability standards.

Siting solid-, hazardous-, and nuclear-waste facilities often involves people in battles both for and against. The coming of age of the generation of young people educated in the principles of sustainability reinforces the activist trend.

WORLD 2 AND 3 NATIONS—
A mixed environmental record

Nations well along the path to industrialization are the new leaders in environmental matters. They turn their attention to the environment as they become wealthier, following the example of the affluent countries. In the interim, their environments suffered, and upped the ultimate cost of cleanup.

The spread of affluent nations' middle-class values to these nations raises expectations and promotes consciousness of the good things in life, such as a healthy environment. As the middle classes become more educated and informed, they are demanding action regarding the environment.

Over the last 35 years, economic growth has generally won out over the environment when the two were in conflict. Short-term economic gains have been easier to justify than long-term environmental preservation. The price is still being paid today and will be for at least the next few decades, as the mess is cleaned up. This will present exciting business opportunities for affluent nations with experience in environmental mitigation, cleanup, restoration, and prevention.

The environmental report card for middle nations below shows the wide range of actions, with grades in the A range for Oceania and Southeast Asia to the D range for Russia and Eastern Europe.

The State of the Middle: An Environmental Report Card

Country & Grade	Description
Middle East **B**	Water has long been the primary concern. Efficient irrigation and desalination projects and protection of limited underground water supplies are underway but not yet complete. In Kuwait, reefs are no longer endangered by oil spills. Prevention technologies, such as double-hulled tankers have been successful. Egypt's low-lying deltas have been flooded periodically over the last 15 years, due in part to global warming. Prevention is still weak.
Russia **D+**	Russia's myriad environmental problems are overshadowed by its high, albeit negative, international profile regarding its long history of mismanaging radioactive waste. The waste was often buried in shallow trenches or simply dumped into waterways. Cleaning up this mess diverts attention from Russia's other problems such as greenhouse emissions and energy inefficiency.

Eastern Europe	D-	Mitigation and cleanup are today's responses to the environmental damage that was the legacy of communism's single-minded industrialism. Air, land, and sea are each highly polluted. There has been some progress in cleanup, but lots of work still needs to be done. These nations suffer from problems long dealt with in advanced nations, such as acid deposition, deforested hillsides, diminished crop yields, rivers used as open sewers, unsafe drinking water, and lower life expectancies.
China	C	Today's primary problems are a legacy from exploding energy use, in particular coal, over the last 35 years. China became the world's number one contributor to global warming in 2003. It continues to endure acid rain and smog in too many urban areas.
Central America	B	Mexico's membership in NAFTA in 1993 pushed the environment to center stage in Central America. Neighboring nations long seeking to join NAFTA are taking steps to get their environmental houses in order. The chief remaining problems are in overpopulated urban areas unable to provide sufficient clean water, air, or sanitation.
South America	C+	Rapid population growth has undermined some positive steps. Urban areas are often overwhelmed. Brazil has been a focal point of environmental observers. Sâo Paolo suffers the air and water quality problems typical of megacities. Land clearing has stopped and started almost in lockstep with international aid. Other problems today include insufficient sewage treatment, water resource depletion, overgrazing, and damage to coastal estuaries.
Southeast Asia	A-	As these nations decide that their industrialization is well in hand, they are committing to cleaning up the environment. Reefs that were in trouble are given immediate attention. Nuclear waste disposal is at the top of R&D agendas, and state-of-the-art technology is being experimented with. Thailand, once the most recalcitrant of these nations, has changed its ways. It has restricted logging, is cutting population growth, reducing energy-related air pollution, and cleaning its extensive canal systems of industrial waste and sewage.
Oceania	A	The farthest along in environmental matters. The ozone hole over parts of the region was eliminated by banning CFC use, water conservation efforts are succeeding, and the Great Barrier Reef is in better shape than most reefs.

Nonindustrial destitute nations that depend on their land for sustenance face a much worse situation. Many of these nations continue to have declining incomes, as they approach their land's carrying capacity.

The story for many destitute nations has been one of collapse and near collapse, with famine and disease drastically increasing deaths and leading to subsequent drops in population growth. This is a typical scenario that, sadly, has been seen over and over again in the last 35 years. The driver of environmental degradation in World 3 is overpopulation.

The degradation takes place first in rural areas, where most of the countries' economic activities, e.g., agriculture, and populations are located and then presses in on already overburdened urban areas. International aid is typically inadequate.

As agricultural practices overwhelm the land's carrying capacity, streams of environmental refugees pour into urban areas. They move into shantytowns around large cities already suffering from smoke from cooking, inadequate sewage and water, and toxic air from industrial pollution or smog. Civil strife is another result. Ethnic and tribal conflict break out as groups seek scapegoats for their severe problems.

As entire nations collapse, refugees streaming toward neighboring countries are turned back at the border. This has led to violent conflict. In regions across the globe, countries including Haiti, Egypt, Bangladesh, Nicaragua, and several African nations have resorted to arms in search of arable land or water that was disappearing in their own countries. The look at Haiti that follows summarizes how a nation can slide into collapse, and how it can be rejuvenated.

It is plausible that the last national collapse has occurred. Strengthened international institutional mechanisms for dealing with these scenarios should prevent future recurrence.

A CASE STUDY—Haiti overshoots and collapses

Haiti was the first nation to be recognized as exceeding its carrying capacity and effectively collapsing. Famine, disease, and rioting were widespread as the land became incapable of supporting the burgeoning population. Civil works were in disrepair, farmlands were ruined, and law enforcement broke down. The exact point at which the collapse occurred is difficult to determine. Some say it occurred in the 1990s, others say only 15 years ago.

Deforestation was at the center of Haiti's environmental collapse. It served as a warning to others. The recipe for disaster was a combination of overpopulation, political corruption and conflict, economic decline, and environmental degradation. These factors fed off of one another. Political incompetence and outright thievery got in the way of confronting the country's bleak economic opportunities. Survival in turn led people to degrade the environment further.

The downward spiral

Overpopulation propelled Haiti toward collapse. Although Haiti was the poorest country in the western hemisphere, its population grew more than 2% a year in the last few decades of the 20th century. The growing numbers of people cleared large amounts of forest for marginal cropland. Using charcoal as the main energy source further contributed to cutting down trees. The deforestation in turn contributed to erosion, which further degraded the already marginal land. It also silted waterways and destroyed reefs. These circumstances created a reinforcing feedback loop until the ecosphere was unable to support sufficient agriculture and fisheries to feed the population.

Well-intentioned aid fails

International aid to save Haiti continued up to and during the collapse. Nations tried to halt the chaos but were unsuccessful. Initially promising programs, such as the 1985 Tropical Forestry Action plan, were overwhelmed by population growth. Reforested areas were simply cut down again. People asked to preserve the environment for their grandchildren balked, as their children at the time were starving. Resettlement failed as well. Haiti's environmental refugees were refused entry across the globe.

Restoring the environment

Haiti was taken over by a UN stewardship in 2009. The UN had hoped to stabilize the situation, build up stronger political institutions, and hold elections within a decade. But the task has proved more difficult than anticipated. Haiti remains a ward of the international community. At least a decade will be required before the transition back to self-government.

Some progress has been made. The population has stabilized. The hundreds of thousands of deaths from famine, disease, and violence reduced the population and decreased fertility rates. Corrupt politicians were prosecuted or fled, and their fortunes seized. UN-coordinated aid efforts are rebuilding the infrastructure and resource base. The economy has been growing at a solid 3% per year since 2015.

In the 20 years since collapse, much international attention has been devoted to restoring the environment of countries like Haiti. Although it was the first, and perhaps the worst case, it was not the only one. Bangladesh, the Maldives, parts of India and Africa collapsed as well. International groups are still learning how to rebuild these nations using the four-step strategies so successful in the affluent nations.

Critical Developments, 1995-2025

Year	Development	Effect
1995	Sustainability principles become part of school curricula nationwide.	A generation knowledgeable and committed to sustainability is nurtured.
1999	Environmental Audit Certification Act passed.	Provided assurance that auditors with the seal met performance standards.
1999	Clear-cutting banned in the United States	Sustainable forestry practices become the norm.
1999	OORTS deployed.	Prevents serious collisions between debris and space vessels.
2000	International Global Warming Federation Forms.	Goal of transferring technologies from affluent to middle and destitute nations.
2001	The EPA is absorbed into the DOTE by a new organic act.	Diffuse responsibility for environmental issues concentrated in one agency.
2002	Selling Sustainabilty campaign.	Public and private commitment, and money for promoting sustainability.
2003	Delaney Clause repealed.	Risk-benefit principles balance safety and economic concerns.
2004	Bradshaw Act passed.	Shifts fiscal and tax policy to encourage business to invest in long-term future.
2004	Banks-Hyde process developed for mixed-hazardous waste problems.	Tackles one facet of waste problem; provides hope that other solutions will come.
2005	Haiti collapses.	First clear case of population overshoot and environmental collapse; warning to others.
2009	75% of Fortune 500 companies have adopt CERES principles.	Corporations publicly proclaim their commitment to sustainability.
2011	International Agreement on the Troposphere.	Established council to oversee climate manipulation.
2012	Convention on Hazardous Waste Export.	Worldwide agreement to ensure safe hazardous-waste disposal.
2019	Massive flooding in Maldives.	Plans to evacuate the island if global warming continues to produce sea-level rise.
2020	Zero Waste Proclamation promulgated by U.S. manufacturers.	Goal of zero waste by 2040.
2028	The United States, Korea, and Japan to inject vitrified nuclear waste into subduction zones.	Success would essentially solve the nuclear waste disposal problem.

Unrealized Hopes and Fears

Event	Potential Effects
Global cooling.	Triggered by a cloud feedback effect from greenhouse effect.
Population growth at replacement levels.	Would have eased the burden on the environment.
Asteroid strike.	Climate disruption or worse.
Planet's feedback mechanisms prevent global warming.	Would have avoided damage from flooding of low-lying areas and coastal areas and the disruption from shifting of the agricultural zones.
Nuclear winter.	Potential to wipe out humanity.
Biotech wipeout of an environment.	A newly created species rages out of control and destroys an ecosphere.
Massive ozone depletion.	Proliferation of skin cancers, crop losses, and ocean biota disrupted.

8

MANAGING THE PLANET

Advances in science and technology are culminating in humanity's growing ability to manage the globe. Managing the globe is taking an active role in influencing, guiding, supervising, or governing the natural and built environments, where human activity continues its expansion underground, underwater, and in space. It also includes managing the social environment, which encompasses the planet's 8.4 billion people and their demographic, economic, and political systems as they grow increasingly complex.

Most people who agree that we are in the early stages of global management question whether we can do it wisely; others question whether we are even doing it at all. Gaining an understanding of one piece of a global system or issue often reveals a much larger scope of implications.

Defining the bounds of global management will continue to be difficult, characterized by overstepping and retrenching. A UN executive, in response to an question about the progress to global management on G-NET (the global news network) said that "we are perhaps 10% of the way towards true global management. It is hard to say what 100% is, but it is clear that we are just scratching the surface. There are many things that we will accomplish over the rest of the century. Managing the globe indicates a shift from reactive to proactive and, increasingly creative approaches to the future."

Although the debate over what constitutes global management continues, it is clear that we are increasingly able to manage complex global systems and issues. Global managing is growing more effective but is still incomplete with regard to environmental issues, population control, war, crime, design and location of business facilities, trade regulation, disease prevention, and business practices.

The benefits of global management include improvements in weather prediction, controlling the effects of natural disasters, instantaneous worldwide communications, and conflict resolution. The trade-off is a more regulated world. For example, free trade has given way to managed trade. In many cases, however, global regulation has rationalized confusing or conflicting national or regional approaches. Regulations and standards worldwide are slowly converging. Many businesses have been bringing overseas operations up to world-class environmental standards over the last 25 years. These companies have not been caught short when global standards were enacted.

Most global management has emerged from incremental expansion of regional and national systems. More systems and issues are managed regionally than globally. Global infrastructure management evolved from grouping national infrastructure management programs like the Federal Infrastructure Administration of 1998. In other cases, new organizations or approaches were devised. The founders of the IGWF, for example, deliberately bypassed existing institutions and their built-in biases so that they could respond immediately to the risks of global warming.

The transition to global management is occurring in bits and pieces and starts and stops. Management initiatives are more advanced in some areas than others. For example, there have been many more successes in settling trade disputes than border disputes. The table below characterizes progress in global management.

Systems/Issues	Grade	Characteristics of global management
Natural environment	C	Progress falls in the middle of the 3 environments; getting non-affluent nations aboard has been more difficult than anticipated.
Environmental monitoring	B-	Remote and in situ sensing networks are well-established; great progress in managing data overload in the last 25 years.
Global commons	C	Solid action regarding global warming but still foot-dragging on nuclear waste; atmospheric programs get most attention, although encouraging steps in ocean management recently.
Natural resources	D+	Lowest score in this sector due to delays in getting programs enacted; too many species lost over the last 35 years.
Built environment	C+	Furthest along of the 3 environments; more and more nations tapping into global systems; trend is away from private management toward greater public management, which should enlist greater participation.
Information	B	Fiber optics traverse the globe supplemented by satellite, cellular, infrastructure personal communicators; the challenge is to expand coverage.
Global logistics	B	Key advance has been streamlining and modularizing cargo shipments to more efficient delivery.
Energy management	C-	Not truly global, due to ocean barriers; regional grids and wheeling are becoming more common.

Global technology	D+	Establishing global oversight of technology has languished; there is still considerable debate over whether it is necessary or desirable.
Macro-engineering	C-	There have been a few successes but still in the early stage of development.
Social environment	C-	Least advanced of the three environments, because it deals with the intricate area of human relationships, values, and ideals.
Population control	C	Success in affluent and middle nations is offset by continuing exponential growth in some destitute nations.
Immigration	C	A code has been regularized, but enforcement is still spotty.
Crime control	C-	Tentative steps toward global control have not gone far enough; weak measures, while they may be portents of stronger ones in the future, also damage credibility.
Public health	D-	Has been totally reactive to date, responding to crisis and diminishing in between.
Trade	B-	Solid progress toward a rational global division of labor in affluent and middle nations, but destitute nations still struggle.
Economic dispute resolution	C+	Has been effective when used, but many parties prefer to traditional legal channels, especially as a stalling tactic.
Intellectual property	C	Another case where regulations are in place, but enforcement is weak.
Financial markets	C+	Growing, but still incomplete understanding of how these incredibly complex systems work.
Countertrade	C	Solid but underutilized mechanism; many established businesses are still reluctant to use it.
Political conflict resolution	D+	Peacemaking has a mixed record; conflict continues to flourish.
Arms control	D-	Bringing reluctant or recalcitrant nations into compliance has been difficult.
Managing natural disasters	B	Great progress in detecting and managing disasters with the aid of information technology; prevention efforts just getting established.

Forces shaping global management

Three primary goals are shaping the move to managing the globe:

- maintaining global economic health and well-being
- preserving the environment
- resolving international conflicts

Although information technology has increased our ability to manage complex global systems and issues, it is not in itself the solution.

The explosion of global economic activity over the last 30 years has often outpaced attempts to manage it. Ties between national and regional economies continue to strengthen, and the lines between economies continue to blur. As a result, it has become increasingly difficult to determine a product's or company's national origins. The growing interconnectedness of economies increases their vulnerability to mismanagement. A severe downturn in Japan, for example, in turn affects the EC and the North American economies. Therefore, preserving global economic health is in most nations' collective interest.

Act globally, act regionally, act locally

The most popular slogan of environmentalists for the last three decades of the 20th century, "Think Globally, Act Locally" adorned bumper stickers and T-shirts. It has been amended to "Act Globally, Act Regionally, Act Locally." The change reflects two factors:

- increased emphasis on regional and global matters

- a shift away from only thinking and problem identification toward acting and problem resolution

---

Sustainability is a fundamental organizing principle for global environmental management. Environmentalists have long recognized that global environmental problems require global solutions. Education campaigns have helped raise awareness and move environmental issues into mainstream global society, but it took the threat of global warming to get people's attention and lead them to press for action.

Obstacles to managing the globe

The more complex an issue, the more stakeholders. Complex situations imply radical moves that go past individual, incremental, or minor actions. The more trouble the complex situation is, the greater the need for radical solutions. But the future has little reality to most of us. Global actions with long-term payoffs are difficult to sell. It is also difficult to take bold action in new situations. The tendency is to go slowly and tentatively. Scientific knowledge and horizons have far exceeded and are out of synchrony with political time frames.

So-called peacemaking (the term's euphemistic sense is often objected to) is the primary advance in international conflict resolution over the last 35 years. *Peacemaking* is the international community's military intervention in a conflict against the will of one or more of the participants. Coercion distinguishes it from *peacekeeping*, in which the involved parties request assistance. Collective action became feasible with the end of the Cold War in the early 1990s. The Persian Gulf War coalition, which formed through the UN in 1990, marked the transition from peacekeeping to peacemaking. It set the precedent of international intervention to redress aggression. The formation of a UN standing army in 2011 institutionalized peacemaking as a widely accepted strategy.

Information technology's coordinating capabilities—information gathering, manipulating, storing, and disseminating—enable global management to transcend time and space barriers that made global management impractical in the past. Massive distributed computing power oversees financial networks. Sensing networks and databases monitor the environment. On-line global surveys, facilitated by automatic language translation, are an important tool for global managers. Macroengineering managers, for example, routinely assess public opinion about proposed projects, so objections can be dealt with in advance. Reliable and timely information is crucial. For example, reports that air pollution is increasing in Mozambique must be reliable in order for the African EMO (Environmental Monitoring Outpost) to coordinate the appropriate response with local authorities.

Why manage the globe?

Many systems or issues, such as global currency and global warming, have long grown beyond the capabilities of individual nations to manage. Growing global linkages and complexity are redressing the paradox aptly characterized by sociologist Daniel Bell in the last century, "government is too big for the small problems of our society and too small for the big ones."

Radicals object to managing the globe

Fringe groups are springing up worldwide to protest global management. Although many object to specific techniques, or to mismanagement, others object to the principle itself. Many groups were started or have their roots last century. Three are:

Earth Firsters: Believe that people have no moral right to tamper with nature and advocate a hands-off approach. Many subscribe to the notion that the Earth is a living organism and that humans are a cancer.

Free Traders: Believe in the sanctity of the free market and that global management is wrong because it interferes with the market. They refuse to acknowledge that free trade has not been free for decades—if ever.

Neo-Orwellians: Believe that international information networks are an inherent invasion of privacy and represent the fulfillment of Orwell's *1984* prophecy. They have cut cables, disrupted communications cells, and even sabotaged a satellite launch.

Global management saves money for governments strapped for cash and for global companies trying to keep costs down. Smart infrastructures, smart management, pooling resources, and reducing redundancy are cost-effective. A key criterion is that the global management has long-term beneficial effects for sizable groups—typically defined as groups of nations or re-gions—while not having adverse effects on others.

Three years ago an international commission was charged with forming "A Vision for the Planet" statement due in 2027. The commission is canvassing world-wide for a framework for humanity's relationship with the planet. Debate has been raging on virtual communities and issue forums.

Who is managing the globe?

Global managing operates on public, private, and mixed levels, for-mally and informally. There is no command center. Responsibility is diffused across many people and organizations. Global management is an evolving enterprise that began without deliberate effort but is moving towards institu-tionalization. A trained cadre of global managers is emerging from universi-ties and other learning centers worldwide, such as the Centers of Excellence, sponsored by the International Institute for Applied Systems Analysis (IIASA).

The global management curriculum

Global management programs began in occasional courses at the turn of the century. They appeared within environmental, civil engineering, systems science, or future studies programs. Lehigh granted the first Master's degree in global management in 2013. Typical core courses include:

- macroengineering
- systems science
- sustainability
- international institutions
- global infrastructures
- macroeconomics
- networks
- social change

The UN leads global management in the public sector

Formal activities combine thousands of international organizations and international regional, and multilateral agreements and treaties. The UN fam-ily of institutions is the dominant organization. There is no single dominant treaty or agreement. The box below shows some important international agree-ments of the last 25 years.

Some international agreements for managing the globe

2000 International Global Warming Federation	transferred technologies to alleviate greenhouse emissions
2002 International Space Agency	part of UN system; provided a negotiating forum
2007 Lima Space Weapons Treaty	preserved space as a weapon-free zone
2007 Geneva Conference on Hunger	established principles for food assistance only in natural disasters
2009 International Energy Agency	nuclear nations submitted to standards after a grandfather clause is added
2009 ISO	established materials characterization standards for recycling and reclamation
2011 International Agreement on Standards in Construction	developed standard measures of performance and capability for materials in construction, appliances, vehicles, and other durable goods
2011 International Agreement on the Troposphere	established a council to oversee climate manipulation
2012 Convention of Hazardous Waste Export	ensured safe hazardous-waste disposal
2016 Bangkok Accords	set up an institution to formalize limited trade protectionism
2017 ISA (International Space Agency) and the Red Cross	formed the International Disaster Tracking Program, which uses space-based monitoring to provide advance warning and help coordinate relief efforts.
2017 International Commission on Animal Care in Agriculture	established approval mechanisms and regulations for transgenic animals in agriculture
2019 International Hypersonic Craft Consortium	formed to commercialize hypersonic craft by 2030
2019 GMAN (Global Manufacturing Network)	upgraded national pre-alliance service networks like FAN in the U.S. to include nations across the globe
2024 Convention on Environmental Balance	got its 30th through 32nd signatories in efforts to make materials and other resource use more sustainable

Global companies drive global management in the private sector

Global companies are the dominant actors in the private sector. They form around the need for natural resources, to ease exports, to expand production, and to enhance competitive position, such as improving market access or taking advantage of local labor pools. Large global companies routinely operate in 70 to 80 countries. They have largely broken free from national control and are global entities in their own right. The similarities of big global companies are becoming more prominent than the differences, as ties to home nations weaken.

Interest groups with global reach continue to proliferate. They range from quasi-governmental political groups like the Greens to nongovernmental organizations (NGOs) to humanitarian groups like the Red Cross and Project Hope to religious groups like the Salvation Army to on-line virtual communities. They heighten international awareness of issues and can mobilize effective actions, including economic pressures or boycotts, quickly. Easy global communications make them welcome allies and dangerous enemies.

How is global management being done?

Global management uses incentives and disincentives. Economic sanctions are a primary disincentive, used mostly against small nations that cannot withstand collective economic action. Creating incentives that work is an ongoing challenge. A rule of thumb is that rewards work better than punishments. The challenge is in devising innovative rewards. Another truism is that global managers must be delicate in matters affecting national sovereignty.

Changing minds

Social change is rarely smooth; it is usually two steps forward and one back. People reformulate their views of the world.

Sociologists today describe a widespread shift in world views over the last 50 years. The world as a precise, reductionist, clocklike machine operating according to natural laws has given way to a world of webs of chaotic, interconnected and interacting holistic systems.

Social change to accepting greater global management has also been engineered by political leaders. Regulations, laws, taxes, and other policy tools are used. People are increasingly willing to accept the need for trade-offs, acknowledging that there are limited resources for unlimited problems. Risk-benefit analyses are standard practice today. Health-care rationing, for example, once unthinkable, is common worldwide.

Many transformationalists have argued that some form of collapse or disaster is needed to motivate radical social change. They argue that an incremental, muddling-through approach falls short of what global management could accomplish if given its due.

But it turns out that complex global systems are more resilient than many thought. Anticipated disasters have not happened, or they did and society adapted to them, as with the stock market crashes of 1987 and 1999. The resilience often comes unexpectedly or unpredictably.

Global management tools can be grouped by how they are enforced. International agreements, laws, and treaties are the top layer, as they are supported by collective economic or military action. Regulations and legal standards are the middle layer. They are more difficult to enforce because they rely on market forces. Voluntary standards or arrangements, such as product specifications, are the bottom layer, because there is no formal enforcement.

Coercion may include economic sanctions, arms embargoes, or communication cutoffs. Blocking access to the Net has been particularly effective in the few cases when it has been used. For example, when China, the world's number one greenhouse gas offender, continued to evade IGWF standards for five years, it was cut off from global communication and financial networks in 2019. Chinese negotiators were back to the table in a week and in compliance within a month.

Regulation has been a growth enterprise over the last 35 years, but this has not necessarily been bad news for business. Moving regulation to a higher level, from national or regional to global, in many cases has simplified situations. Rather than dozens of national regulations, businesses can deal with a single code. Rationalizing the thicket of building and housing codes in the United States last century improved efficiency within these industries. It also enabled companies to more easily expand into new markets.

Regulations have often made economic competition fairer. Seeking competitive advantage by relocating from nations with strict regulations to those with loose ones, has become much more difficult and costly as loopholes and havens are being closed. Last century, NAFTA, for example, was ratified by the U.S. Congress only after guarantees were made that Mexican environmental standards would be brought up to U.S. levels—many feared that companies would relocate to avoid tough U.S. environmental laws.

Standards are imposed by governments or negotiated and adhered to voluntarily through markets. They provide for control, such as fuel economy standards, a guarantee of quality, such as product classification, and for processes, such as bills of lading or EDI (Electronic Data Interchange).

The United States had some difficulty with standards setting in the last and early part of this century. U.S. companies were used to being market leaders with the clout to set de facto standards. They were less skilled in negotiating standards, which has become the norm since the United States began sharing world economic leadership with the European and Asian trading blocs in the 1990s. The U.S. government had a hands-off approach. As the need for government help was acknowledged, there were turf battles within government agencies. The Government Standards Bureau, formed in 2001, emerged as the government's chief negotiator for standards. It works closely with private groups like the American National Standards Institute (ANSI).

What are the effects of managing the globe?

Global management is generally stronger in World 1 and weaker in Worlds 2 and 3. Most affluent and many middle-income nations are substantially committed to global management. The remainder of nations participate selectively. A common objection is the alleged hypocrisy of global environmental regulations. World 2 and 3 countries assert the right to pollute as affluent nations did on their road to economic development. They typically participate when the economic benefits outweigh the cultural costs. A small group of nations, including Algeria and Myanmar, refuse to participate at all, citing cultural and religious objections.

There have been some failures. There are cases in which the international community has refused to take responsibility for cleaning up an environmental problem or mess or to intervene in a bloody conflict. Last century, Yugoslavia broke apart while actions were deliberated. Just 15 years ago, the on-again, off-again civil war in Sri Lanka culminated in what many have termed genocide. Sometimes intervention fails—witness the tragedy of over 30 years of civil war in the Sudan.

Technical failures have also occurred sporadically. Portions of the global telecommunications network have been out of service for varying lengths of time. A single-day blackout in most of Europe in 2016 led to billions of lost dollars of data and output. Energy grids have failed across areas larger than possible last century—NAPNET was down for a week in 2005. It was estimated that less than half of businesses and homes had portable backup systems.

THE NATURAL ENVIRONMENT—Managing global environmental issues

Environmental issues have been prominent on international agendas for the last 30 years. The global community has been mitigating and cleaning up the environmental wreckage of the past and today is moving more strongly into prevention and restoration. A continuing challenge is enforcement of existing regulation. Last century, for example, the former Soviet Union had the strictest environmental laws in the world, but they were not enforced. As the situation was opened to public scrutiny, it was clear that severe environmental problems existed.

Five forces provide the impetus for global management of the environment:

- risk response: global warming, nuclear-waste disposal, limits to carrying capacity

- value drivers: sustainability as a global value; caring for the global commons, e.g., Antarctica, space

- institutional drivers: Rio Summit, IGWF

- technological drivers: global sensing and monitoring capabilities tied into massive databases

- economic drivers: markets in environmental technologies and pollution control

Global warming continues to be the single most important driver of global environmental management. The threat of it last century, until its confirmation in 2010, was seen as serious enough to require international action. It established a precedent for global action that is relied on today. The well-reasoned and successful actions of the IGWF defused opponent's arguments that global actions were a new imperialist scheme. Cooperation extends beyond simply pooling resources to dealing with threats. Weather prediction, for example, has provided a model. Sharing data improved common understanding. Today, however, experiments with climate control are resisted by some.

Global environmental agreements proliferate

International agreements cover ozone holes, global warming, deforestation, soil erosion, overfishing, disaster monitoring, and the 3Rs. Examples include:

Orbital Object Radar Tracking (1999)

International Global Warming Federation (2000)

International Agreement on the Troposphere (2011)

Convention on Hazardous Waste Export (2012)

Convention on Environmental Balance (2024)

Sustainability has emerged as a core global value. Sustainable societies, especially in World 1, rely on growing capabilities to manage global environmental issues. Sustainability principles are part of school curricula worldwide. The Rio Summit in June 1992, sponsored by the United Nations Environment Program (UNEP), is today hailed as a milestone. Its Agenda 21 report is the framework for much of global environmental management today. It proved to be a key stimulus for further institutionalization of environmental actions. The early paper tiger organizations that Rio spawned have matured into effective action bodies today. The box above highlights some of the major international agreements that have in large part emanated from these groups.

2025

Complete around-the-clock global monitoring of the environment has been considerably refined over the last 25 years. A primary challenge at the turn of the century was making better use of the masses of data that were pouring in while the international archive of remote sensing data bulged with unprocessed data. Pricing issues have finally been worked out. Fine-tuning global monitoring technologies makes this possible. Fuzzy logic programs, neural networks, and expert systems are helping to sort data. Global positioning systems bring the essential dimension of location to data gathered on the Earth or from Earth orbit. Geographic information systems also help collate, organize, and interpret the data.

Businesses have capitalized on this new partnership with the environment—20% of U.S. GDP growth has been devoted to environmental cleanup over the last 30 years. The numbers are similar in other World 1 nations. Environmental groups present the environment as an opportunity for businesses today. National governments continue to sponsor environmental technology research as well as cleanup and restoration programs. Pollution control is a multibillion-dollar industry.

Monitoring the environment around-the-clock and around-the-globe

Remote and in situ sensors feed into networks of databases, which in turn relay instructions for mechatronic systems to carry out. The systems are moving from self-monitoring to self-diagnosing to self-repairing. The Resource Tracking Model developed in 2001 with the assistance of the EOS is an integral part of international efforts to assess the planet's carrying capacity.

Managing the commons: oceans

Oceanographers can measure just about everything in the ocean today. Satellites, ships, interactive real-time observation, automated smart instruments, detailed maps, and supercomputer-based models provide the tools. Understanding how oceans work is an important element to understanding climate change. The confirmation of global warming in 2010 spurred intense interest in oceans and hydrology that continues today.

Collaborative research on coastal issues is ongoing. For example, coastal over-development has created environmental problems worldwide. Growing shares of populations live near coastal zones. In the United States, for example, 75% of the population lives within an hour's drive of the coast, up from 50% last century. Overdevelopment harms fish populations and drinking water and often overwhelms civil works like sewage systems.

Tending the global commons

The global commons—the atmosphere, oceans, land, and space—as well as the particular global-scale issues such as global warming, ozone depletion, and nuclear waste are managed internationally. There are several cross-border arrangements to manage water power and water supplies, telecommunication and transportation infrastructures, and even waste disposal. The Global Commons Agreement on Infrastructure in 2013 established the principle that national infrastructures should strive for integration with the emerging global infrastructure by adhering to construction and other standards.

Managing the globe: the atmosphere

Climatologists use state-of-the-art global data collection, satellite and remote observations, high-performance computing, and data visualization and analysis, as part of their systems approach to understanding climate and the atmosphere. The growing body of knowledge informs global management.

The U.S. Clean Air Act Amendments last century set the precedent for the International Agreement on the Troposphere this century. They were one of the last major national acts dealing with the atmosphere, as global oversight is now predominant.

Trading blocs have proven to be useful bases for managing regional projects. Linking regional projects, where it makes sense, has been and will increasingly be, a logical next step.

Conserving natural resources

Biodiversity and deforestation are primary global resource management issues. Debt-for-nature swaps are slowing the extinction of species in rainforests, although substantial damage has been done. International authorities are managing portions of rainforests in Madagascar and New Guinea. Plans are being laid for a multibillion-dollar reforestation project in the Sahara in the next decade. It will be based on artificial tree technology used to reclaim several deserts in Australia.

Managing the global food supply is in its embryonic stages. The UN's WHO has been the lead agency in food distribution as well as public health, reflecting the close links between the two. The second Green Revolution from 1995-2025, based on bioengineered crops and automation, has kept the food supply, but not its distribution, ahead of population growth.

THE BUILT ENVIRONMENT—Managing global infrastructures and technologies

The built environment is humanity's technological intermediary with nature. It adjusts the natural environment to better accommodate the social one. Increasing technological capabilities have raised some projects to the global scale, which in turn has led to global management of them.

Three types of global management are in place regarding the built environment:

- oversight of technological infrastructures

- cooperative R&D and regulation of enabling technologies

- macroengineering projects

Historically, these technological systems have been local in scale. Over the last 35 years they have been moving to national, regional, and today, global capacities. The primary technological infrastructures being managed globally are information, transportation, and energy. More traditional business and industrial infrastructures, such as manufacturing facilities, chemical plants, and electric generating facilities, are not globally managed yet, but they are part of managed systems and subsystems.

The information infrastructure: the backbone of global management

Information technology is so pervasive that it is largely invisible. Today's networks are widely available and easy to use, melding computing into the background. The information infrastructure is taken for granted as electricity became last century.

A global network of information technologies is the backbone of the global move to information- or knowledge-based economies. The Internet has merged into a larger information infrastructure, which in turn has expanded into a worldwide broadband network of networks—the GlobeNet. They are framed on fiber optics and supplemented with satellites, cellular, personal communicators, and microwave arrays as ancillary.

The breakup of monopolies and subsequent globalization of national telecommunications led to hundreds of cross-border acquisitions that sped up the transition to global networks. The networks are moving toward universal standards. In some nonaffluent nations, networks are still bridged by software protocols that allow incompatible networks to communicate with one another.

Milestones on the way to the global net

1990 analog and digital
1995 64K ISDN
2005 multigigabite ISDN
2010 terabit broadband ISDN
2015 total fiber nets*

*in most affluent nations and in up-and-coming middle nations

---

Millions of computers are linked into the global network. Groupware and vidoeconferencing are important business collaboration tools today. They enable work teams spread across the globe to work together less constrained by time and space barriers.

INTELSAT continues to be a benchmark of international cooperation on telecommunications. This long-standing arrangement to coordinate access to satellites has become especially important as low Earth orbit fills up. Access to LEO must be carefully regulated to avoid multimillion-dollar accidents. Another successful international agreement is the International Telecommunications Union, whose forerunner the International Telegraph Union goes back to 1865. It is considered by most scholars to be the oldest function-ing international organization today.

Security on the global net

Global management of the Net centers on guaranteeing security and reliability of transmis-sions. Because so much of human enterprise is dependent on the information infrastructure, nations are willing to pay for oversight. Fees vary based on the levels of privacy, encryption, and guaranteed delivery.

Security and reliability standards negotiation is ongoing. To keep up with technological ad-vances, which often outpaced standards-setting in the past, flexible, scalable standards are continually updated. EDI last century was the initial area of cooperation. Spectrum allocation and reallocation have also been successfully negotiated.

Motorola's Iridium satellite network deployed in 1998 has significantly enhanced global positioning capabilities. It has improved logistics and navi-gation, and supplements remote sensing satellites monitoring the environ-ment. Linking remote sensing and mechatronics has created smart elements and monitoring systems.

The sensing network is also supported by data storage capabilities that have been growing exponentially for years. Intelligent, self-learning databases update themselves autonomously, absorbing new information and deleting obsolete information. Global modeling simulation capabilities, increasingly using virtual reality, are valuable tools for global management.

A principle of collective response to attempts to interfere with global infrastructure systems was established by the international police action that followed the cutting of trans-Atlantic fiber-optic cables by terrorists in 2011.

Moving goods and people: managing global logistics

The widespread move to just-in-time and just-when-needed econo-mies over the last 35 years has required a substantial boost in logistics capa-bilities and the physical transportation infrastructure. Air, land, and sea traffic control has been enhanced by information technology. Cooperation built from simple steps such as sharing weather data.

Adopting the standards in place today has eliminated redundancy and improved coordination on the global scale. Cooperative global agreements are becoming increasingly common; for example, the IHCC that formed in 2019 hopes to commercialize hypersonic craft by 2030.

Travel survives the infotech onslaught

Many experts predicted the demise of the travel industry with the advent of information technologies such as networking, groupware, and videoconferencing. Why travel when you could meet remotely? Most expected a drop in travel. Others felt information technology, by expanding the number of contacts that people could make, would actually stimulate increased travel. The verdict today, based on the travel numbers, and the continued position of tourism as the world's number one industry, supports the latter. Travel has been spurred, not reduced, by infotech.

National systems expanded into continental ones. The EC maglev system was an early continental project. NAFTA and SAFTA (South American Free Trade Agreement) later built continental highways as well.

Sea traffic control has been using the GPS since last century. The fully linked modular cargo networks now used by most nations greatly facilitate international trade.

Global energy management: sharing grids, guarding impacts

Global management has emerged in support of utilities' wheeling and dealing. Utilities across the globe are being made common carriers, as power generation is separated from transmission and distribution. NAPNET became the largest electric grid in the world 10 years ago with the connection of Mexico to the United States and Canada. As superconducting transmission line costs fall and reliability rises, they will lead to growing continental grids.

The IGWF has coordinated and aided shifts away from fossil fuel power. Renewable technologies are in place and their use is poised to expand. This shift is crucial, since nonaffluent nations, typically lagging World 1 nations in shifts to advanced energy technologies, now account for 60% of the world's energy use. In many cases, however, these nations pioneered renewable technologies, when the technologies were cheap and obviated the need for a supporting grid or infrastructure. Portable photovoltaics and fuel cells, for example, were popular in many nations that lacked the funds for an infrastructure. Global management of energy in most cases has been confined to the affluent nations and selected middle nations who could afford access to regional grids, usually with the assistance of affluent trading partners.

Nuclear power is globally regulated. Damage from the last incident, a meltdown induced by sabotage in Pakistan, was well contained, but still raised international alarm. The International Energy Agency (IEA) in 2009 finally got all nuclear nations within its jurisdiction, i.e., subject to its inspections and standards.

Cooperation on global technologies: R&D, standards, and regulations

Some technologies require global oversight due to their centrality to economic activity or to regulate their possible global impacts. Biotechnology and materials are technologies with worldwide consequences whose prospects have been advanced by international cooperation on precompetitive R&D. In addition, biotechnology's novel effects warrant global oversight.

International genome research programs built on the early national programs. Procedures for setting regulatory lifetimes and for updating and changing regulation have been standardized. The U.S. biotechnology regulatory structure evolved from a tangle of differing jurisdictions to a harmonized new federal Biotechnology Regulatory Agency formed in 2004. It is addressing long-standing public safety concerns and industry's competitive concerns.

The biotechnology industry was the prime mover behind the evolution of international regulatory structures. The industry argued that uncertainty about the future and a growing hodgepodge of biotechnology regulations was slowing expansion. It pushed for the International Biotechnology Commission organized by UNEP three years ago. The commission is charged with developing a comprehensive vision of the implications of biotechnology for global society over the next 25 years. Its report should be out within the decade.

Risk analysis becomes standard practice

Risk analyses accompany global technology projects. In addition to anticipating and monitoring health and safety risks, social, political, environmental, and institutional dimensions of risk are within its scope. It is standard business practice and required for government projects today. A key enabler of risk analysis is improved modeling, including virtual reality simulations. Risk analysis is integrated with decision analysis and technology assessment.

In parts of World 1, explicit programs have begun for the aggregate enhancement of populations' physical and mental abilities (as opposed to disease prevention) based on the understanding of human genetics. International sports have been debating whether or not to admit genetically enhanced athletes. Some bioengineering of people, animals, crops, trees, ecosystems, microorganisms, chemicals, and materials is regulated internationally. For example the ICACA, established in 2017, set up approval mechanisms and regulations for transgenic animals in agriculture. Other agreements cover releasing new organisms into the environment.

For materials, the International Agreement on Standards in Construction of 2011 developed standard measures of performance and capability for materials, including their energy efficiency and insulative capacity and established guidelines for use of standardized materials and parts in construction, appliances, vehicles, and other durable goods.

Macroengineering projects gaining credibility

More macroengineering projects are in the planning than building stages. They are global scale by definition. They are too expensive for a single country and require international cooperation. They have substantial, measurable, lasting impacts on the environment with cross-border social, political, or economic consequences. There is not yet a formal institution for the oversight or funding of macroengineering projects. The private Global Society of Civil Engineers plays an important advisory role. The Mitsubishi Research Institute in Tokyo's Global Infrastructure Fund begun in 1977 has helped fund projects. The key is not so much the know-how, but the politics and economics.

Reclaiming the desert

Planting natural and artificial trees has been used to reclaim deserts in Australia, taking advantage of weather and climate manipulation to stimulate and increase rainfall. Artificial palm trees, for example, cause rain to fall. The trees have perforated plastic and foam trunks, branches, and leaves, and polyurethane roots which, injected as a liquid into perforated steel tubes will percolate into the ground and solidify to form roots. By shading the ground and through evaporative cooling, cool fronts form and cause rain to fall. Gradually natural trees replace the artificial ones. This project was funded internationally as a test bed for further large-scale application in Africa.

The reversal of the Ob and Yenesi rivers in 2018 in Russia is an example of a macroengineering project. The project redirected the rivers from their natural drainage into the Arctic to the Aral and Caspian Seas in order to water the central plain. Other projects include the solar power satellite deployed to provide power for the Moon base, and a Saudi-led Middle East consortium that is towing icebergs to cities and deserts for water.

New water and power sources were made available five years ago with the Great Replenishment and Northern Development (GRAND) Canal, funded by public and private contributions and coordinated by the NAFTA Board of Governors. The states involved were essentially bought off by promises to boost infrastructure spending in the affected areas. The Great Lakes were converted into a water storage and distribution reservoir. A 160-kilometer dike across the southern end of James Bay in Canada captures the inflow of fresh water and pumps it southward to the Great Lakes along new channels and existing rivers. With water management systems it provides fresh water for some of the United States and Mexico as well as providing hydroelectric power. The cost was about $100 billion. As with most hydroelectric projects this century, there were environmental protests, but modification of the projects after the environmental impact statement satisfied all but a few extremist groups.

THE SOCIAL ENVIRONMENT—Managing global society

The primary issues in the social environment today center around managing population, social, economic, and political issues. Population issues include keeping the planet, regions, and nations within their carrying capacities. A related global issue is managing immigration and refugees. Social issues being addressed on a global scale include crimes of terrorism, drugs, and financial, and public health issues, namely infectious diseases. Substantial global oversight of economic issues includes trade blocs, economic dispute resolution, intellectual property rights, financial networks, and countertrade. The difficult area of political issues includes conflict resolution and peace-making, arms regulation, and disaster relief.

Managing population issues

Population growth in the nonaffluent countries has pushed the world total to 8.4 billion today. Ninety-four percent of population growth has been in middle-income and destitute nations. World 1 nations have completed or are well along in the demographic transition to replacement fertility levels. Ten nations have fallen below replacement level and have shrinking populations, whereas others—the United States, for example—are still growing. Middle-income nations are in varying stages of the transition. India's population is still growing 2% per year. It passed China as the most populated country five years ago. Thailand has made remarkable progress over the last 35 years. Its growth rate is under 0.9% today. Some World 3 nations are still growing exponentially. Nigeria's population, for example, has almost tripled over the last 35 years.

Crowding into cities

A companion trend of increasing populations is urbanization. More than 60% of people live in urban areas, up from less than 50% in the 1990s. There are now 25 cities with populations of 10 million or more.

Heavy urban concentrations have overwhelmed supporting infrastructures. Eighty-five percent of Mexicans, for example, are urban. Shantytowns have sprung up worldwide as a result. Estimates of carrying capacities have been fairly accurate in the cases of small nations or regions. These figures are no longer easily dismissed as doomsaying.

The effects of overpopulation are primarily local, but they have spillover effects worldwide. Overpopulation is a primary driver of environmental problems, hence reducing it is an international priority. Expanding energy demand from 1.5 billion people, for example, led China to burn more and more coal, which causes acid rain in some Japanese coastal areas and adds to global warming. China is the world's largest contributor to global warming today.

Conflicting values have stalled global attempts to manage population. Different nations have different values about human life. Attempts to impose one culture's standards on another have consistently failed. The United States, for example, often refused to assist population control programs of nations that allowed abortions in the 1980s and 1990s. This proved especially damaging in the northern Asian republics of the Commonwealth of Independent

States. Other U.S. assistance programs to these republics were ineffective due to the overriding effects of runaway population growth.

Progress has been through large, high-profile activities like improving women's education as well as market forces. Making morning-after pills and multiyear implants available and affordable has helped manage population—in tandem with hundreds of international population control programs. The male contraceptive pill plays a role in affluent nations with educated males. The lesson of the demographic transition is that economics and education, especially of women, are the key factors. Addressing the reasons that people have large families, such as fears that some children will die or to assure that children will take care of them in their old age, has led to success. Programs for old age security, such as social security, pensions, or long-term health care, have been proven to reduce birth rates.

Weak sanctions, such as reduced foreign aid, have been levied on nations that have done little to curb population growth. Economic development aid to India was halted three years ago, as population growth rates climbed back up. The Geneva Conference on Hunger in 2007 established the principle of food assistance only in natural disasters. It has been difficult to enforce the principles, since global media coverage of people starving has led to pressure for governments to intervene. Over the last 35 years, however, the balance of global opinion is weighing towards aid for long-term development, rather than stopgap measures. There is growing impatience with nations that are unwilling or incapable of helping themselves.

Tight borders slow immigration and refugees

Immigration is being managed globally today. The emphasis has been coordinating policies and tightening controls of illegal immigration. The UN High Commission on Refugees, which expanded to include immigration in 2005, passed the Convention on Immigration of 2019. It regularized national and regional policies into a coherent single approach. Regional solutions were insufficient as weak policies and enforcement in some areas led to problems for all.

The U.S. melting pot or assimilation model collapsed and made U.S. leaders willing to join in an international agreement. The convention has provided a cover for the United States and other national governments. When Mexican leaders object to deportations, for example, United States leaders refer to the convention.

Immigration problems are similar and therefore conducive to global coordination. Nations using guest workers to solve short-term workforce needs found themselves with the long-term problem of supporting the worker's children and relatives. Key clauses of the convention closed loopholes that automatically naturalized the children of illegal aliens or guest workers, or by marriage. There is now a 10-year limit by which time noncitizens must return to their country of origin, which cannot refuse reentry.

Illegal immigration continues to be a problem. There are large smuggling rings worldwide. Although information networks and identification cards have tightened security, innovative conspirators have found ways to evade laws.

The long-term strategy for discouraging immigration is to improve conditions in the originating countries. There are smaller-scale examples. Germany integrated East Germany after the collapse of communism, and South Korea integrated the North in this century. Although initially expensive, they were politically desirable and they provide a model for integrating people from faltering economies.

Immigration is still possible for those with money, desirable work skills, or for education. For the less fortunate, there are few options. The convention, however, provides for recruiting migrant labor. The massive project to reforest the Sahara, for example, was supported by migrant laborers who were granted citizenship in any of the participating nations after the project if they wanted it.

Some social groups have great mobility. There is an expanding group of global citizens with global passports. They are the business and political leaders for whom global citizenship is a reward for hard work or good service.

Managing social issues

Social issues being addressed on a global scale today include terrorism, drug trafficking, financial crimes, and preventing the spread of infectious disease.

Policing the globe

Terrorism has been aided by improved transportation as the perpetrators and their materials are easily shipped worldwide. Old institutions, like Interpol, have not been able to expand the scope of their activities beyond information sharing or their reach beyond the affluent nations, North America and Europe. Cultural differences have proven particularly difficult to overcome regarding terrorism. A key problem is that one person's criminal or terrorist is another's patriot or potential martyr.

International sports model social cooperation

The Olympics are a symbol of internal cooperation and competition. Cooperation in this arena provides a model for another. The games have also been used to make political statements, or symbolic acts, through boycotts or in siting. Harmonizing sports rules has provided useful lessons for setting voluntary global trade or product standards.

The United States has led the so-far unsuccessful push for global news blackouts of terrorist incidents. Terrorists are aware that one well-publicized attack can induce widespread fear. Proponents of blackouts feel that by denying terrorists publicity, they will reduce their effectiveness. Coverage is officially prohibited in the United States, but bootleg disks circulate unofficially. One way that terrorists thwart news blackouts is by targeting communications equipment and facilities. Bringing down information networks for a short time has more devastating effects than blowing up a building or shopping mall—favorite targets of the past.

Advances in combatting the drug trade have come from internal improvements in drug education and treatment programs. On the global scale, laundering drug money has become more difficult because of increased oversight of financial transactions. The Hague Accord of 2016 was an important advance in the global war on drugs. It established World Court jurisdiction for trans-border shipments of drugs. It has proven politically feasible to turn drug suspects over to international bodies, in contrast to the extradition struggles of the past.

Computer crime has accompanied the spread of global financial networks. While oversight and security measures are continually improving, the strategies and tactics of the criminals have kept pace. Wiping out financial crime is still a distant dream. As long as the stakes are so high—the equivalent of the world GNP passes through financial networks each day—criminals will attempt to skim a percentage.

Global management of crime is proceeding slowly. Yet there is growing support for an autonomous international police force. The last vote in the UN General Assembly fell only a dozen votes short. Current enforcement officials are held in check by national governments. There is a fair degree of cooperation and sharing of data for international crimes, and there has been cooperation for decades on financial, or white-collar, crime. Extradition agreements have moved beyond national to regional agreements, but global extradition is probably a decade away according to most experts. Establishing a venue for trials is often tricky, especially in the case of global companies without an easily confirmed home base. In the 2010s, the Ribowka Co., based in 77 countries, was able to put off a trial for more than a decade, relying on objections to improper venue.

Protecting global public health

Few public health issues are truly global, but many cover sizable regions. Global public health programs, anchored by the WHO, have centered in destitute and some middle-income nations. Sanitation programs in affluent nations have brought most public health concerns under control, marred only by occasional outbreaks of disease in poorer urban areas. Public attention and dollars have shifted to the diseases of aging and mental health. Just 10 years after the introduction of an AIDS vaccine in 2000, over 50% of people in the United States with the illness were cured.

The glory days of global public health

Public health officials like to cite the case of smallpox. This killer disease was eliminated worldwide in the 1970s through vaccination programs run by organizations like the WHO. A similar global killer has yet to emerge.

The lesson still being learned in destitute regions, however, is that in cases of highly contagious diseases, prevention efforts must be widespread to be useful. The reappearance of infectious diseases like tuberculosis, the worldwide spread of AIDS last century, and the cholera epidemics in destitute areas this century are cases in point. But it was the devastating spread of hantaviruses in the early 2000s that put public health into the global spotlight. The viruses began in rural areas and were spread to urban areas by rats. They triggered an epidemic in Calcutta which ripped through south Asia, the Middle East, and even found its way into Africa. Public health programs that were languishing received higher budget appropriations after the devastation caused by the hantaviruses revealed their weakened state.

Managing economic issues

Global economic management continually lags the rapid expansion of global economic activity. Ties between nations have strengthened over the last 30 years. Supra-national oversight is taking root in trade blocs, economic dispute resolution, intellectual property rights, around-the-clock financial markets, and countertrade.

Blocs help expand global trade

Trading blocs harmonize trade practices and standards within designated regions. The blocs that began forming in the 1990s differed from their predecessors in that they were not formed to be exclusionary, but to improve internal access to markets. Their primary advantage is removing trade barriers within the bloc. In the EC, for example, German chemical companies improved their access to burgeoning Spanish markets.

Helping themselves by helping others

The EC has been focusing attention on integrating the former communist Eastern European nations. EC parliamentary leaders decided it would make more sense to help them rebuild, then to deal with migrant and refugee problems that would result from having moribund economies on their borders.

Since the blocs are not exclusionary, they are stepping stones to a single global market. Differences in trading practices are diminishing. They will likely be precedents for a single global market.

The biggest three blocs are NAFTA, the EC, and the east Asian bloc. Smaller blocs have formed in South America (SAFTA), a West African bloc led by Nigeria, a south Asian bloc led by India, and an Oceanic bloc of Australia and New Zealand. There have been negotiations to merge blocs, with NAFTA and SAFTA being the likeliest candidates. At the same time, a Chinese faction within the east Asian bloc has been advocating a split from Japan and Korea.

Trading between blocs continues to thrive. The completion of EC economic union in 2010, for example, did not significantly affect trade with outside nations. Trading outside the bloc dipped only slightly or not at all. Access to the EC for non-members changed little. The EC trade and tariff policies for outsider's selling to their market fell in between the least restrictive United Kingdom policies and the most restrictive French ones.

Avoiding the courtroom: economic dispute resolution prospers

Economic dispute resolution is an important part of global management. Alliances are standard practice. A recent on-line survey of 1,000 20-50 person companies by Business Inc. Newsforum found that on average 63% of the revenue of these small companies derived in part from alliances. Alliances often lead to disputes, which can have global ramifications. When U.S. marshals padlocked a Toyota plant in 2007 in a controversy over nonpayment of taxes, it set off an investor panic that lead to a near-collapse of financial markets. Only immediate intervention by the World Trade Organization group got the doors unlocked, the parties back to the bargaining table, and confidence restored to investors.

Arbitration, mediation, and conciliation are common tools for resolving international economic disputes today. After the turn of the century, it became evident that all parties lost in trips to the courtroom. Alternate means of conflict resolution are continually experimented with. Regional contract standards have avoided many disputes that resulted from sloppy agreements. Expert systems rigorously analyze contracts as well and have further preempted disputes.

Securing intellectual property

Intellectual property includes patents, copyrights, trademarks, and trade secrets. The Intellectual Property Act of 2014 provides a framework to sort these issues out. It has been a difficult issue due to different cultural views about property. In the United States, for example, scientific discoveries are viewed as the private property of the discoverer, whereas in other cultures, for example Korea, such discoveries are viewed as public goods.

Earning by giving it away: encrypting software

Software piracy is an area of intellectual property rights where global management is making progress.

Most software in affluent and some World 2 countries today is downloaded from the Net. Users do not buy their own software, but simply call it up on the Net and pay a royalty based on use. There are alternatives to direct usage fees, in which one pays a flat fee for access to software or software packages.

Software on the Net cannot be easily pirated, although hackers occasionally beat the system. Some users still use old-fashioned software packages, but they are becoming obsolete because they only run on older systems.

Many World 1 pharmaceutical companies were severely damaged by clones. Software and video piracy also cut substantially into revenues. Cloning and piracy also harmed the countries from which they originated. Domestic software industries in nations where piracy was rampant, for example, were unable to flourish due to this piracy.

Piracy in the affluent nations was considered technology transfer in World 2 and 3 nations. World 1 nations agreed to ease access to their discoveries, when faced with the threat of united action by the rest of the world. One strategy has been to encourage other countries to develop their own software industries, hence raising their stake in protecting it from theft. They followed the example of the video industry last century, which found that video piracy dropped when joint ventures or distribution companies were set up in the offending nations. The combined market power of China, India, Brazil, and other nations prevailed. Patent lifetimes have been shortened, and compulsory licensing arrangements were agreed to.

Money and finance go electronic

Financial markets are 24-hour, around-the-clock operations that are complex practically beyond human understanding. They often behave in ways that overseers are surprised by. The increasing complexity requires global oversight. There have been failures, but disaster has been avoided. The system has shown unexpected resilience. Many experts predict universal monitoring of financial and business transactions within a decade.

A de facto global currency, in which national currencies are pegged to a global standard, is in use. There is no Global Reserve Bank, but an agglomeration of regional and national institutions and controls. Twenty-four markets redistribute financial power. Chicago, for example, once controlled about 75% of futures and options contract trading. It now has just 21%. Access to capital has been leveling out the global economic playing field, by reducing the competitive advantage of easy access to capital. Clearing and settlement times have gone from two weeks to less than three days at the turn of the century to real-time today.

Computer-based crimes have been a growth enterprise over the last 35 years. International regimes, such as the fraud division of the International Telecommunications Union and the private sector Prevention of International Fraud and Forgery (PIFF) have been set up. Their security measures include voiceprint validation and encryption. Criminals are often quick to circumvent security. There is an intellectual arms race between network criminals and security.

Many private networks have sprung up to exploit gaps in regulatory structures and avoid broker's fees. There has also been growth in off-exchange trading, which is a step away from global management. These arrangements have typically been small-scale and have not drawn significant traffic away from global networks. Industries have established global information exchange networks, such as the global manufacturing network (GMAN) in 2019 that upgraded national prealliance service networks like the Factory America Network (FAN) in the United States to include nations across the globe.

Countertrade fills in gaps in global trade

Countertrade accounts for about one quarter of world trade today. These arrangements are extremely complex—a single transaction can involve a dozen nations. It is a primary tool for trading with currency-short countries. It was originally widely used by affluent nations trading with middle and destitute ones. Its use has expanded to middle and destitute nations trading with one another. Brazil and Mexico, for example, have a flourishing countertrade in vehicles and agricultural products. Computer matchmaking networks for barter, such as TRANET, have helped rationalize previously chaotic arrangements of barter, counterpurchase, offsets, buy-backs, and switch trading.

Underground economies flourish

Black and gray economies still flourish despite improved global economic management. Many businesses feel they cannot abide by the regulation that is intrinsic to global management, with its attendant costs and complexity. Uprooting underground economies has proven more elusive than many global management proponents anticipated. Underground economies cut across destitute, middle, and affluent nations in different forms. It may be in the form of street peddling, hidden sweatshop production, or servants. It often goes hand in hand with illegal immigration.

Managing political issues

Progress in managing political issues has come slowly. True global governance is decades away. The progress that has been made is in peacemaking, arms regulation, and disaster relief.

Conflict resolution: making the peace

Conflict has been a growth enterprise over the last 30 years. It involves terrorism, insurrection, civil unrest, ethnic and racial violence, border conflict, and balkanizing and irredentist movements. The end of the Cold War blew the lid off many smaller conflicts that had long been simmering. Hesitant, tentative arrangements have gradually been strengthened since then, although funding questions are still hampering peacemaking.

Divisions along religious, ethnic, tribal, and economic or resource lines have been played out in battle. The number of nations in the world today appears to have stabilized at 210, up 30% from the 162 of 1990. Most of the new nations come from Africa, where borders based on colonial convenience were reshaped along tribal and ethnic lines, creating more nations. The United Nations has moved firmly into peacemaking with the establishment of its standing army—the international equivalent of a foreign legion—in 2011. The North Atlantic Treat Organization (NATO) and Warsaw pact forces combined with troops from other nations to form a UN army. Recruits from less well off nations continue to be eager to join the force, as the troops are well-trained and well-paid.

Amnesty on-line

A perhaps unlikely beneficiary of the global information infrastructure was Amnesty On-line AO (previously Amnesty International). It became more difficult for nations to control information flows into and out of the country. Beginning in the 1990s, stories of people who were imprisoned secretly or unjustly were posted on the Net anonymously. AO would investigate the charges and followed up when it was merited.

In addition to making it more difficult for nations to hide their bad behavior, the information infrastructure also made it possible to bring more pressure to bear on renegade nations. AO chapters sprang up across the world. The AO news forums were hotbeds of activity. Millions of supporters could be organized to flame nations, or to influence governments or people dealing with the renegade nation.

For example, AO organized a boycott of Wallace & Maxwell Industries products in 2011, because they were doing substantial business in Pakistan. The Pakistani government was found to have been systematically violating political prisoner's rights. W&M threatened to move their operations out of Pakistan if the government did not comply. The government complied in this case, but in cases where governments refused to give in, companies were harmed or ruined.

There have been many battles over the lines between national and international sovereignty. Many demagogues have tied their political fortunes to bashing global invaders. Growing interconnectedness has made economic sanctions more effective than in the past, but nations with sufficient economic clout have been able evade them. China, for example, has often acted contrary to global initiatives, especially in arms trade. Its economy is too big to be harmed by sanctions. In some cases, there is little international bodies can do. Although this revives references back to the ineffectiveness of the UN in its early days, and even the League of Nations, these cases are the exception rather than the rule.

International receivership has been necessary for Haiti, Bangladesh, and Djibouti. Others have been or are on the precipice. The total collapse of these governments led to peacemaking troops taking control, without the invitation or authorization of the country involved.

Nonproliferation struggles

The overall assessment of nonproliferation, including conventional and nuclear weapons, over the past 30 years is dismal. Most nations claim that this area warrants greater priority on the global agenda, but actions are limited. China has been resistant to stop its arms trade, which has led other nations to refuse to abandon theirs.

The Lima Space Weapons Treaty of 2007 set up space as a weapon-free zone. It has remained that way. Another significant achievement was North Korea's reluctant compliance with the Nuclear Nonproliferation Treaty in 1999, as part of Korean unification.

Predicting, easing, and responding to natural disasters

Natural disaster monitoring networks have been built around environmental monitoring networks. Prevention infrastructure is emerging as the next step in managing nature. Earthquake and flood prevention works are particularly advanced. In 2017 the ISA and the Red Cross formed the International Disaster Tracking Program. It uses space-based monitoring to provide advance warning of disaster and help coordinate relief efforts.

A growing understanding of the physics of the Earth, particularly patterns of crust, mantle, and core boundaries, and the overall evolution of the crust, hydrosphere, and atmosphere has been critical. Earthquake prediction now provides warnings weeks, days, or hours before the event, taking advantage of well-established precursors. The Parkfield segment of the San Andreas was the first fault zone equipped with prevention systems. Experiments with earthquake control, such as fluid injection, are part of ongoing global activities.

What lies ahead for global management

There has been progress in global management, but few systems are complete or issues resolved. It is likely that international institutions will increasingly gain jurisdiction over realms now under nation-state sovereignty. International armies, and in some cases crime task forces, are already exempt from national jurisdictions.

An emerging and likely candidate for global management in the next 10 years is coordination of recycling, reclamation, and remanufacturing programs springing up worldwide. Proposals abound, but experience with the

3Rs is still scant, and disparities in progress are wide. Some nations, such as the United States and Germany, have been pioneers, while many middle and destitute nations have no formal 3Rs programs.

Another key future challenge will be to bridge the rich-poor gap, which continues to fuel most international difficulties. Taking advantage of affluent nations' search for new markets is a likely path. The interconnectedness of the planet is bringing home Ben Franklin's axiom that "we must all hang together, or assuredly we shall all hang separately."

Critical Developments, 2000-2025

Year	Development	Effect
2000	IGWF forms.	Transfers technologies in response to global warming.
2007	Lima Space Weapons Treaty.	Preserves space as a weapon-free zone.
2009	All nuclear nations under IEA standards.	International negotiations on nuclear safety standards begin.
2011	International police action responds to network sabotage.	Sets precedent of global police responses.
2011	UN standing army forms.	Regularizes international peacemaking.
2013	Lehigh University offers the first Master's degree in global management.	Begins training of a cadre of global managers.
2013	Global Commons Agreement on Infrastructure.	Establishes principle that national infrastructures should adopt common standards.
2014	Intellectual Property Act.	Sorts out the culturally-based differences on patents, copyrights, trademarks, and trade secrets.
2016	Bangkok Accords.	Sets up institutions to formalize limited trade protections.
2016	Hague Accord.	Established World Court jurisdiction for trans-border drug cases.
2019	Convention on Immigration.	Regularization of national and regional immigration policies into a coherent approach.
2019	China cut off from communication and financial networks.	Cutoff becomes a credible threat, as China quickly backs off.
2022	Vision for the Planet commission forms.	The vision statement will frame global management strategies and actions for the future.

2024	Convention on Environmental Balance.	Gets 32nd signature in support of sustainability standards.
2025	World population hits 8.4 million.	Population is well within planetary carrying capacity, but some regional and national capacities still in danger.

Unrealized Hopes and Fears

Event	Potential Effects
Global government.	Rationalize global management by concentrating power at global level.
Global coup d'etat.	One nation or group of nations seizes control of the globe under the cloak of global management activities.
Economic collapse.	Interconnected economies lead to a crisis in one region bringing the rest of the global economy down with it.
Closing the rich and poor gap.	Global management leads to sharing of resources, transfer of wealth, and brings nonaffluent nations up to affluent standards.
Systems failure or sabotage of global infrastructure.	The information infrastructure collapses and is not reconstructed.

9

PUTTING SPACE TO WORK

Space activities are moving ahead on many fronts. Steady gains in understanding planetary processes, technology transfer for social applications, and better space science add up to a winning investment for the nations and corporations involved, as well as the planet and the society.

Big space projects, since 2000				
International Space Station	Space Plane	Moon Base	Human Expedition to Phobos	Human Expedition to Mars
2013	2017	2020	2024	2030

The big space projects started back with the 400-foot international space station in low Earth orbit completed in 2013. The station provided the stimulus for the R&D that is bearing fruit today. Similarly, it triggered interest and investment in the Moon base, a mostly underground complex of research labs, warehouses, commercial leasing space, and an observatory, which was finished in 2020. The snowball effect continued with a human expedition to Mars' moon Phobos last year, and a planned expedition to Mars in 2030 is being rehearsed on the Moon base. It will follow six robotic missions.

Benefits from space activities are streaming in. Nations no longer use space as a playground for national prestige, but seek concrete scientific and commercial gains from it. Space activities aimed at Earth have provided the most attractive gains. New commercial ventures are forming, new scientific discoveries are being made, and new space technologies are being applied on Earth. The scope of space activities continues to expand. Workers involved in space activities worldwide number in the hundreds of thousands, an order of magnitude increase from the late 20th century. The graph below shows the steady increase in space launches. It has evolved from a specialty to a commodity operation.

World Space Launches: All Earth to Orbit Trips

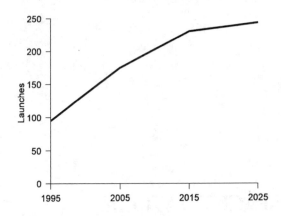

The overriding feature of the space program, and a exciting side effect, at this quarter century mark is the unprecedented degree of international co-operation. The International Space Agency (ISA) was formed in 2002 as part of the United Nations (UN) family and provides an institutional mechanism for negotiating public and private multinational space projects.

Dealing with global environmental problems set the stage for international space activities. The Montreal Accords on the ozone hole back in 1988 were followed by a ban on ocean dumping in 1998 and international global warming research in 2001. The protocols of these agreements were later adopted by the space community.

A second significant development in space is the flourishing of the commercial sector. Slow progress in commercial space in the 1980s and 1990s dampened hopes that space commerce would ever become a viable industry. At the beginning of the century, with guidance and funding from governments, commercial successes were achieved. The time line below highlights some key commercial events in space.

Timeline of commercial space activities	
1997	Earth Observation System
1998	Iridium Satellite Network
2001	Resource Tracking Model
2002	Detection of phytoplankton depletion
2004	Private weather forecasting proliferates
2009	AI robots construct space factory
2015	Biocompatible blood vessels developed
2016	Cancer fighting drug developed
2018	Solar mirrors for light in Alaska
2020	Moon mining begins
2021	Alloy for engines developed
2022	IVHS with global postioning satellite
2024	Solar power satellite for Moon base

International cooperation

Today's cooperation arose from economic necessity. Space activities often lost out to other budget priorities through the 1990s and early 2000s. But international computer and communications networks enabled collaboration among the world's space community and enhanced awareness of the wastefulness of developing redundant programs. These redundancies were addressed by ISA in the 2000s, and the stage was set for multinational projects.

The debate over the U.S. space station *Freedom* in the 1990s provided a clear example of the scope of big space projects being beyond the capabilities of a single nation. Even after scaling down expectations for the station considerably, it was not able to go forward until the project was internationalized. It took the resources of several countries, including all the primary space players—the United States, Japan, Europe, Russia, and China—to get the station in place in 2010. After that accomplishment, discussions about the planned Moon base began with the assumption of international control and cooperation, which enabled it to be constructed relatively quickly. Rather than three programs developing their own shuttles, for example, they collaborated on one and developed other projects with the money saved. It has become clear that the scale and sophistication of space activities is beyond the capabilities of any one nation. The European Space Agency (ESA) model of multiyear funding was adopted by the international space program in 2003. The decline of military missions and transfer of budgets to civilian purposes provided much of the initial funding. Restrictions on international ownership stakes of companies benefiting from the National Aeronautics and Space Administration (NASA) programs were eased. The table below summarizes the strength and weaknesses of the world leaders in space activities. It differs little from the situation 30 years ago, with the exception of the emergence of China as an important member of the space community.

Profile of World Leaders in Space

Countries	Strength	Weakness
United States	Reputation; large, trained, space workforce; thriving commercial sector	Deciding on priorities; reluctant and demanding international player
Russia	Developing scientists; rocket propulsion; space power plants	Still rebuilding the economy
Europe	Launch vehicles	Getting consensus
Japan	Commercial spinoffs for space activities; space robotics; satellites	Commercial interests in space dominate
China	Inexpensive launch capabilities; R&D into new propulsion technologies	Still catching up with the rest of the space community

Competition for the limited space to deploy satellites in low Earth orbit (LEO) also had to be worked out. The problem has been less severe than once forecast, as the size of satellites and other space structures have sharply decreased with continued technological advances in miniaturization, and the recognition that large multimillion- dollar satellites increased the risk of financial ruin. Smaller satellites are easier to launch and present less financial risk. The successful deployment of the Iridium III network of 66 LEO satellites completed in 1998 gave confidence to others that small satellites were as capable as big ones. Extended-lifetime Iridium III satellites are joined by hundreds of satellites in orbit today.

United States

The U.S. space program has regained the credibility it lost with the international space community beginning back before the turn of the century. Political problems hamstrung the U.S. space program. The Cold War mentality of competition and jealous guarding of secrets prevailed a decade after its obvious rationale had expired. The eventual easing of export controls and technology transfer restrictions after the turn of the century was important for big international space projects like the space station, which was able to use technologies once restricted for military reasons.

The eventual warming of the United States to international space cooperation came more from necessity than good will. NASA's spotty ability to deliver on promises led nations to shop elsewhere for their space needs or meet them domestically. This enabled other nations to close the technology gap with the U.S. program around 2000. The United States today is first among equals in the space community.

High levels of safety that were required following the *Challenger* explosion in 1986 hamstrung rapid progress. Eventually a decision was made, after an extensive series of risk analyses, to make a greater commitment to robotic missions. These missions served as test cases. They proved that the presumed need for multiple redundant systems was excessive. Redundancy was able to be reduced when human spaceflight resumed with testing on the Moon base in preparation for the Mars expedition. The Moon base was built almost totally without human labor, although technicians were at the site.

Russia

Russia's space program is only now slowly recovering from its fire sale of the 1990s and early 2000s. Desperate for hard currency, the government auctioned off a significant portion of the space program, including hardware and technical experts. The space program was split into different jurisdictions with the dissolution of the Soviet Union, and cooperation among them was difficult. Economic troubles further exacerbated the decline. Although the sale of Russia's space program was often under the guise of international co-operation, desperation was the driver. And the international community took full advantage. Now that relative stability prevails, one can expect a renewed commitment to space by Russia. The government is hinging a great deal of its credibility on the ability to regain prestige through space activities.

Europe

Grandiose plans for European unification in 1992 faltered and only slowly regained momentum following the turn of the century. European Space Agency programs, once a model for cooperation, were a victim of this failure to get along. Dealing with post-Cold War Eastern Europe since the 1990s further diverted attention from space activities. Integrating some of those battered economies into the EC, or in the case of Germany, reunification, took time and money away from space activities.

Europe's position in the space community, however, declined significantly. Even though internal wounds healed, and ESA got back on its feet, much ground had been lost. Europe clearly became, and remains today, a junior partner in international projects. As unification gains momentum, however, many space experts feel it will rise to the top.

Agreements with Japan, which provided much of the funding to revamp ESA, shifted the direction of ESA from developing its own space capability to sharing its expertise and developing niches required by the international community. ESA, for example, remains the world leader in launch capability and has concentrated efforts in that area.

Japan

The Japanese joined the international community's space efforts after much internal political wrangling. Those who argued for Japan becoming the world leader in space technology were only narrowly defeated by those favoring cooperation. The debate over participating in the international space station beginning in 2000 was the turning point. Japan finally signed on in 2005.

Japanese leaders came to the realization that their drive for global technological supremacy had extended beyond the country's limited resources and that Japan simply could not be the world leader in every technological area. Because Japan's position in the space race was far from the top, its leaders decided not to aim for technological supremacy in space, but instead to draw on the resources of other programs that were further advanced. Yet the Japanese have become leaders in space technology areas such as robotics, and Japan is a valuable partner to the international space community.

The Japanese are particularly interested in commercial spinoffs from space ventures. Private industry dictates the government's commitment to space to a much larger extent than in the United States. The nation has been working towards becoming the world leader in the information-based economy for decades. Satellite communications have played an integral part in developing their information infrastructure. Although communications has been the primary driver for Japan's participation in space, research into materials processing, manufacturing, biotechnology, and medicine have been of great interest.

Rather than take on the United States and Europe, partnerships that precluded redundancy and improved overall efficiency were formed, especially after the 2005 decision to join the international space station. The partnership is wary. There is concern in the United States that Japan will take what it has learned from the United States and try to take over the market. On Japan's side, there are questions about the political will and economic capability of the United States to hold up its end of the bargain.

Japan's primary contribution is in space robotics. Autonomous guided vehicles that were developed for factory use have been adapted for navigating the Moon and Mars. Robotic systems repair and maintain space structures, such as platforms, the space station, and probes destined for distant planets. They also perform more mundane housekeeping tasks on space vessels. Expert systems play an important role here. They have developed to the point where they mimic, and in many cases surpass, human capabilities. Low-level learning capabilities are emerging.

Other countries

China, the best of the rest, continues to attract launch business by charging significantly less for commercial satellite launches than is charged by its international counterparts. There are no other significant international or commercial players. Some countries continue to develop domestic space programs for reasons of national prestige, but these are minor programs. Some equatorial countries rent launch sites, because launches are easier near the equator, and others house tracking stations for space activities.

National pride

While the international community plans to go to Mars, Indonesia is now in orbit, having just launched its first domestically manufactured and operated agricultural sensing satellite. Despite the availability of international satellite networks, some countries invest national prestige in developing their own satellite or space program.

Most countries participate in space activities through leasing space aboard international vessels, stations, and bases. Some countries, and, increasingly, large multinational corporations, pay for small pieces of the big international space projects.

Military applications

Military applications in space are evolving from conflict prevention to conflict resolution. Military prospects in space declined as ideas such as the Strategic Defense Initiative (SDI) or Star Wars lost momentum. Funding based on the Cold War fell sharply in the 1990s. The Cold War mentality had set up space as the next battlefield, but today space provides the capability to monitor troop and equipment movements, whereas enforcement takes place on Earth. The 2007 Lima Space Weapons Treaty outlawed weapons in space.

Reigning in international outlaws

The case of Johan Sabini, the international outcast leader of Zaire, appears in textbooks across the world. Despite repeated warnings from UN peacemakers, Sabini invaded neighboring Zambia. Reconnaissance satellites detected his troop movements. Peacemaking troops stopped the invasion, and Sabini was arrested, tried, convicted, and removed from power by the World Court.

The scope of reconnaissance activities is beyond the capacity of one nation. The UN nominally coordinates the multinational peacekeeping observation system. Individual countries own the satellites, and their participation depends on their perception of benefits from drawing on the resources of others. Secrecy concerns have not been satisfactorily resolved. The United States continues to threaten to withdraw from international space projects, citing lax security. These threats have typically been accommodated with granting a greater voice to the United States in decision making.

Satellite reconnaissance provides real-time updates of target or troop movements. Areas with a high probability for conflict are carefully watched. If the reconnaissance detects abnormal patterns, the parties involved are quickly warned. Failure to comply with warnings lead to preventative measures, typically the deployment of UN peacemaking troops.

Reconnaissance satellites have been adapted for nonmilitary operations as well, such as monitoring natural or environmental disasters or other emergency situations. Timely recording of events surrounding the detonation of a nuclear device by terrorists in the Middle East in 2020 helped mitigate the effects of fallout and helped the cleanup.

Technology transfer from military space programs, such as the old SDI was a boon to peaceful applications because it emphasized surveillance and tracking. Spy satellite data was made available for commercial and scientific research. Law enforcement has benefited from surveillance and tracking technologies. The proper balance between monitoring needs and privacy needs, however, is still an issue for governments and the World Court.

Commercialization

Today's thriving commercial space industry has its roots in former U.S. president Ronald Reagan's decision, following the explosion of the space shuttle *Challenger* in 1986, to bar the shuttles from carrying commercial satellites. This encouraged private industry to get into the launch business. Europe's Ariane competed with U.S. companies and won the lion's share of the business in the 1980s and 1990s. Russia and Japan also became competitive. By the 2010s, China made significant inroads by offering cut-rate prices.

Commercial space activities, 1990 and 2025*	
1990	**2025**
satellite communications	satellite communications
launch services	Earth observations
Earth observations	launch services
microgravity research	space robotics
orbital facilities/infrastructure	materials processing
*In order of importance	

The competition was fierce in the early stages, as launching capacity far exceeded payloads. Following a shakeout and a gradual growth in demand for launches, a strong industry began to emerge around 2000. Today, governments focus on the big programs, such as space stations, the Moon base, and the manned Mars expedition. Private industry continues to provide launches in addition to activities such as processing, interpreting, and selling data; satellite communications; remote sensing; and materials processing. Private contractors are responsible for carrying out government plans with its oversight.

Profile of U.S. commercial space players		
Player	**Description**	**% of market**
The space establishment	Origins as NASA and/or military contractors, e.g., Hughes Dynamics	48%
The entrepreneurs	Small start-up venture capitalists that gamble on high-risk, high-potential projects	32%
Business consumers	Companies established within their particular industry seeking to take advantage of potential payoffs from space, but space is not their primary activity	20%

A reasonable balance has been struck between industry risk and reliance on tax dollars for space activities that indirectly benefit society. Since 2010, most governments have turned management of some space activities over to private enterprises. In the United States, for example, government spending has declined slightly since 2010, and private spending over the same period has more than doubled.

What is commercial space?

A commercial space activity in contrast to contract work is defined as a venture where private capital is at risk, where there are existing, or potential, nongovernment customers, where the commercial market ultimately determines the viability of the enterprise, and where primary responsibility and management initiative are with the private sector.

Governments provided the initial investment that stimulated the private sector to get involved. Government commitments to space activities, in the form of start-up incentives, persuaded companies that the risk was worth taking. Government-owned, contractor-operated, joint ventures and consortia arrangements were stepping stones. Joint Endeavor Agreements, which beginning in the late 1980s allowed companies to fund space experiments with free transportation on the shuttle in exchange for sharing the research results, became popular only about 15 years ago. The figure below compares public and private spending over the last 30 years.

Public and Private Space Spending

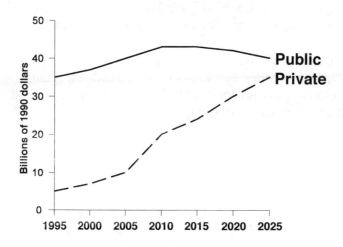

The remaining seven (down from 17 in the 1990s) Centers for Commercial Development of Space (CCDS), with hundreds of affiliates from industry, academic institutions, and state and federal agencies, received large cash grants of millions of dollars to leverage cash and other kinds of investment. NASA paid the transportation costs to space. When the program began in the 1980s, it was hoped that it would take five years for it to become self-financing. But it was not until 2018 that the centers turned a profit.

Communications satellites, back in the 1980s, were the first space industry to be completely owned and operated by the private sector. Arrangements similar to COMSAT, a private organization to spur the satellite industry, were a model for other commercial sectors. Initially, governments reserved their needs for space launches and other services to native companies. An international agreement in 2003 opened the market, mirroring the global trend toward freer trade.

Partnership between the two sectors is comfortable. Dissemination of data from remote sensing has been private, with government subsidy, since the 1980s, despite its unprofitability until the last few decades. The private sector makes widespread use of the data. Private industry continues to launch the big government programs. The inter-space freight industry, which includes shipping among and between space facilities and specializes in orbital transfer vehicles, docking capabilities, and satellite servicing, continues to grow.

In the 1990s, big players in space diversified into other industries. They believed being totally dependent on space for their existence was too risky. This opened the playing field. There was an influx of military personnel into space occupations around the turn of the century, whose positive impacts have only recently been felt. Small, nimble, entrepreneurial companies foresaw the potential to fill niches and are doing so. These companies came from all across the globe. Some established companies in other industries, such as construction, adapted quickly to being players in the building of the Moon base. Insurance companies are heavy investors in satellite technology. Weather prediction and natural disaster monitoring, for example, can help save them money, so they fund research in this area.

What follows

The benefits of space activities are in three categories: planetary, social, and space science. The table below lists the topics within each category.

Applications of Space Activities

Planetary

pollution tracking and control
natural disaster monitoring
atmospheric research
managing the environment

Space science

advances in astronomy
SETI (the search for extraterrestrial
   intelligence)
interspace travel
space station
Moon base
planned Mars expedition
prospects for colonies

Social

satellite communications
weather forecasting
crime surveillance and tracking
entertainment opportunities
robots for hazardous duty
energy technology transfer
coordinating transportation
insights into human physiology and
   psychology
mining
manufacturing and materials processing
   modules
health, medicine, and pharmaceuticals
crop and fisheries monitoring
solar lighting experiments
systems engineering

Planetary

Space activities continue to improve our understanding of the Earth. Space observations demonstrated the impact of human activities on the atmosphere and the ozone layer in the 1990s. These observations enabled the creation of a strategy to deal with the ozone holes. Corrective actions first involved the banning of chemicals harmful to the ozone layer, completed by 2000. Further research focused on and led to the development of plans to patch the holes by releasing "adhesive" chemicals and synthetic ozone into the atmosphere. Similarly, the identification and beginning of corrective actions for global warming are underway.

Remote sensing applications, such as the now full-blown Earth Observation System (EOS), monitor hundreds of planetary processes such as photosynthetic activity and tides and wind currents and enable an integrated study of the Earth as a planet. Last year, for example, a UN-sponsored program to measure the world's oxygen supply began. International cooperation in caring for the physical planet is a model for cooperation in other areas. It is indicative of the trend to a totally managed environment. Water, coral reefs, the atmosphere (including global warming and the ozone holes), the polar ice caps, biodiversity, minerals, hazardous waste, landfills, deforestation, and soil are some of the myriad ecosystems of the Earth being monitored from space.

2025

Planetary applications of space activities

- pollution tracking and control
- natural-disaster monitoring
- atmospheric research
- managing the environment

Pollution tracking and control

Pollution tracking and control is a robust industry, accounting for 20% of GDP growth this century. The trend is toward a greater emphasis on control. Knowledge of pollution and its effects has advanced rapidly as a result of years of data collection and interpretation. Analysis is giving way to corrective actions, such as emissions limits, banning of harmful chemicals, and heavy government investment in R&D for alternatives.

Natural-disaster monitoring

Advances in remote sensing and geology enable the prediction of natural disasters, such as earthquakes and hurricanes, weeks, days, or hours in advance. Remote sensing detects ground motion (seismic activity) and maps and monitors undersea faults, greatly improving prediction accuracy. Many natural disasters are now effectively managed, controlled, or prevented. Damage from the Great San Francisco Earthquake of 2018, for example, was held to a minimum thanks to advanced preparations enabled by long- and short-range forecasts that were acted on by government and business.

Policy question

What do government leaders do when informed of a pending natural disaster that they are powerless to do anything about? Do they inform people anyway? Do they say nothing? Leaders have been following a mixed strategy.

Atmospheric research

Climate change drives a great deal of planetary research. Understanding of upper atmosphere composition and dynamics benefits from this attention. Definitive proof of global warming, derived largely from EOS observations a decade ago, directs lots of funding into enacting plans that have long been sitting on shelves. Interaction between the atmosphere and the oceans has enriched our understanding of hydrology. Satellites and earth probes are supplemented with land observatories.

Managing the environment

Satellites interact with ground-based sensors and microprocessors to create a smart environment and dramatically increase understanding of the planet. Changes can be measured instantaneously. Genetics adds to the capability to manipulate the environment directly or those responses to it. Crops resisting local pests, for example, have been genetically engineered. The effects of genetic manipulation on the environment are closely monitored by environmental groups.

Approaching the Limits: World Population to 1990 to 2025

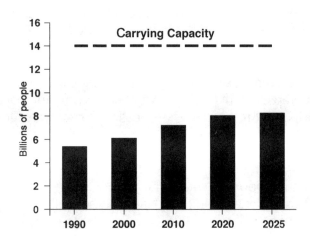

Diagnosis of the planet's health and monitoring functions are precursors of restoration, maintenance, and planning projects. Environmental engineering is a booming industry. An annual assessment of the health of the planet, first popularized back in the last century by the WorldWatch Institute's *State of the World* reports, is now an international undertaking. WorldWatch now puts out *Restoring the Earth.* The United Nation's *Earth Development Report* draws on a massive database made possible by the EOS a constellation of sensing networks under UN supervision.

WorldWatch Institute
Restoring the Earth, 2023

Table of Contents

1. The Cleanup Continues
2. Ocean Dumping Slows
3. A Steady State for Coral Reefs
4. More Demographic Transitions
5. Brownouts in Developing Countries
6. Patching the Ozone
7. Coastal Zones Fear Sea-Level Rise
8. Intelligent Vehicles Assessed
9. Global Environmental Governance

2025

This database is the foundation for global environmental management. It is eagerly tapped by the public and private sectors, as well as individual citizens. For example, what urban planner today would not take advantage of the mass of knowledge about potential hazards, estimated carrying capacity, and estimated growth potential?

An interesting development in planetary management was the flourishing of the 3Rs (recycling, reclamation, and remanufacturing). Their evolution can be traced to the institutionalization of the sustainability concept in the 1990s. This triggered discussion of 3Rs practices. Recycling received the early emphasis. Gradually, the three came to be considered together. Today 40% of materials in the United States are recycled, 30% are reclaimed or remanufactured, and 30% are either burned or buried.

The primary enabling capability for the 3Rs was the assembly of the Resource Tracking Model, an offshoot of the EOS, which provided the first reliable quantified estimates of the planet's carrying capacity. Although they are incredibly complex, early findings were that we are approaching depletion of many important resources. The so-called doomsayers, going back to 1972's *Limits to Growth*, now have evidence to back their arguments in favor of sustainability. Armed with this data, their arguments found an audience, and 3Rs programs sprang up primarily at the local level. Regional programs, with some national support, are now well underway. An international agreement on the 3Rs is currently being negotiated at the Melbourne Waste Minimization Conference.

Social

A second area of space applications includes those applications that directly benefit society, in contrast to planetary applications with indirect benefits. Markets range from communications to the fishing industry, which locates fish by identifying phytoplankton, to shipping, ocean drilling, and industrial environmental monitoring groups.

Satellite communications

Satellite communications networks continue to be the primary economic payoff from space activities. They complement the worldwide, broadband network of networks based on fiber optics. They are particularly useful for remote or sparsely populated areas, where the economics for laying fiber optics are unfavorable. People are able to talk, see, or send information to almost anyone in the world. LEO satellites supplement terrestrial cells to form personal communications networks. Direct broadcast satellites broadcast television programs in remote areas to very small aperture terminals (VSATs) which are small rooftop antennas. They provide personal communications services as well.

Today's satellites use higher frequencies, have more onboard power and processing capabilities, are fully digital, weigh less, are compatible with ground-based broadband digital networks, and use ion propulsion, which is weak on Earth, but is sufficient in frictionless vacuum of space. Intersatellite links, in which information is exchanged by transmitting modulated laser beams rather than the microwaves that were used in the past, are growing. Intelsat, Eutelsat, and Inmarsat are connected in space by laser intersatellite links.

Weather forecasting

Weather satellites combine with aircraft, balloons, ground-based observations, and telecommunications linkages to supercomputer-based modeling, smart databases, real-time data, and graphic displays to deliver unprecedented forecasting accuracy. Today's forecasts have longer time horizons (months and seasons rather than weeks) and shorter time horizons (minutes rather than hours) than those of the previous century.

The limits to forecasting

Incomplete understanding of chaotic systems limits further improvement in weather forecasting accuracy. Chaos theory tells us that even a tiny change in initial conditions can have dramatic effects on the behavior of a chaotic system. Until we learn to map initial conditions better, weather forecasting will remain an imprecise science.

Better characterization of initial conditions of the atmosphere since the 2000s has improved short-term weather forecasts. Private forecasting services delivering detailed, tailored forecasts to corporations have proliferated since then. Advances in meteorology enable the prediction of specific storms with a useful lead time and a finer ability to distinguish among types of weather, such as rain versus hail. Forecasting solar storms, which produce solar flares capable of damaging satellites or other space equipment, has been a growth enterprise. Industries heavily influenced by the weather, such as tourism and agriculture, finance many of the projects that have extended the forecasting envelope.

Crime surveillance and tracking

Satellite tracking and monitoring capabilities, including the ability to detect electronic vehicle identity tags from space has been bad news for criminals. Car theft, once a major crime, has virtually disappeared. Fugitives are more readily tracked down. This is not meant to imply that crime no longer exists, but it has been significantly curtailed.

Many are calling for the use of solar mirrors to blanket high-crime areas at night with light, therefore discouraging criminal activity or aiding law enforcement officials pursuing potential lawbreakers. Opposition to this proposal on civil liberties grounds has been equally vociferous, and it has not yet been tried.

Entertainment opportunities

Rapidly growing interest in space-related entertainment suggests that it will be the fastest-growing segment of the commercial market for space activities over the next few decades. Entrepreneurs have seized on the soaring global growth of the entertainment industry and tapped into space as a novel medium. Movies continue to borrow space themes. Space tours are projected to be a thriving industry within the next 50 years. Plans are already being laid for a globe-circling sightseeing space ship. A huge market awaits virtual reality participation in exploration. Virtual reality spinoffs include training in addition to recreation. It is a routine simulation tool. There are hundreds of space simulation games on computer networks across the world. Space educates as it entertains. Interest in science among school children has risen with each new space achievement.

Robots for hazardous duty

Robots and telerobots performing hazardous duty in space have been adapted for Earth duty such as monitoring fusion experiments, running fission plants, overseeing hazardous waste disposal and conversion, and other dangerous duties. Robots and other automated machinery are commonplace inside and outside the factory in agriculture, building construction, underseas activities, as well as in space.

Breakthroughs in machine sensing and vision after 2000 enabled a much wider range of applications for robots and telerobots. Equally critical have been developments in artificial intelligence (AI), which improve robots' onboard information processing capabilities. As advances in AI have progressed since 2010, use of teleoperation has diminished and fully autonomous AI robotic systems have increased.

Robots are used for satellite retrieval; servicing and maintenance; deploying or assembling large structures such as space platforms, space stations, large antennas, and solar power stations; rescue operations; and in situ exploration and analysis of lunar and planetary terrain. Robots perform more space operations than humans. But the versatility of astronauts and long-term plans for space settlement keep human space exploration alive. The key technological advance has been in sensors, particularly machine vision, as well as onboard and teleoperated intelligence. Computer integration of cells of robots continues to improve. Robots are capable of receiving communications, understanding their environment, formulating and executing plans, and monitoring their operation.

Energy technology transfer

Perhaps the most exciting development in space activities is the experiments with solar power satellites (SPSs), which collect solar energy with large arrays in orbit and convert it to radio waves, which are then beamed to collecting antennas. SPSs are exposed to the sun 99% of the year and receive 10 times as much energy as an earthbound solar station. The rising costs of Earth-based energy stimulated a revisiting of SPS research. The development of supporting infrastructure was crucial. The space station and Moon base provided the platforms for the necessary construction and manufacturing.

Earth energy technologies in space

Photovoltaic efficiencies continue to improve. Today's 50% efficiency is much higher than was thought possible 25 years ago when efficiencies seem stalled in the 30% range. Space experiments are responsible for this breakthrough.

Fuel cells are a supplement to SPSs. They produce water in addition to electricity, invaluable to the Moon base, given the absence of water on the Moon. During the day, solar power breaks down water into hydrogen and oxygen which is stored, and then recombined in fuel cells to produce energy when solar energy is unavailable.

Experiments with nuclear fission in space continue. Nuclear thermal propulsion is promising in that it produces a larger amount of energy from a smaller propellant mass. The disadvantage, at least when human crews are involved, is the mass of shielding required to protect from radiation. Space disposal of radioactive wastes has been proved technologically feasible but is still too expensive and politically unpalatable.

An SPS has been used successfully for power on the Moon base. Advocates are confident that the transition to providing power to Earth on an experimental basis will be made within a decade. Opponents deride the proposal as science fiction fantasy, citing the harmful effects of a powerful energy beam on the environment. Further study will be certainly required before a trial run is allowed.

This new energy source could be critical as there has been a substantial increase in energy demand on Earth. Although *per capita* energy consumption has been declining in World 1, it has risen dramatically in the rest of the world.

Public utilities use remote sensing to monitor energy efficiency of users. Wasteful energy practices are easily detected with heat-sensing infrared devices. Large users are routinely monitored, while plans are being laid to extend coverage to all users.

Cryogenic fuel storage of liquid hydrogen continues to lower boil-off levels. This technology has been proposed for the Mars expeditions, which will require large amounts of fuel storage.

Coordinating transportation

Satellites play a role in monitoring transportation infrastructures as well. The safety of aviation, rail, and highway infrastructures is maintained with the assistance of sensors interacting with satellites. Intelligent vehicle highway systems now dot the landscapes in the United States, Japan, and Europe. Initially, IVHS relied on cellular stations for positioning but now take advantage of the greater accuracy that satellite global positioning systems offer.

Global positioning uses ground-based devices, detecting beacons or ranging signals to triangulate a position, as well as three-dimensional sensing for collision avoidance. Positioning is accurate to within centimeters.

Satellite mapping techniques are continually increasing the real-time information available for applications such as crop or forest status, environmental assessment, energy production, geographic information systems, disaster assessment, and land use planning.

Transportation has improved not only through advances in navigation technology enabled by satellites, but also in technology transfer to avionics. Hypersonic travel, at five times the speed of sound, is now commonplace. Space planes developed in 2020, which can take off horizontally and fly to low Earth orbit, and single-stage, heavy lift rockets are used for launching satellites and other payloads. Proposals to use space planes for rapid intercontinental travel—getting from Europe to the United States in less than two hours—have been stalled by high costs.

Insights into human physiology and psychology

The international space station and Moon base are test beds for the physical and psychological effects of space on humans. Extensive life-support systems for extended stays in space have been developed. As the current human expedition to Mars so far indicates, extended stays in space are possible. The full effects will not be known until the mission returns and detailed debriefings take place.

Understanding of human biology continues to grow with assistance from space research. Microgravity effects, including effects at varying gravity, are much better understood. Conception in microgravity has been successful, although the first birth in space awaits further research. Autonomous and/or robotic systems augment human capabilities in space. Cell research studies, such as aging, anemia, bioprocessing, cell secretion, diabetes, drug delivery, hormone production, immune function, light effects, muscular atrophy, and osteoporosis, are ongoing.

Artificial gravity, created by revolving the space structure, is an effective means to avoid the effects of microgravity. It is now the preferred means of space stays, with microgravity experiments ongoing, but of lesser priority.

The study of human psychology in novel environments continues to be aided by research from the study of astronauts. Critics argue, with justification, that astronauts are not representative of the general population and, therefore, the results are questionable. The rigorous selection process for astronauts, and the intensive training they undergo indeed prepares them in such a way that it is hard to extrapolate the findings to the rest of the population.

Still the findings form a basis from which further study can proceed. As more people become space travelers, findings should become richer. Human factors such as productivity in space and teamwork in extreme environments and conditions have been adapted to hiring practices; cognitive/personality screening; stress measurement, evaluation, and management; and interfaces with sophisticated equipment here on Earth.

Genetic testing determines who is suitable for space travel. Some of the criteria used are response to g-force, heart and metabolic rates, and predisposition to claustrophobia. Genetic manipulation, namely augmentation to better withstand the space environment, is being planned for the next Mars mission.

Mining operations

Mining operations include solar concentrators to melt lunar soil to create glass and other building materials. Mining ilmenite, a mineral made of iron, titanium, and oxygen for rocket fuel on the moon is scheduled to begin next year. Active exploration of mining Helium 3, an abundant source estimated to be capable of providing over 10 times the energy equivalent of fossil fuels remaining on Earth, is ongoing.

Mining advances and lessons from the Earth aid lunar mining and vice versa. Petroleum engineering technology has been modified for lunar mining. Remote sensing and expert systems lend precision to the targeting of lunar resources.

A graduate degree in space geology has become a valuable commodity. It includes mining as well as sample analysis from the Moon, Mars, Venus, and eventually other planetary specimens. Experience in lunar mining will provide the basis for future asteroid harvesting.

Manufacturing and materials processing

Manufacturing in microgravity offers one of the most practical applications of space activity. Its importance is diminishing, however, as new technologies developed on Earth make the microgravity environment less attractive. Space manufacturing modules still dot low Earth orbits. The primary products are those that could not be made on Earth, or could be manufactured to a quality that would be prohibitively expensive to achieve on Earth. Manufacturing techniques on Earth and in space are transferred to and from one another. Automated materials handling and automated storage and retrieval systems have lessons directly applicable to remote operations in space, such as docking vehicles and transferring materials from launch vehicles to a space platform or station or Moon base.

Abundant materials opportunities

Knowledge of how materials interact in microgravity are applied to materials processing on Earth. Processing pure materials such as crystals and microscopic latex spheres for calibration are important for pharmaceuticals and other biologicals, biotechnology, chemicals, and electronics industries. Advanced materials processing, such as powder metallurgy; crystal shaping for electronic optics, detectors, and separators; advanced coating and composites tooling; radiation-resistant materials; fault and damage tolerant, self-healing materials and structures; lightweight and self-assembling structures; and adaptive smart materials, surface coatings, electrodeposition, and polymer projects are routinely employed. The vacuum in space allows for the controlled deposition of thin films by epitaxial growth, which is now a powerful technique for electronic, superconducting, and magnetic materials and devices.

Space provides new materials, as well as new means of materials processing. Materials advances improve space structures. Research is enabling the development of materials in space for space structures. A planned addition to the Moon base, for example, will be made almost entirely of materials indigenous to the Moon. Ceramics are able to withstand extremely high temperatures. Advanced composites are light, strong, stable, temperature-resistant and long-lived. Materials for all space structures are stronger, more tolerant of space conditions, and more intelligent, as they are armed with mechatronic devices and integrated with information and control systems. Lighter launch vehicles increase energy efficiency. Photovoltaic grid materials have boosted energy efficiency to 50%.

Zero gravity offers intriguing possibilities for combining materials which cannot be combined on Earth. Liquids that cannot be mixed due to the effects of gravity can be mixed in microgravity to form new alloys. Applications could include aircraft or even motor vehicle parts that are highly heat resistant and longer-lasting. Commercial ventures are researching the possibilities. The Centers for Commercial Development of Space, with help from NASA, spurred research in this area, until it became self-sustaining. CCDS's first important contribution was in 1986, to the development of the first superconducting materials at temperatures above liquid nitrogen, 77 Kelvin. Experiments in this area led to improved materials for radiation detectors.

Recoverable launch capsules, or commercial experiment transporters, called COMETs, were developed in the 1990s for low Earth orbit experiments. They parachute back inside capsules to Earth. Spacehab, Inc., modules inside the space shuttle, and reusable external launch vehicle tanks boosted into orbit, were used until the new international space station was constructed in 2010.

Astronomical technologies offering markets and generating spinoffs include glasses and lightweight materials, high-tolerance machining and polishing, reflective coatings, wavefront control and intelligent optics, and large-scale computer simulation.

All materials are, of course, designed for recycling, reclamation, and remanufacturing.

Health, medicine, and pharmaceuticals

Space medicine research competes with biotech medical research for R&D dollars. Niche applications for Earth predominate. Most space medicine research focuses on treatments needed for those in space.

The health effects of microgravity have improved our understanding of how the body works. Macromolecular crystallography, the primary technique for determining the three-dimensional atomic arrangements within complicated biological macromolecules, has been essential to understanding the fundamental structure/function relationships that govern biological systems.

Primary applications are in pharmaceuticals—such as the delicate separation of complex, nearly identical substances—drug design, protein engineering, and in the chemical and biotechnology industries. Experiments have helped determine enzymes targeted by acquired immunodeficiency syndrome (AIDS) and cancer researchers.

Cancer treatment from space

A new treatment for cancer resulted from protein crystal growth experiments performed in a microgravity space laboratory that identified structures triggering the destructive growth of cancerous cells and enabled the development of drugs to block them.

Bioprocessing experiments, which form and manipulate biomaterials in microgravity, have been instrumental in developing advanced biocompatible materials formed from the self-assembly and polymerization of proteins and other macromolecules. Applications may include artificial body parts, such as skin, tendons, blood vessels, and corneas, as well as advanced membrane technologies.

Crop and fisheries monitoring

Soil science experiments, as precursors to terraforming, have made significant contributions to developing erosion-resistant soils. These synthetic soils, designed for specific surroundings, are used to restore terrain as well as enhance agriculture. Sensing technology enables the alignment of forestry practices with environmental objectives of forest ecology. In agriculture, production is fine-tuned to meet soil and weather conditions. Remote sensing detects plankton and monitors fish population to prevent disruption of the food chain and overfishing.

Lessons from growing food in difficult space environments have been transferred to growing food in harsh environments on Earth. The use of artificial lighting and new growing media combine with careful genetic manipulation to produce food practically anywhere. Automated plant growth facilities have been developed to grow food in space. They provide oxygen as well as food, remove carbon dioxide, and purify water. These facilities are in use at the space station and the Moon base.

Solar lighting experiments

Solar mirrors bring light to regions once darkened for seasons at a time. Alaska and the Scandinavian countries are pioneering this technology. The effect is similar to a bright moonlit night. Though still experimental—the first large-scale trial was just five years ago—initial response is encouraging. Surveys of the local populations indicate that a majority favors continuing the trials. Some complained of trouble sleeping and that their biological clocks were disrupted. Entrepreneurs are investigating using multiple mirrors in tourist areas to extend the day.

Systems engineering lessons

Space programs continue to pioneer the practice of systems engineering for managing extremely complex systems. Huge computing and data storage needs have been met with advances in computer engineering such as parallel processing, artificial intelligence, neural networks, fuzzy logic, and routing theory. Data storage problems have been abating only within the last five years. Exponential increases in data volumes from the Earth Observing System overwhelmed capacity. The trillion bits of data required 10,000 reels of the old magnetic storage tape. Optical storage technology and magnetic and semiconducting storage, still used today, filled the gap. The long-awaited magnetic bubble or holographic storage technologies are finally being used, albeit on a small scale so far. Three-dimensional interactive images are integrated into computer and information systems, mixing video, graphics, text, sounds, and voice.

Space science

Long the primary purpose of space exploration, in recent years the focus has shifted to more tangible applications as the terrestrial agenda continually squeezed budgets. Yet, space science has prospered despite the shift in emphasis.

Space science applications

- advances in astronomy
- search for extraterrestrial intelligence
- interspace travel
- Moon base
- planned Mars expedition
- prospects for colonies

Creating a separate category does not imply that there are no planetary and social benefits from space science. Our growing understanding of the universe, in particular neighboring planets, improves knowledge about our own planet. Certainly the protective measures we have taken to avoid an asteroid collision are practical.

Advances in the pure science aspects of physiology, materials processing, geology, archeology, robotics, automation, virtual reality, cryogenics, energy, materials, computer and software engineering, information and control systems, and sensors in turn advance commercial prospects. The practical applications of these sciences are discussed in the social applications section of this chapter.

A critical management strategy was the shift from custom-designed space structures and vessels to standard models. This innovation, adopted by the international space community, saved money in times of tight budgets and increased the prospects for interaction among different space programs. It was accomplished only after years of excruciating negotiations.

Advances in astronomy

We know that the universe is roughly 18 billion years old. The origin and evolution of the our solar system is far clearer today than at the turn of the century. In the 1990s the Hubble Telescope, despite its flaws, narrowed the error in the universe's age to +/- two billion years. The new generation of orbital telescopes emerging after 2015 further narrowed the range of error to hundreds of millions of years.

Advances in telescopes

Telescopes are now constructed of space-durable, lighter-weight radiation-resistant materials and coatings with greater thermal dynamic and structural stability. They continue to improve by orders of magnitude each generation, roughly every 20 years. Large, rigid, single mirrors have been replaced by segmented mirrors, multiple-mirror telescopes and interferometers, instruments in which an acoustic, optical, or microwave interference pattern of fringes is formed and used to make precision measurements. Space-based interferometry offers resolutions 10,000 times greater than individual telescopes. Adaptive optics, which correct for atmospheric distortion, are enabling large gains in resolving power and sensitivity.

Astronomy today benefits considerably from being largely outside the Earth's atmosphere, with its haze, fog, and clouds shielding some of the radiation emitted from celestial bodies.

Permanent orbiting observatories are now in various orbits. The profusion of space probes is too large to detail here. Intelligent probes are studying planetary systems, comets, asteroids, quasars, and black holes. Results include thousands of asteroids being identified. They are made of rock, metal alloys, and carbon compounds. The carbon-based ones may be useful for mining. (Space entrepreneurs are already expressing interest.) They could be used to make oil, food, or have water extracted from them. The nature of black holes and quasars has been clarified.

The search for extraterrestrial intelligence

SETI is now an international program built around the hodgepodge effort begun in the United States last century. The name was retained for sentimental reasons. Confirmation of the existence of planets around other stars in 1996 boosted the credibility of SETI researchers arguing for funding. This finding dramatically increases the prospects for detecting intelligent life.

False signal

For a brief week in 2007, the world drew together in marveling over the seeming discovery of intelligent life in the universe. Scientists cautioned against overreaction, but most became so caught up with the possibility that caution was abandoned. A common cause for humanity drew people together. Daily life paused, much as it does in wartime, until scientists identified the signal as false and ruled out intelligent life for now.

Interspace travel

Liquid and solid rocket fuels continue to predominate, despite an extensive menu of alternative launch technologies. Single-stage-to-orbit rockets have predominated since 2000. The alternative technologies are presented in the table below, followed by a more detailed discussion.

Propulsion Technologies

Type	Year Proved	Prospects
Fission: thermal propulsion	2019	Shield technology for manned expeditions has been improved enough to be considered for the Mars expedition
Laser	R&D	R&D in this area has lost out; small-scale research only
Tethers	2001	Experiments ongoing; more practical applications needed
Rail guns	2014	Are used to launch small satellites into low Earth orbit
Ion drives	R&D	Losing favor to solar sails
Solar sails	R&D	Significant long-term potential
Plasma Thrusters	2023	Prototypes are being developed
Antimatter	R&D	Costs still too high

Fission thermal propulsion is under consideration for robotic missions, but the problem of shielding human expeditions has not been resolved. Prospects for fusion are dimming, as long-awaited breakthroughs in fusion energy research have failed to materialize. The old maxim that "commercial fusion is always 50 years away" probably still holds true. Helium 3, which is below the surface of the Moon, is a likely fusion fuel source, if and when the propulsion technology arrives.

Protests by environmental groups have impeded research into laser propulsion. The effects of generating the powerful beam that would be required are not fully understood. Environmental groups have opposed any testing until safety can be assured. This presents a chicken and egg problem, as tests are needed for greater understanding.

Tethers, which conduct electromagnetic energy between two spacecraft (usually from a large vessel to a small probe) got off to an inauspicious start back in the last century, when a NASA tether failed to unravel properly. The feasibility of tethers was demonstrated in 2001, but in this case the technology awaits a market.

An electromagnetic rail gun was deployed on Mauna Kea, Hawaii's highest mountain in 2019, and in the Ural mountains in Russia last year. In the United States, a bitter struggle over environmental effects postponed the construction of the rail gun for five years. Use so far has been limited, and will be for at least the next few years, to small payloads, such as satellites, and to low Earth orbits. R&D continues into possibilities for expanding use. For instance, plans are underway to deploy a rail gun on the Moon for transporting mining materials back to Earth.

Ion drives and solar sails are exotic technologies competing for the same market. To date, solar sails have received more attention. They are thought to be effective close to the sun, not much further out than Mars. Research favors ion drives for travel beyond Mars, but getting beyond Mars has been a lower priority to date.

Plasma thruster prototypes are under development. They use electrical and magnetic fields to force propellant out of the engine. They are an excellent transition technology, since they are not far removed from liquid and solid fuel rockets currently being used.

The concept of antimatter propulsion, in which the antimatter converts all of its mass to energy, has been proven, but the cost of producing the antimatter is still prohibitive.

Space planes are beginning to find commercial practical applications. Lessons from the composites required for entering and leaving orbit, as well as novel aerodynamic techniques, are being applied in everyday aviation. Space planes take off from conventional runways and exit the atmosphere, rather than being launched. Originally developed for military missions, they are now being proposed for very fast intercontinental travel. A flight from the United States to Japan or Europe could take under two hours on a space plane.

Space travel and space-walking technologies are advancing. Extravehicular activity suits have done away with the requirement that astronauts breathe pure oxygen for hours before walking in space. Robotic orbital transfer vehicles, space tugs, and in-space assembly and construction technologies have reduced the need for human expeditions and space walks.

Space sanitation companies have emerged over the past 10 years to clean up the growing volume of space debris. They are paid by satellite and other space-structure operators to prevent debris from colliding with their equipment. The companies have also taken advantage of salvage opportunities.

Space station

The international space station in low Earth orbit endured lengthy battles over its utility before its completion in 2013. Opponents argued that the cost of the station, now estimated to be $120 billion over its proposed 30-year lifetime, did not justify its expected yields. They argued that smaller-scale projects from which funds were diverted ultimately would have been more useful.

Although commercial benefits clearly did not recoup the investment, it established a space infrastructure with electric power, thermal control, warehousing space, communications, fuel storage, docking capabilities, and interspace transportation. The flight telerobotic servicer for assisting in operations and maintenance of space structures had its trial run on the space station.

These capabilities have led most to conclude that building the station was, on net balance, positive.

The station evolved from being a U.S. facility with some international participation to an international facility with U.S. participation. It supports eight full-time astronauts. The four research modules are divided between the main partners—the United States, Europe, Japan, China, and Russia—and countries or corporations leasing space. The leasing funds pay a significant portion of keeping the station afloat. Optimists expected the station to be self-supporting by now, but it is not, and it does not seem likely to be for at least the next decade.

Preparing for the Moon base and Mars expedition became the primary rationale for its construction. Commercial benefits took a back seat. Plans for an observatory and an Earth observation post were abandoned in the initial design to keep costs down. The observatory was built on the Moon base instead. An Earth observation post was added to the space station in 2022. The space station's previous role as the hub of space activity has been down-graded somewhat as the Moon base assumes that role. It remains, however, an important way station. People visiting the moon base, for example, launch to the space station and transfer into an interorbital vehicle to finish the trip. The station has long been a satellite launch point and service and mainte-nance center. It refuels space vehicles bound for longer journeys and has been a test bed for experimentation with new forms of space-vehicle propul-sion systems. It will have direct responsibility for maintaining the solar power satellites, if they ever produce power for Earth.

Moon base

Scientific experiments, astronomy, and preparations for the expedition to Mars are the base's most important functions. The Moon is being thor-oughly probed and sampled to determine its origins and for potential mining operations. Small-scale mining operations began with the completion of the initial stage of the base in 2020. Astronomy from the Moon sharply increases resolution, which is inhibited by the atmosphere on Earth. The effects of microgravity on astronauts for extended periods of time have been studied in preparation for the Mars expedition. The microgravity environment is also used for materials processing research and development.

The Moon base consists of mostly underground modules, which protects inhabitants from the radiation that hits the atmosphereless surface. Fission and photovoltaics supply energy, supplemented by recently deployed solar power satellites. Solar mirrors generate artificial light that eases the monotony of month-long day-night cycles.

Designs for building the base were solicited from the world's leading architects. Civil engineers assist with supporting infrastructure, and a long-term goal is to use moon materials and terraforming experiments to make the base self-sufficient. Although the initial modules were constructed on Earth and launched to the Moon, scientists expect to be able to add on a module with indigenous materials within the next decade.

The Moon base is an ideal way station for deep space travel, as its gravity is one-sixth that of Earth. The expeditions to Mars will include a refueling on the Moon base. The base has been a rehearsal studio for the Mars expedition. The base is supplied from space platforms as well as the international space station.

Planned Mars expedition

The planned expedition to Mars is expected to last two years. The international participants have been debating the means for getting to Mars for the past five years and have yet to decide. Japan, Europe, and China have argued in favor of a novel approach, in which a robotic return vehicle and fuel-processing facility would be launched from the Moon base and precede the astronauts. The astronauts would then arrive in a one-way launch vehicle. The fuel-processing facility would take advantage of the carbon dioxide in Mars's atmosphere to make the fuel necessary for the return trip, which would be loaded into the return vehicle. The United States and Russia prefer nuclear thermal propulsion, which they feel is a more proven approach.

The astronauts will spend months on the surface conducting experiments. A goldmine of scientific information has been gathered. Balloons will be deployed in orbit, to test, and deploy instruments, and two Mars rovers will take soil samples and map the terrain.

A robotic mission last year to Phobos, a Mars moon, has not returned yet but promises to provide valuable data about surface and atmosphere of Mars. Two more robotic missions to Phobos, followed by three to Mars itself, will gather information and lay the groundwork for manned expedition. Robotic rovers provided samples, and other scientific and engineering information useful to the manned expedition.

Prospects for colonies

Plans for space colonies are already being laid by private entrepreneurs. Colonies are not a solution to the population problem at this time or for the foreseeable future. They could be useful as test beds for long-term prospects, but at this time they could be no more than novelties.

Terraforming experiments, which modify conditions to enable human survival, have been undertaken on the international space station. Primitive experiments with creating ecospheres began on Earth in the various biosphere experiments beginning in the 1990s. Although the science was sometimes questionable, and the line between science and entertainment was often breached, the results were helpful. Small-scale terraforming experiments are scheduled on the Moon.

Critical Developments, 1994-2025

Year	Development	Effect
1998	U.S. space station *Freedom* project scrapped.	Paved way for U.S. participation in international space station.
2000	Chemicals harmful to ozone banned worldwide.	Remote sensing monitors compliance.
2001	Resource Tracking Model developed with assistance of Earth Observation System.	Part of international efforts to assess the planet's carrying capacity.
2002	International Space Agency formed.	Part of the UN system; a forum for negotiations.
2003	European Space Agency multi-year funding model adopted by ISA.	Boost to long-term planning efforts.
2003	International trade agreement opens launch markets.	National space programs no longer reserve launches for native industry.
2005	Japan signs on to international space station project.	Japan decides in favor of international cooperation rather than competition in space.
2007	Lima Space Weapons Treaty signed.	Space is preserved as a weapon-free zone.
2010	U.S. government spending on space levels off; private spending continues to surge.	Commercial ventures becoming increasingly viable.
2013	China launches most payloads of any nation.	China joins the world's space leaders.
2017	ISA and International Red Cross form the International Disaster Tracking Program.	Space-based monitoring to provide advance warning and help coordinate relief efforts.
2025	Pollution tracking and control accounts for 20% of world GDP growth from 2000 to 2025.	Space-based tracking efforts an important commercial application.

Unrealized Hopes and Fears

Event	Potential Effects
SETI discovers intelligent life.	Alters humanity's conception of its role in the universe and boosts the importance of the international space program.
International political movement against space activities emphasizes the need to take care of problems on Earth first.	Loss of political support for space exploration.
Nanotechnology enables nanomachines for application such as flushing astronauts' systems from the effects of microgravity, terraforming, or developing a space suit managed by billions of mechanical nanocomputers.	Enhances technological capabilities available for space activities.
A hostile nation develops space weaponry.	Shifts the use of space from peaceful to military purposes.
Meteoroid demolishes the space station or moon base.	Questions wisdom of manned exploration and leading to greater emphasis on robotic missions.
Economic depression.	Loss of funding for space activities.
UN votes down the use of space surveillance technology, citing invasion of national sovereignty.	Reduces effectiveness of peacemaking operations.
UN collapses.	Disintegration of international cooperation; reemergence of competitive national space programs.
Rocket or satellite crashes to Earth, killing or injuring people.	Loss of support for space activities.
World Court fight over ownership of space as countries lay claim to space above them.	Assigning areas of space becomes divisive, much like disputes over water rights.
Environmental impact statements required for space.	Slowing of space development as impacts are studied.

10

OUR BUILT WORLD

Social and economic advancement requires an adequate infrastructure. A society can barely survive and will not thrive without the means to transport goods and people, produce and transmit energy, and, most important, communicate. Sustainable growth and development require complex systems to manage the supply and use of resources and the environmental effects of human activity.

A new concept of infrastructure

Infrastructure today means central and distributed systems established to manage the environment, the movement of goods and people, communications, resources, and wastes. The infrastructure is in space, on land, underwater, and underground.

Infrastructures were once thought of as local, regional, and national. Today, we recognize the importance of the global infrastructure, in managing natural disasters such as floods and earthquakes, weather modification, and ecosystem reclamation.

Today there is a developing global infrastructure made up of the evolving infrastructures of each country and of emerging international systems including, air, rail, and communications networks. The world's infrastructure development is gradual, punctuated by revolutionary events such as the Iridium Satellite Network and continental-scale high-speed rail. In the coming decades, an international and eventually global infrastructure to manage natural and human systems on a world scale will emerge. The table below compares infrastructure and construction in 1990 and today.

Infrastructure and Construction Then and Now

1995	Today
INFRASTRUCTURE	
Heavy steel and concrete bridges, overpasses, rail trestles, etc.	Lightweight, dynamic structures of composite materials and other advanced materials; superconcrete composites.
Large-scale airports with 300-meter runways serve metropolitan areas	Small airport complexes for super-quiet VTOL/STOL craft, tilt rotors, and helicopters, serving urban cores and individual satellite communities.
Steel-rail diesel and electric power railroads	Magnetic levitation and other high-speed rail lines, monorails.
Steel-reinforced concrete structures for commercial buildings, roadbeds, dams	Polymer and composite and ultralightweight metals in most construction, along with rammed and treated soils for dams, roadbeds.
Steel and concrete pipelines for water, sewage, and commodity materials	Smart sewers that have composite pipelines, are self-healing, and are constructed by automated systems; smarter, smaller robots called "smart pigs" live in pipelines to monitor and maintain them.
Little experience with macroengineering projects	Growing recognition that global management can be achieved through macroengineering and large distributed systems such as those taping polar ice for fresh water and controlling fault zones for earthquake prevention.
Limited conception of infrastructure, covering transportation, energy transmission, communications, and civil and industrial works	Expanded definition of infrastructure, from undersea and underground to outer space, covering the traditional categories plus more resource conservation and use, and environmental interventions and management.
CONSTRUCTION	
Computer-aided design relying on printed standards, and legal codes, and human expertise	Human concept/aesthetic design, using expert system engineering and structural design, component specifications, and materials ordering. Virtual model building. Designers and engineers simulate visits to structures, experimenting with aesthetic, structural, and functional possibilities.
Site preparation using human-piloted earth moving, excavation, and grading equipment	Automated site preparation equipment, laser-guided and computer-directed. Teleoperated machinery for some customized tasks
Site construction of most structures, with some prefabricated components	Majority of effort and investment are in factory component production and assembly
Manual labor and heavy machinery dominate	Automated equipment, robots, and teleoperated machinery dominate
Optical and steel tape measurement. Lots of on-site measurement trimming, fitting, and customization	Laser measuring/plumbing/squaring tools, most components manufactured to close tolerances for immediate fit
Traditional paradigms: structures based on strength and tension	New paradigm: dynamic structures that responde to changing uses and stresses

It is possible to characterize the world's affluent, middle, and destitute societies according to the relative conditions of their infrastructures:

- *The affluent* are endowed with well-integrated, technologically advanced infrastructures. Great efficiency in operating their infrastructures and coordinating them as a resource is routine. World 1 nations rely on infrastructures for continued economic and social well-being. At the same time, World 1 societies benefit from automation and other construction technology for cheaper, safer, and aesthetically satisfying structures.

- *The middle* is a more varied story. Many World 2 countries have a mix of outdated and new infrastructure.

  — Many struggle to update their infrastructures and avoid slipping down the economic scale to disaster and destitution.

  — Others have successfully built infrastructures that are a boon to their economies and that raise their standards of living.

  — The middle-income societies have had partial success at innovations in construction, including factory preparation of structural modules, and some automation in field operations.

- *The destitute* have aging and failing infrastructures, which characterize blighted societies such as those of sub-Saharan Africa and southern Asia that are operating on the brink of starvation and suffering episodes of famine. Outdated, failing, or nonexistent infrastructures are characteristic of their continuing struggle.

  — Most large-scale new infrastructure projects undertaken since the turn of the century have been either technological or financial failures.

  — Advanced construction methods are occasionally found in the destitute countries in new construction materials in aid-supported factory housing and in the wealthy enclaves.

The table below is an infrastructure score card based on ratings issued by several groups in 2024 and 2025. It compares the middle and destitute countries with several World 1 countries.

An Infrastructure Report Card

The ratings below use a point scale of 1 to 5 with 5 being the highest quality/efficiency rating for an infrastructure element.

	Roads/Highways	Rail	Airports	Water Transport	Telecommunications	Environmental	Waste & Water	COMMENTS
United States	5	3	5	4	5	4	5	Excellence in telecommunications; continued emphasis on autonomous motor vehicle transport at the expense of high-speed rail; pressures of sustainability promote advanced water, waste, and environmental management.
EC	3	5	5	4	5	4	3	Coalescence of the EC promoted first-rate high-speed rail, air connections, and telecommunications networks; strong in ports, ship channels, and environmental management infrastructures.
Japan	3	5	5	5	5	4	3	Leads the world in integrated transportation infrastructures based on rail, air, and water links; first-rate telecommunications system, linking to global network.
China	2	3	3	3	3	1	2	Some success in rail, water transport based on upgrading of old systems; now building a complete fiber communications network, but service is still spotty; development at the expense of resource and waste management leaves those infrastructures shoddy.
World 2	3	2	2	4	2	1	1	Economies rely on adequate rail and road links and good ports; catching up on environmental and resource management.
World 3	1	1	1	2	2	1	1	Decrepit or nonexistent infrastructures, except where foreign money has built ports and telecommunications links.

Source: Based on Global Business Alliance, June 2025, and *Places Rated*, October 2024, ratings for infrastructure.

WORLD 1 NATIONS—FINE-TUNING THE HIGH-TECH INFRASTRUCTURE

Technology has solved many affluent society infrastructure problems. Today, those societies are fine-tuning their infrastructures for even more efficient and better services.

New York City, for example, with an aging water supply infrastructure, revamped its water system by relining existing pipelines and using smart controls and other technologies, with a net water savings of 46%, comparing water use in 1998 and 2016. New systems are designed for modification and alteration 20 or 30 years from now, as new technologies become available.

Japan's Alice (in Wonderland) cities move people underground

The underground city complexes of Japan are the realization of plans and visions from the last century of solving that country's chronic space problems. The underground mixed-use complexes, the largest of which is Taisei Corporation's Lifeplex complex in Kyoto (2019), houses as many as 20,000, with office space, retail, and commercial space.

The Alice Cities take advantage of year-round comfortable underground temperatures, can have direct connections to subways and other underground transit systems, and are less prone to damage from earthquakes. They free the land surface for parkland, greenspace, and kitchen gardens in some cases. In other cases, such as in Tokyo, they underlie urban areas that preexisted the underground cities. There are now seven Alice City complexes in Japan, with three under construction. Elsewhere, there are resorts and other recreation facilities built entirely underground, including the Club Subterrainee resorts in Tokyo, Akureyri, Iceland, and St. Paul, Minnesota.

Meanwhile, some of the long-dreamed-of visions of engineers are under consideration or construction. These are macroengineering projects to tap resources, reverse environmental degradation, provide living space, and construct intercontinental and global-scale transportation and information linkages. These projects will be instrumental over the next several decades in extending global management. Most basic elements of the infrastructure, and the systems that coordinate them, are designed and managed to operate sustainably, with little negative and as much positive impact on the environment as possible. Global management infrastructure components such as environmental monitoring networks and restorative ecology projects will support the development of sustainable lifeways.

In construction, the affluent countries have achieved unprecedented levels of quality, sophistication, and flexibility. Safety, durability, and environmental harmony are fundamental goals, along with function, of any construction project.

In looking back to the year 1995, the most remarkable change in construction technology has been greater automation and factory production of modules and, occasionally, entire structures. From design to assembly, information technology and computing make it possible to coordinate all processes. Thirty years ago, although computers were enabling many advances, there was little coordination between the steps in the construction process and among the people and institutions involved. Today, information systems are the equivalent of the general contractor of 30 years ago. They coordinate other systems in the push for greater efficiency, reliability, and highest quality.

A U.S. CASE STUDY—Approaching total integration of infrastructure elements

Information technologies are making it possible to integrate all components of the national infrastructure. In the United States, the greatest successes so far have come in the transportation and telecommunications infrastructures, enabled by information technology. Establishing a fully integrated infrastructure was a central goal of the Second National Infrastructure Initiative begun in 2017.

Putting it all together

Thirty years ago, the planning, management, and use of the U.S. infrastructure were poorly coordinated or not coordinated at all. Today, people can move from rail to PRT to road to air, counting on smooth and timely connections. Goods can move in nearly unbroken paths under automated or human piloting along roadways, slurry lines, and pneumatic tubes in aircraft and ships, loading and unloading in a timed, coordinated logistics network. These and other infrastructure elements are designed and operated for maximum efficiency and integration. Routing optimization and travel timing are calculated and readjusted continuously in transportation systems and communications. Fixed infrastructure elements such as waste and water treatment plants and dams are coordinated with linked systems, energy, and resource supplies.

There will be unprecedented infrastructure integration and efficiency in the coming decades as obsolete systems are replaced and networks are fine-tuned. Information is the critical element across the infrastructure, whereas, in the last century, concrete and engineering principles were dominant.

A wider definition of infrastructure has accompanied greater integration and coordination. New systems for new purposes have expanded that definition since the 1990s. The U.S. infrastructure today includes systems to control natural processes, such as groundwater management systems and the proposed Earthquake Control System for the San Andreas Fault Line. It includes networks of communications and navigation satellites. Integral to today's infrastructure are the telematic systems that control and coordinate transportation and communication routing and other complex infrastructure processes. Distributed energy systems such as small-scale hydro power, biomass, and photovoltaics arrays join large energy plants as part of the infrastructure. The diagram below shows the relative growth and shrinkage of components of the U.S. infrastructure from 1990 to the present.

The Changing Emphases of U.S. Infrastructure Spending
Index of Government and Private Infrastructure Spending:
1990 to 2025

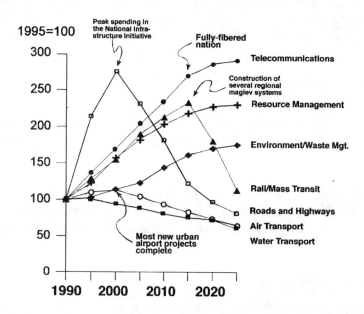

Integrating the infrastructure requires it to be international. For example, the airline industry has long been internationally coordinated. Since 2015, the power and water grids in North America are fully coordinated in a cooperative alliance of Canada, Mexico, and the United States. The highway and rail systems are less thoroughly coordinated, because of lingering disputes over controlling contraband and illegal immigration.

Paying for the infrastructure

In the early 1990s, policymakers forecast that it would cost $1 trillion to $2 trillion to fix the existing public works infrastructure of the United States. In fact, the United States has spent $3.2 trillion to do that and more in the past 30 years. Largely public money rebuilt the traditional infrastructure of the United States as part of the First National Infrastructure Initiative (1998-2010). The Second National Infrastructure Initiative (2017) recognized the role of private-sector funding and market-driven infrastructure development. Private money added complex systems for telecommunications, environmental and resource management, and other infrastructures.

In the 1970s, government spending for infrastructure was 2% of GDP. Today it is 1.1%. Private spending more than makes up for the decrease. Over the past 30 years businesses invested in the infrastructure where they used it, because it raised their productivity. In the 1980s and 1990s, federal, state, and local governments took up user fees to solve their budget problems and changed the public infrastructure financing paradigm in the United States.

Meanwhile, for-profit infrastructure investments, such as the national data highway, interurban high-speed rail, and downtown PRT systems flourished, paying for their investments and making billions of dollars from user fees. With the nearly ubiquitous phase I intelligent vehicle highway systems, advertising on the IVHS pays much of the bill.

Integrating the infrastructure

Twenty years ago, the United States recognized the need to take a total systems approach to its infrastructure. In 2001, the U.S. government set up a coordinating agency, the Federal Infrastructure Administration (FIA) to accomplish better what was needed. Networks of distributed computing and monitoring systems oversee routing, system use, traffic flows, and interconnections. The utilidor, now in use in 57 cities and thousands of neighborhoods became popular late last century. Utilidors streamline the construction and maintenance of utility lines—water, sewer, electric, and communications—by putting them in a common corridor, usually buried.

Sensing and monitoring in infrastructure coordination

Today there is nearly universal use of sensing to coordinate infrastructure. Technologies for remote and proximate sensing make much of the integration, management, and maintenance of infrastructure possible. Mechatronics allows more elements of the infrastructure to be self-diagnosing and monitoring. Dynamic structures respond to localized stresses and changing uses, and feed back status information to coordinating systems. Remote sensing allows coordination of systems that are not wired into other systems.

Infrastructure goes to sea

The sea has absorbed some of society's need for additional commercial, residential, and transportation space. Historic examples of land claimed from the sea include Japan's urban airports, and Veracruz, Mexico's Puerto Morales community. The Zuider Zee is an example of land being given back to the sea, with 54% of the historic reclaimed area now restored to tidal marsh.

More recently, developers have built several floating communities for the United States, jokingly called "maritime subdivisions" for cities with expanding populations. In contrast to their success in Japan, floating communities have not been readily accepted because of problems with storm protection, environmental impact questions, and financing.

Smaller-scale structures, including floating highways and bridges, have had greater success in the United States. Hailed as having fewer negative environmental impacts, floating roadways and bridges have replaced steel and concrete structures in dozens of U.S. areas, especially on rivers, lakes, and other water where boat passage is not necessary. On bays exposed to ocean storms and high surf, the structures would too often be closed for bad weather, and would require drawbridge passage for ships and boats.

Ice highways in Alaska have proved a boon to mineral extraction above the Arctic Circle. The highways are cheap to build and maintain, requiring only seasonal regrading. Most environmental scientists consider them a reasonable compromise over the construction of paved highways and railbeds. A recent proposal for a more-engineered ice highway would use repulped waste paper to create a super durable ice material. The highway would be sprayed into place, freezing on contact in the Arctic temperatures of eastern Siberia. Similar techniques have been used to build ice structures, including living space, in Antarctica.

Infrastructure moves underwater

Thirty years ago, about the only underwater structures were pipelines and cables under the sea. Today, underwater infrastructure includes:

*   "bridges"—actually vehicle tunnels that float at maintained depths under the sea

*   systems for exploiting sea-floor resources

*   ocean thermal-energy conversion

*   cable and pipelines

*   habitable space and work space

Infrastructure in space

The United States also depends on the space infrastructure. The space infrastructure is defined by the growing number of communications satellites and the emerging space-based manufacturing and its associated transport docking facilities. Analysts expect a full-fledged space infrastructure in the future to deal with space debris, govern interplanetary communication and research, and serve space transportation.

Failure amid success

Not all of the U.S. infrastructure story for the past three decades is about success. Economic and engineering failures punctuated the last 30 years of infrastructure development. There will continue to be failures amid success. The proposed Grand Forks, North Dakota, city dome exemplifies one kind of failure.

In 2011, developers and city officials in Grand Forks decided to bolster their city's economy by making the community more attractive to outside business investment and travelers, despite consistent subzero winter temperatures. This winter city, with the backing and technical expertise of Trimar Corp., proposed and began construction of a dome of unprecedented size to cover the central core of the city. The geodesic dome was to be built by robots. It was to be constructed of a lightweight, high-strength polymer, with elaborate airflow controls built into the shell. Heating elements would melt away snowfall. The dome would cover just under 2 square kilometers of land in the downtown area, and would reach 1,000 meters in height. After four years of feasibility study, a bond issue, and $255 million spent in planning and research, the project was deemed an engineering impracticality.

Other infrastructure failures were purely financial. Private sector developers of information linkages, transportation projects, and other projects often went bankrupt on ultimately profitable systems. Government bailouts and private buyouts ensured the completion or continued operation of the projects.

Logistics

Innovations in goods transport transformed the urban landscape in 15 large U.S. cities over the past 30 years. The large cities have networks of underground pipelines carrying retail products, food supplies, and packages for delivery, with all such goods traveling via pneumatic and hydraulic capsules.

Some networks use slurry transport for solid wastes and industrial raw materials and fuels, including silica, bauxite, and coal. In several cities, engineers over the past two decades converted aging water mains and storm sewers for this kind of transport. They developed robots that traveled the pipelines to inspect them and install tubing with precise tolerances for the new cargo systems. They replaced water mains with high pressure composite pipelines, often in the same tunnels. Dynamic controls help assure that breaks in the high pressure lines are contained quickly.

Putting much of goods transportation underground has lengthened the lives of surface roads, lessened traffic, and made streets safer and cleaner for other uses. Urban and industrial zone noise levels are reduced. Some cities, including downtown Chicago, bar traffic from certain or all streets, and bar truck traffic entirely.

Personal mobility

Moving people has involved key shifts in transportation technology. Those shifts are, in turn, beginning to reshape communities, aligning their centers along public transit lines where those lines have proved efficient and accepted by communities. In several large metropolitan areas, property value patterns show the reorientation, though new development and new commercial enterprises are just beginning to reflect the changes.

PRT systems are a model for greater transportation infrastructure management

Analysts expect personal rapid transit systems, such as those in use in Chicago and Boston, and planned for El Paso, Tulsa, and Atlanta, to be the model for future metropolitan personal mobility. PRT systems are a boon to urban aesthetics because they remove heavy vehicle traffic from densely populated areas. They also free riders for leisure in transit. The systems save on energy and infrastructure maintenance costs and have proved popular with urban riders, though less so with those traveling from suburb to suburb and among satellite cities. Chicago banned non-PRT traffic in its downtown area in favor of PRT vehicles. That allowed an almost complete reconfiguration of the city streets to improve pedestrian use and to bar truck traffic.

The PRT systems in Chicago (2017) and Boston (2021) are the entering wedge for that technology. A half dozen other cities have plans in the works for PRT systems. The system under design for Charlotte will use on-the-road/on-the-rail dual purpose vehicles that use automatic guideways and can be driven on roadways.

Aircraft technology and new airport forms are reconfiguring air travel

Vertical takeoff and landing aircraft and short takeoff and landing aircraft were fundamental in allowing new airports to be constructed close to or in urban areas. The key to their use was the development of whisper engines in the 2010, solving the problem of excessive noise. The smaller facilities made possible by VTOLs/STOLs cost as little as a tenth that of airports constructed in the 1990s to serve growing metropolises in the United States.

Two airports built in the 2010s, Chicago's New Midway and Atlanta's North Metro, innovated with completely underground passenger, baggage, parking, public transit, and ticketing facilities, saving as much as 60 hectares of space. North Metro is a combined airport and underground high speed rail facility. The Interurban maglev lines operated by Southern Crescent converge there.

High-speed rail, IVHS, and other transport systems ease mobility

The United States is witnessing the development of a national maglev system. Today's regional and interregional links are to be part of an integrated national system. Its capstone will be the continental line expected to be completed in 2037.

IVHS systems of some form now serve most of the federal highway system, and continue to be constructed for smaller roadways. Coupled with other personal and mass transit links, these systems bring unprecedented mobility to the U.S. worker and leisure traveler.

Waste management

Waste management in most U.S. communities is a different game than it was a generation ago. Starting in the late 1980s, municipalities recognized the twin threats of too much solid waste and too much sewage. Technological responses to each problem involved reducing the sources of waste and revamping waste handling and processing.

Mining waste heaps

Thirty years' experience in mining landfills and other waste sources has proved the practice to be lucrative for those doing it, and beneficial to the environment. The practice took off when some U.S. recycling facilities began to run low on recyclable materials.

The company called Waste Management in 1993 renamed itself WMX Inc. Then, in 1998 it took yet another new name—WMX Resource Corp.— reflecting its greater emphasis on waste as a resource. WMX and others have made billions of dollars mining landfills and other waste sources and have made the commitment to take the bad with the good. In exchange for tax relief, they recover and dispose of hazardous materials safely as they remove reusable and saleable materials. This has meant many historic landfills are no longer environmental threats.

Three things may be recovered from landfills: energy, materials, and land. Successful operations promise and deliver all three, because it is inefficient to work the landfills for only one purpose at a time.

Typical solid-waste management systems use the Dallas-Martin process, available since 2011, in decentralized facilities in urban areas to preprocess material and then transport the material by slurry to facilities for materials separation and recycling, toxics removal, etc. Intensive recycling and reclamation efforts meant that solid wastes began to have value and became commodities. Instead of simply being hauled to incinerators or landfills, as they were in the last century, waste material came to be sorted at the source and at collection facilities into reclaimable and recyclable fractions. The miscellany left over could be treated by processes to reclaim further materials. For example, in one process, wastes are dumped into a bath of molten metal at extreme temperatures. Their component fractions are separated and reclaimed. This process is especially important today for dealing with pre-2010s goods that were not designed for dismantlement and reclamation.

Smart sewers manage waste

Several dozen U.S. cities today have smart sewer systems to manage their liquid wastes. The sewers use sensors and actuating devices to track and direct their contents. Membrane extraction preprocesses some wastes and separates materials before they reach processing facilities. Usable fractions, such as organic materials useful as fertilizer, are extracted for garden and agricultural use. Fresh water is extracted and recycled.

Sewage in most communities is treated biologically, turning it into a material useful in agriculture or disposing of it harmlessly on land or in the sea. In several dozen communities, including some multiunit housing complexes, subdivisions, and several entire cities, decentralized sewage processing is undergoing experimentation.

Biological waste-processing systems serving from 40 to 2,000 households are located in neighborhoods, fully contained underground. They take household sewage, including food and human wastes. They produce clean water that is cycled back to the area's water supply system. They also produce processed sludge collected by pipeline or "honey" trucks that take it for use in agriculture. Toxic materials, typically only a tiny fraction of the plant's output, are collected and processed for safe disposal.

Resource systems

The traditional problem of many natural resources is that they are in abundance in some places and in chronic or acute shortage in others. Integrated resource systems make it possible to overcome these inequities. Central to these systems are automated information exchange forums that broker the supply and demand for resources.

The integrated water supply system

Mimicking the electric power grid of the last century, North America's water supply has slowly become an interconnected supply grid, allowing local water companies to wheel supplies—shunt them through the North American network—as needed. The system does not yet cross the Rockies throughout the continent. It is most fully developed in the western part of North America, where water supplies reached crisis points earliest. A similar network, not as complete, exists for nonpotable fresh water for agricultural and industrial uses. The system has reduced agricultural losses to drought by about 75% since 2000.

Traditional and innovative infrastructure systems are making it possible to more effectively move these materials to where they are needed. For example, slurry pipelines, rail cargo links, and unpiloted river barge towing can move resources to industrial centers and speed the exchange of resources from areas of abundance to areas of scarcity.

Water supplies are centralized and decentralized

In addition to recycled water from sewage and storm sewer processing, traditional potable water systems continue to serve most communities. In several notable cases, groundwater has been made safe for use after decades or as much as a century of unreliability. The formula for taking groundwater is based on increasingly accurate estimates of rates of natural replenishment. Artificial replenishment and runoff collection equalize aquifer water levels.

Interventions to repair and clean groundwater aquifers have proved successful in the Imperial Valley of California, in coastal New Jersey, and in other communities. In the Atlantic City area and in other coastal regions, the overuse of groundwater was causing saline water to infiltrate the region's aquifer.

By the 2010s, state natural resources authorities took a double-edged strategy to cope with the problem. They installed monitoring systems to guide withdrawals. They also installed impermeable barriers in underground areas that had become the main channel for infiltrating sea water.

The Atlantic City barrier involved injections of a polymer into the soils at about 75 meters depth. Over a 2.7 kilometer stretch, technicians made about 500 injections. This strategy is under consideration for other coastal aquifers and for several aquifers threatened by tainted groundwater.

Energy supplies

Electrical energy is similarly centralized and decentralized. Household and neighborhood power generation from mixed sources is integrated with the central power generation and distribution systems—the traditional power utilities. This means that at the household level, there are smart connections between any generating source in the immediate vicinity and the local, regional, and national grids. The household or commercial user relies on automated decision and buying systems to negotiate for and purchase power and sell excess capacity from on-site generation.

Wheeling power throughout the continent

An intercontinental power grid serves the United States, Canada, and Mexico through a computerized network of power buying and selling. Annual agreements permit regional power utilities to sell and buy power as needed. The system is enabled by an international information network linking central power production and some distributed systems. Buried superconductive trunk lines move the power long distances with minimal wattage attenuation.

Communication

The telecommunications network that evolved over the past century is a complex web of wired, fibered, and fiberless connections. Almost a thousand satellites, transponders, up and down links, repeaters, and optical switching centers are distributed across the globe and in orbit. They are coordinated by widely distributed computing/switching/routing centers and end-use devices for optimizing transmission.

Fiber communication

Eight-and-a-half-million kilometers of fiber-optic networks traverse the United States. Many were installed from 1980 to 2000, using often cumbersome methods of routing the lines underground or through existing tunnels. Today, engineers install new lines using automated tunneling equipment and robotic machinery that can lay lines in storm sewers and other tunnels. By 2008, telecom companies were using automated robots to travel underground cable lines, laying fiber-optic cable alongside metal cable, or exchanging fiber for metal cable. Telecoms have installed about 190,000 kilometers of optical cable in this way.

Fiberless communication

The extended Iridium system of satellites is the core of the global fiberless communications network. In conjunction with the global fiber network, it provides a personal communication service (PCS), connects machines to people, machines to machines, and connects sensors to analytical and management systems.

The satellite network freed people and devices from being tethered by wires or fiber cable to other devices. Since the 1990s, it has enabled a flowering and spread of electronic devices above, across, and into the landscape. We have:

- geological sensors thousands of meters down in the Earth's crust

- sensors aloft, traveling the globe's jet stream and other air currents

- sensors installed in sea animals, land animals, and plants for ecological monitoring, research, and agriculture

- actuating devices in the land, on the sea, and in structures, taking command from people, machines, and their own onboard intelligence chips

The net effect of wireless communication is the movement of computer-chip based smarts to more things in more places than ever before. A key advance relying on this capability is environmental management, discussed below.

Environmental management

Perhaps the biggest changes on the U.S. landscape, though they are often invisible, are the networks and systems devoted to maintaining, protecting, monitoring, and restoring the environment. These systems result from a new paradigm in environmental science—the acknowledgment that sustainable existence on Earth requires people to manage the environment to balance its protection with human needs.

> ## Technology lets us manage nature and people
>
> *Managing Nature*
>
> - Earthquake control systems in the field
> - Earthquake monitoring
> - Iceberg docking and processing facilities for towed-in freshwater
> - Deep underground pipelines and nuclear-waste storage
> - Shallow underground living spaces, storage, etc.
> - First explorations of deep-mantle energy resources
> - Climate control infrastructure
> - Mass-scale floodplain/watershed controls
> - In-sea barriers, established to modify/manage the marine environment
> - Recapture of river silts from delta bottoms
> - Permanent installations to control/manage aspects of nature, restore ecology
> - Mobile, robotic ecological restoration equipment
>
> *Coordinating nature and human activities*
>
> - Greater protection and integration of soils into civil works design and functions
> - Coordination of ocean sensing and monitoring with dynamic infrastructure elements
> - Coordination of weather sensing and monitoring with dynamic infrastructure elements
> - Coordinated pest control associated with infrastructure integration, e.g., potable water systems

Environmental management involves, at a minimum, networks of fiber-connected and fiberless sensors and monitoring equipment. Often, devices to manipulate the environment are distributed across the landscape. One case in point is the complex control facilities of the Mississippi River Valley, described in the box below.

Conquering the Mississippi watershed

By the late 1990s, it became undeniable that the United States had to rethink the traditional schemes for managing large rivers and their tributaries. Urban and agricultural runoff continued to disrupt the biology of rivers. Seasonal fluctuations, floods, and storms, disrupted human activities on shore.

The traditional way to manage rivers in the United States was to build complexes of dams, revetments, culverts, and levees to manage the flow of water and restrict that flow to where people wanted it. Designers took into account caprices of nature and designed things to withstand flooding, at least to 100-year flood levels.

Episodes of failure in those systems, such as the floods of 1993 and 2011, as well as agricultural problems and the drive to sustainable agriculture, changed how people viewed rivers. A new paradigm emerged, often called "let it flood."

The new paradigm flew in the face of traditional practices of governments, particularly the U.S. Army Corps of Engineers, which has become the Federal Infrastructure Management Agency.

The emerging complex is one in which dynamic structures, including flood gates and levees, protect existing cities, but community leaders relocate smaller towns and industrial facilities to higher elevations. Small communities can no longer get insurance and loans to build and rebuild in certain flood zones. At the same time, the flows of the Mississippi and its main tributaries are coming under control through dams and cataracts. Runoff management systems gather excessive flow from heavy rains in flood lakes, tens or hundreds of acres in size. Some of them are constructed as riverside wetlands. There the water is available for agricultural use, or to maintain river flows in times of less rainfall or to be directed into underground aquifers.

The "let it flood" notion comes from a further strategy that is emerging. Much of the agricultural bottomlands in the Mississippi Valley historically flooded as a natural seasonal process. That flooding can rejuvenate agricultural lands, and is seen as a net positive for sustainable farming. Today's bottomland farming corporations are making seasonal flooding a part of their integrated agricultural practices, much like the ancient Nile.

CONSTRUCTION TECHNOLOGY

A new structural-design paradigm is transforming structural engineering and design. Throughout human history, construction rested on two paradigms; tension and compression. The resulting structures were strong and functioned well as long as their substrates remained stable and their materials remained strong. Today, engineers build structures that rely on dynamism. The structures can adjust themselves structurally to stresses from people and nature. This is done by simulating how living things respond to stresses: sinews, muscles, and joints react flexibly to the stresses put on them. Surfaces adjust to temperature, moisture, and pressure, especially wind pressure, to preserve a structure's contents and integrity. Smart materials, sensors, and actuators, embedded and distributed throughout the structure make all this possible.

Dynamic structures proliferate

- *The Barrow Tollway* in San Francisco is an earthquake-proof elevated roadway with continuous dynamic adjustment to substrate shifts and lateral and vertical shaking.

- *The Hitachi Building* in Seattle is Hitachi's 270-story U.S. headquarters, built as a demonstration project to show how interior space could be freed from much of the structural support required by traditional skyscrapers.

- *Twin Forks Dam* in Montana spans 255 meters. Unstable substrates and variable water loads required a dam that could dynamically respond to compression.

- *The New York-Boston Maglev* is a partially elevated high-speed rail line that reacts dynamically to vehicle loads and lateral tensions, as well as relies on dynamism for its basic structural integrity.

- *The Transierra Viaduct* in California depends on dynamic trestles, particularly as insurance against shifting substrates in earthquake conditions.

Construction technology supports infrastructure development and management. The construction process has been revolutionized from project conception and design through construction, finishing, repair and maintenance, through end use and demolition. Some areas in construction lag the available technology for social and institutional reasons. For example, the construction industry took several decades to accept the emerging premier role of factory construction of buildings.

Design

The design phase of construction, whether it is infrastructure, commercial buildings, or housing, is the most critical phase of any project. In the last century, two elements governed structural design: (1) the materials and techniques of construction and (2) aesthetic choices that could be made using those materials and techniques.

Simulated tactile planning and design

Using artificial reality tools, designers and their clients can enter the structures they are planning. They can have the simulated experience of walking around in the structures. They can examine the heft, textures, and look of materials, and consider engineering alternatives suggested by their computer design systems.

Today, a greater repertoire of construction techniques and a wider choice of materials mean that functional design, economic choices, environmental impacts, and more aesthetic possibilities are parts of the design process. Powerful designing computers assist architects, engineers, manufacturers, and aesthetic designers, while giving them unprecedented choice of the final design outcome. Virtual walk-through-design contests, with judges and designers exploring a structure in a virtual environment, are increasingly popular.

Site preparation, excavation, construction/assembly

Robotics and other automation pervade the next three phases of construction. Autonomous robots, such as grading and excavating machines, are used for site preparation, excavation, and construction. More difficult problems can be solved with teleoperated machinery that allows human operators to work safely at a distance from the machine.

Finishing

. There is relatively little structural finishing to do on site. Construction members typically are finished in the factory. Custom construction still requires finish work, some patching and seam covering on site. There, powerful tools enable workers to quickly finish a building.

Where in 1990 it might take weeks to complete building interiors, factory production of components and today's interior finish materials make it possible to do the work in a day or two. For a Cape Cod model home, it takes one worker two days to complete interior finishing; for a Tudor style, 2.5 days. Many modern-style homes require only about two worker hours for final site finishing. Some products that make home, office, and other commercial finishing easier are:

- Factory-applied coatings and finishes on modular components
- Seamless construction with on-site composite laser welding of pre-finished panels
- Automated coating and painting equipment that navigates using electronic floor plans
- Self-healing/self-joining finish materials that close seams automatically when installed
- Near net shape component casting, e.g., cast wood that needs no trimming and sanding
- High tolerances in factory-produced modules, meaning little or no fitting on site

Repair and maintenance

Caring for structures efficiently depends in large part on their initial designs. On top of other design criteria, design for maintenance is essential to the economics of a structure. Planners have more accurate economic tools today to forecast the maintenance costs of a proposed structure than did planners 30 years ago.

Structure life history records start at design stage

The growing practice of recording each structure's life history became a U.S. law in 2013. Today, the electronic cornerstone of a structure contains a full record of its design, construction and repair histories, use, energy consumption, ownership, and other particulars. Visual images, including full motion, are stored with other continuously updated data from sensors and monitoring devices. Any authorized user can query the record from the building or remotely. The record is especially important to rescue workers, renovators, regulators, energy auditors, and owners.

The ideal structure is designed for regular maintenance and has built-in flexibility. Modularity helps. Modular add-on units for commercial structures and homes are planned in the initial design of most structures. Modules are also designed to be replaced in situ. Internal building modules use a fold-down principle so that disassemblers can collapse, remove, and replace them with modules expanded into place or assembled from several components inside a structure.

Demolition

Today's buildings are designed for demolition and materials reuse, whereas before 2005, most demolition materials were wasted—burned or put in landfills. The central goal in demolishing structures is to do so safely and cleanly and to recapture materials for reuse. Because full-fledged design for reclamation construction is a recent development, few large-scale demolition projects on such structures have been completed. The emphasis today is on demolition and materials recovery for old buildings not designed for efficient reclamation. The showpiece of the demolition industry was the dismantling of New York City's Olympic Towers in 2022. The project took 29 days. Had the structure been designed for dismantling, it is estimated that it could have been dismantled in four days.

Three decades ago, by contrast, there was almost no scavenging of structural components, and most materials went to landfills. Modular construction and high value-added materials like composites and metal alloys mean that the parts of structures 20 years old or younger have ready markets for reuse. Older structures pose a more difficult problem. Some of their parts and materials are wanted for historic restoration of other buildings and for some artful materials reuse. However, much of the material is waste that is recycled back to basic mineral components. So far, only a few dozen buildings in the United States have undergone this process in full. However, home demolition wastes have been one of the commodity wastes since early this century, mined for useful recyclables.

Prototype reclamation machinery has continued to fail in the field because of the complexity of sorting and grading wastes. To preserve the value and condition of certain materials, such as concrete composites, such machinery must be virtually flawless in its operation. Teleoperation and automation dominate the field.

Where construction debris can be hauled to processing facilities, the record for automation is better. Controlled conditions make it possible to sort and grade material with little or no human labor.

INFRASTRUCTURE AND CONSTRUCTION IN WORLD 2—Success, failure, and trying to catch up

Infrastructure makes it possible to preserve and advance quality of life in the middle-income countries. World 2 countries that are advancing are doing so because of their investments in infrastructure. Those that are backsliding to economic depression and poverty have let their infrastructures become outdated and decrepit.

Middle countries on the way up are working to drive their development with investment in their infrastructures and to make them integrated and compatible with the emerging global infrastructure. The more progressive among them are trying to leapfrog to the latest infrastructure technologies. Many World 2 countries have been building first-class infrastructures usually based on last-century technologies.

The infrastructure efforts of the successfully developing middle countries were expensive, but usually worth it. International investment, extractive resource exports, and successful foreign trade paid the way.

Playing catch-up has sometimes compromised a resource base or the natural environment. China fueled its growth by burning cheap, high-sulfur coal. Brunei exhausted its petroleum and then its timber supplies. World 2 countries also sometimes financed their infrastructure development with loans and bonds, extractive resources industries, and through economic growth based on cheap energy and cheap labor. In most cases, the scars of precipitous, wrongheaded development strategies are still visible.

Other middle-income countries are in decline, losing the edge in an extractive industry or manufacturing industry or losing ground during civil strife or prolonged economic problems. In those countries, there are failing, worn-out infrastructures with few prospects for improvement.

Collectively, the middle-income countries have a mixed record in infrastructure development. Their infrastructures are a patchwork of the old and the new. Some analysts speculate that World 2 countries' economic success in the 1990s through the 2010s is a pale version of what might have been with more efficient, fully modern infrastructures. As it worked out, many of those countries saw healthy economic growth rates of 3% to 5.5% over the period 1990 to 2020.

Debt crises and politics reduced the investments made in infrastructure in the first part of this century. Politicians seeking reelection often would not let utilities or other state-run or state-regulated firms charge enough for infrastructure services. Those firms reduced investment in new infrastructure and often deferred repairs and maintenance to a crisis stage.

The failure of national infrastructure systems in the 2000s and 2010s was widespread. Fast-rising populations and growing commerce in developing regions led to overuse of roads, ports, railways, and airports. Other systems, such as waste handling, water supplies, and communications were over-burdened from runaway use.

Often, corporate and domestic users turned to alternatives. They built what some call "islands of modernity" amid inadequate or failing infrastructures. A case in point is the Venezuelan port of Santa Marta, built by the newly privatized Petróleos de Venezuela in 2014. Also, developing-world businesses and households often turned to cellular telephony and photovoltaic power sources when national systems failed or were inadequate. Private energy companies grew up in some countries, providing off-line power. Satellite-based communications supplanted wired communications for many people as they became available in the 2000s.

Aging, inadequate, and failing transportation infrastructures were harder to circumvent in the first part of this century. Though some firms built their own ports, airports, roads, and railways, they more typically suffered with everyone else in using their national infrastructures.

Political and economic forces also held back automation and other advances in infrastructure. The political clout of labor often preserves labor's role in systems that should have long since been automated, such as port transshipment facilities and waste handling. In 1993, the port handling costs for steel in Recife, Brazil, were five times those in Rotterdam. Today, they are still 3.4 times as high. Rotterdam advanced its automation, while Brazil gained little ground.

The following case study examines the state of infrastructure and construction in a typical middle-country situation—Venezuela.

A CASE STUDY: Venezuela struggles to build a 2020s infrastructure

Venezuela is a typical World 2 country. Despite its rich resource base and generally high levels of social development, the country squandered its potential for greater affluence from the 1980s through the early part of this century. Repeated episodes of political unrest, overreliance on petroleum revenues, and several ill-conceived infrastructure projects held the country back from becoming one of the affluent stars of South America.

Venezuela's infrastructure is typical of middle countries—a mix of the old and the new. The country has high-speed rail connections east and west, a state-of-the-art metropolitan transit system in Caracas, and well-integrated road and rail connections to the inland port city of Ciudad Bolivar on the Transamazonia Corridor. The high-tech systems are superimposed on a rudimentary system of rural roads and railways, a faltering telecommunications system, incomplete fiber-optic networks, a troubled power grid, and the nearly complete lack of efficient metropolitan transit systems outside Caracas.

As it has been for 40 years, Venezuela is well-positioned to become one of the world's affluent countries, but cannot make the transition. Poverty and income inequality continue, and the spotty infrastructure holds back better distributed development.

Successes and failures

In the mid-1990s, Venezuela had a government that could not maintain or improve public services. For example, in Caracas, water sometimes would be shut off to parts of the city for days. Some infrastructure and equipment in use then was 20 years old, built during the country's 1970s oil boom. Governments of Venezuela through 2000 routinely got away with poor attention to infrastructure and economic development because of a steady stream of oil revenue. Even in years when oil prices were low, the country could count on tens of billions of dollars in oil revenue. Other resources, including gold, iron, and other minerals, nursed Venezuela along economically as well.

2025

Venezuela has a highly urban population, and the concentrations of the country's population in its several large metropolitan areas are still rising. Venezuela's population grew from 18.9 million in 1992 to 27.3 million in 2010 to 34.8 million in 2025. The population increased from 84% urban in 1992 to 93% urban today.

Metropolitanization combined with the steep and rolling terrain around its capital and biggest city, Caracas, has made infrastructure expansion difficult. Caracas has a subway system that is almost 50 years old. The system underwent expansion in the late 1990s, and the 2010s, but remains inadequate for the city's 4 million population. Traversing the steep hillsides surrounding the city are a mix of cable cars, funiculars, and footpaths. Buses and jitneys connect remote suburbs to the ends of the subway system.

Private infrastructure develops

The port of Santa Marta, near Maracaibo in Venezuela, was built entirely with private funds and bond issues in 2010-2014. Petróleos de Venezuela, then a newly privatized oil firm, decided the port facilities in Maracaibo were inadequate and that labor unions and other forces were in the way of modernizing the port. So the company built its own, state-of-the-art facility. Today, Santa Marta is a successful operation, serving 24% of Venezuela's trade.

A proposed circular monorail will connect the radial subway lines to better serve the growing satellite communities around Caracas and give more people access to the system. The Venezuelan government will attempt to sell development bonds to pay for the publicly-operated system. Earlier development of the Venezuelan system depended on development loans from the Interamerican Development Bank and the World Bank. The country also borrowed money from the EC in exchange for hiring Italian and Swedish firms to do the construction and build the subway cars. Venezuela has never defaulted on its bonds or loans.

Connections south and west

In the 1990s and 2000s, Latin American trade began to flourish. Venezuela continued to supply the continent with much of its petroleum and petroleum-based products. It developed other extractive and manufacturing industries over the past 30 years as well. Parallel developments occurred in Colombia, Brazil, Argentina, Uruguay, and Chile. The growth of inter-Latin American trade made it economical to build new infrastructure in the more developed of Latin America's countries.

In 2006, Brazil and Venezuela began work on the Transamazonia Corridor, linking, São Paulo, Belo Horizonte, Manaus, Ciudad Bolivar, and Caracas. The parts of the corridor that cross the Amazon rainforests involve innovative schemes to protect the neotropical ecology.

Floating highways and railways protect rainforest ecosystems and resources

The principles of so-called floating transportation corridors are the same for rail and highway. They do not actually float, but are constructed above the ground in sensitive environmental zones to ensure minimum impact on those environments. The structures sit on pilings that enter the soil and extend to depths where they can be stabilized without excavation, foundations, or footers. This permits the passage of groundwater, and the maintenance of other natural processes on the forest floor.

The resulting highway sits at about mid-canopy, and requires only that tree limbs and other growth be kept away from about 10 meters height and 15 meters width surrounding the rail or road bed. The pilings are designed to be no more a factor in the ecology than tree trunks, and in most cases are less so.

The rail and roadbeds built on the Transamazonia Corridor are dynamic structures designed to adjust to shifting subsoils and bedrock. Sensors in the pilings and roadbed trigger continuous adjustment to keep the structures level and structurally sound.

Automated hydraulic pile drivers made the construction of the Transamazonia Corridor practical and affordable. The project cost $120 billion and was completed in 2014.

From the ecologist's point of view, the principles of construction for the corridor are "uninterrupted biological, hydrological, carbon cycle, and social processes." In other words, construction designs and execution sought to "float" the rail and highway links several meters above the rainforest floor to avoid harm to the natural exchange of water, soil and plant processes, animal life, and the lives of native people in several aboriginal reservations in the region.

INFRASTRUCTURE AND CONSTRUCTION IN WORLD 3 COUNTRIES—Struggling with failing, outdated infrastructures

Government inaction, corruption, and wrongheadedness in the destitute countries keep their infrastructures from getting adequate attention and investment. This problem is integral to holding those failing societies back. The few infrastructure success stories found usually involve tens of billions of dollars in foreign aid money or the successful investment of native or foreign corporate interests. Otherwise, there is no money for infrastructure.

Often where there is an extractive industry, such as minerals, oil, timber, or a lucrative agricultural sector, private interests or the occasional savvy government invests in fully up-to-date infrastructure systems. These usually are transport links for cargo, including rail and ports. They generally do little for the social and economic development of the rest of the country but make the country's GDP look healthy.

Superconcrete provides hope for destitute countries

The great boon to impoverished world infrastructure comes from an unexpected source: the development of an easily prepared and worked superconcrete, sold under such names as Duracrete and Thinsculpt.

The concrete's strength is 8.4 times that of reinforced portland cement. Chief among its advantages is the ability to build stronger but thinner-walled houses, roadways, airstrips, industrial facilities, grain silos, and other structures. The material promises to help at least some destitute countries firm up their infrastructures and begin to advance their economies.

The net result of the failure of the poor countries to invest in adequate infrastructure is that they have inadequate patchwork systems for water, power, waste management, transportation, and communication.

Aggravating the problems of the destitute is the rapid shift of their populations from rural areas to totally unprepared urban areas. As a result, only a minority of World 3's urban dwellers have basic public services continuously available. Critical infrastructure systems for destitute country cities include:

- *Transportation, especially for getting to work*—These are nearly universally inadequate.

- *Food distribution logistics systems*—Trucking and food storage are at the center of this need. Labor unrest, civil strife, and poor storage facilities threaten food distribution.

- *Sewage and water treatment systems*—The absence of or failure of urban waste and water systems have stepped up the rate of deaths due to infectious disease.

- *Energy*—The availability of energy for cooking, transportation, and basic services is usually adequate but suffers sporadic outages and occasional sabotage.

Construction technology in the destitute world is, with few exceptions, rudimentary. Although heavy machinery is often available for earth moving and heavy lifting, construction processes, materials, and design ideas are vintage 1950 or 1990, at best. Stick scaffolding, hand excavation, and portland cement still dominate.

A CASE STUDY: Mozambique—A war-torn pauper state holds out hope for infrastructure-led development

Mozambique entered the century as the poorest country in the world, measured by *per capita* GDP. On all measures of poverty and despair, Mozambicans suffered as much or more than their African counterparts and other destitute peoples around the world. On top of that, their country was stressed by years of civil war.

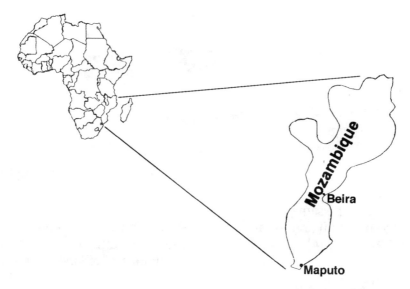

The scars show today. Mozambique is still the fifth poorest country, of those with 20 million or more inhabitants, measured in *per capita* GDP. None of its often-cited potential, based on good ports, hydroelectric power potential, oil, and minerals has been realized. Its location adjacent to the successful states of South Africa and its increasingly successful neighbors Zimbabwe and Botswana have not paid off, despite favorable trade possibilities.

Half Mozambique's GDP today is foreign aid or foreign investment, a situation little changed since the 1990s. Maputo businesspeople rank projects less by their viability than by their eligibility for aid or foreign investment. Chasing foreign money has long been a game played with zeal by Mozambican businesspeople and foreign salespeople pushing heavy equipment and infrastructure projects.

Mozambique's Rising Population Grows Urban
Population in Millions Showing Urban Share

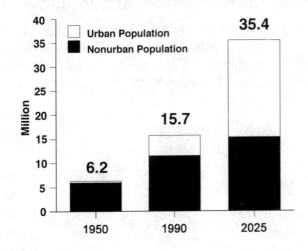

A decrepit infrastructure remains Mozambique's demon. The civil wars the country endured until 2002 left much of the infrastructure damaged and inoperative. At the same time, the country almost completely lacks the skills needed to maintain and expand the infrastructure. By one estimate, 57% of Mozambique's paved roads are in poor condition. Some intercountry transportation companies, such as South Africa's Johannesburg Line, have taken it upon themselves to regrade and repave miles of highway between Maputo and Beira.

The wars also forestalled much of the international investment and lending that might have helped Mozambique keep up or expand its transportation, communications, and public services.

Mozambique has seen rapid urbanization over the past 30 years and more than a doubling of its population since 1990. The population outgrew an already inadequate infrastructure by 1998. Episodes of famine are traceable to the sporadic breakdown of food transport, though international aid shipments of grain have been regular. The country's own food production has lost ground continually since it was first assessed in the immediate post-colonial years in the 1970s. Food production is at 106% of its 1980 level. However, *per capita* food production has fallen to about 50% of its 1980 levels.

Rail connections to Dar es Salaam and other cities to the north were so unreliable that the Cone of Africa Trade Organization (CATO) countries decided in 2006 to build rail links that circumvented Mozambique, assuming that Mozambique should end its warfare, could link into the system at Harare, Zimbabwe, or Lusaka, Zambia. Mozambique railways still do not connect with the CATO Inter-African Rail Link, and their rail gauge is different.

Mozambican trade is served by truck and rail routes to border "inland ports" and by the seaports of Maputo and Beira. Those ports are maintained as adequate but not particularly up-to-date facilities by business interests in Mozambique.

Road and rail modernization became an aid-agency policy in the 2000s. Studies by then had shown that two capital investment keys were essential for successful economic development: (1) modern, efficient transportation and (2) up-to-date telecommunications. Aid packages since 2000 have required attention to those parts of an aid recipient's infrastructure. Aid to Mozambique since about 2005 has been contingent on parallel investment in roads, railways, and communications infrastructure.

Contracts let for many infrastructure projects went to local and regional construction companies who took shortcuts and used inferior materials for their work. The National Highway, completed in 2013, was rutted and potholed five years later because of too much sand in the concrete, improper site preparation, and inadequate use of reinforcing materials. The National Highway was to be an integral part of the southern cone of Africa's transportation network, a key element in the CATO plan to build the economies of the region.

International construction firms, contracted to build several bridges and hydroelectric dams in the 2010s, in several cases quit work and were driven from the country by violent insurgency. There are three unfinished hydroelectric projects in Mozambique today that are awaiting either new financing or the cessation of tribal violence.

Disaster response and failing to keep up

Natural disasters have led to dozens of cases of damaged or destroyed infrastructure over the last three decades in Mozambique. More often than not, the roads, bridges, dams, and other systems damaged are either patched back together haphazardly or not repaired at all. This is true of the Maputo-Chicaulacuala rail line, destroyed in 2017 by a mudslide. Today, rail traffic between the two cities goes through Zimbabwe instead.

Critical Developments 1990-2025

Year	Development	Effect
Early 1990s	Estimates of the cost to repair the U.S. infrastructure at $1 trillion to $2 trillion.	Intensified public- and private-sector concern with the need to improve the infrastructure for competitive advantage and further economic growth.
1999	National Infrastructure Initiative (United States).	Intensification of infrastructure spending on highways, roadways, bridges, and conventional rail.
2001	Federal Infrastructure Agency (United States).	Rising out of the National Infrastructure Initiative, this agency took over the coordination of all parts of the U.S. infrastructure.
2010	First of Japan's Alice City complexes completed.	Demonstration that mass-scale mixed-use structures underground could be technologically and financially viable.
2011	Dallas-Martin process patented.	Capability to efficiently process wastes into recyclable fractions.
2013	Global Commons Agreement on Infrastructure.	Established the principle that national infrastructures should strive for integration with the emerging global infrastructure by adhering to construction and other standards.
2014	Transamazonia Corridor completed.	Demonstration of the floating roadway concept on an intercountry scale.
2015	Mexican, United States, and Canadian water grids fully coordinated.	Annual agreements govern the sale of water supplies throughout the continent, enabled by complex control of the pipeline network.
2017	Second National Infrastructure Initiative (United States).	Renewal of the national program, with a new focus on telecommunications, resources, and environmental management.
2022	Demolition of the Olympic Towers, New York.	First demonstration of efficient construction materials reclamation on a skyscraper scale.

Unrealized Hopes and Fears

Event	Potential Effects
Elimination of the autonomous motor vehicle.	Full acceptance of public transportation systems, elimination of traffic problems and environmental impacts from motor vehicles, dismantling of highways and roads
Dominance of macroengineering projects to reengineer the planet.	Full control of Earthly processes, including human and natural processes, resource use, climate, geophysical events
Transoceanic and transcontinental rocket trains, aka Planetran.	Unprecedented transportation speed for travel between Europe and North America, and across North America
Free energy at the point of use.	Fully distributed energy production from biomass, photovoltaics, or another process removes the need for any power grid
Completely wireless communications based on satellites, microwave, and cellular transmission.	Frees the landscape of transmission lines, makes people completely mobile—untethered by communications lines

11

PEOPLE AND THINGS ON THE MOVE

Transportation systems and technologies today are tightly integrated into local, continental, and global networks. Information technology, a core enabling technology, has rationalized transportation into a coherent system. Intelligent vehicle highway systems (IVHS), personal rapid transit (PRT), hovercraft, ships, and planes are coordinated. Information technology tracks and monitors the transit of goods and people; collects fees, tolls, fares, and tariffs automatically; enforces regulations; governs automated piloting systems and automated cargo handling; and weighs trucks while they are in motion. Other primary innovations over the past 35 years are:

- a cluster of high speed and automated transportation forms, including maglev trains, hovercraft, robotic vehicles, supersonic aircraft, and faster ships

- global positioning and related electronic navigation and logistics

- new fuels and propulsion technologies, especially electric vehicles

- materials technology, contributing to smarter, lighter, more durable vessels and infrastructure

- integrated scheduling programs

Middle and destitute nations benefit from new transportation technologies. Their transportation systems combine travel by foot and bicycle with high-speed rail, ship, and air travel.

Integration has enabled great progress toward the perennial goals of transportation planners: more efficient travel, and improved safety, energy efficiency, and reliability.

2025

Integration is the key to faster travel

New technologies have improved the raw speed of many transportation modes. Nuclear-powered ships, automated IVHS travel, and supersonic aircraft are faster than their predecessors—for example, today's supersonic aircraft are twice as fast as those of 2005. But the most striking advances in travel speed have come from better coordination and interchange among systems enabled by computerized logistics. In air travel, for example, logjams used to result from overwhelmed air traffic control systems and inefficient loading and unloading of passengers and goods. Goods movement is automated and people still move at their own pace. Upgrading control systems has reduced trip times more than increased speeds have.

Just-in-time principles are standard in transportation today. Vehicles, ships, or planes are available where they are most needed when they are most needed. Traffic control systems, covering land, air, and sea, maximize the capabilities of the available craft.

The integrated travel experience

I am happy to report that Department of Transportation's ongoing efforts to keep the transportation system running smoothly are succeeding admirably. I submit my inspection travel log for the committee to review.

5:30 am	There was little traffic on Route 66 to Dulles Airport. I switched my electric hybrid to automatic pilot in the high-speed lane and made the 19 kilometer trip from my home in just 15 minutes. My onboard computer found the nearest parking spot, and I proceeded to gate 7 for my commuter flight to New York on Mid-Atlantic Air.
6:03 am	My 6:00 flight was three minutes late (we still need work here). Boarding was smooth. I got aboard as soon as I reached the gate and had time for the robot attendant to bring me a coffee. I logged on to my daily news report by dialing up my home computer with my seat module.
6:41 am	We arrived on time, making up the three minutes. I had my luggage sent ahead one leg to the Park Plaza hotel in Boston, to test Mid-Atlantic cargo service. I took a people mover to the subway and connected to the New York leg of the maglev.
7:55 am	I boarded the maglev without delay, stopping for breakfast en route. I ordered a taxi from the subway and had its computer hunt for the fastest route to Sam's Breakfast & Deli, which I never miss when in New York.
11:17 am	I arrived in Boston on time once again. The maglev has cut the trip time to just over three hours. Only 10 years ago, it took over four hours. I participated in a videoconference on the train concerning another project I am involved in. I connected to the PRT stop at South Station and toured the city. I chose a four-person car, since I wanted to ask some questions about rider satisfaction with the system. We were on the guidebeam, so I relaxed and asked questions. The common theme is that people are happy to be rid of the "crazy" Boston drivers that once terrorized the city.

1:29 pm	I decided to walk the three kilometers to the hydrofoil after picking up my luggage at the hotel (kudos to Mid-Atlantic Air). It was a pleasant stroll, although I must report that a cyclist ran a traffic light and could have injured me. We still need to beef up enforcement of bicycling regulations. The hydrofoil service, which runs every 7 minutes, whisked me over to Logan airport in ten minutes. The ocean air was refreshing.
1:45 pm	I was back on Mid-Atlantic Air, en route to Dulles. I wrapped up this report onboard the plane.
Source:	Testimony by Henry Walker, Inspector, U.S. Department of Transportation, before the Subcommittee on Transportation and Travel, U.S. House of Representatives, June 3, 2024.

Safety improves across the board

Safety has been improved by reducing the potential for human error. Expert system assistants, such as Driveaid[tm], communicate with extensive databases to get a full picture of the travel route and plan the safest, most direct route. Collision avoidance technologies have been standard equipment in all new transportation devices since 2019. They detect oncoming traffic and quickly locate accident scenes or breakdowns and adjust their courses accordingly. Sobriety detectors, standard since 2010, disable a vehicle if alcohol is detected on the driver's breath. This innovation alone has saved tens of thousands of lives.

Automation for safer driving		
Safety hazard	**1990**	**2025**
Drunk drivers in fatal accidents	20,000	<1,000
All drivers		
accidents	20 million	0.9 million
injuries	2 million	0.09 million
deaths	50,000	2,250
theft	1.6 million	0.4 million

Transportation is more energy efficient

The global move to more sustainable use of energy has had a big impact on transportation. In the United States, the Clean Air Act Amendments in the early 1990s and legislation supporting the "Selling Sustainability" campaign set strict energy efficiency standards and encouraged a transition to more efficient alternatives.

The move to efficiency

Item	1990	2025
Gas costs, per liter	$0.39	$1.02
Average fuel economy	11 kpl	18 kpl
Gas, kilometers per dollar	28	18
Kilometers per electric charge	120	360
Electric, kilometers per dollar	n/a	28
Hydrogen, kilometers per dollar	n/a	10
Availability (% of total)		
Gas stations	100%	43%
Electric charge stations	0%	55%
Hydrogen stations	0%	2%

Standard kilometers per liter went up from 11 to 18 by the turn of the century. Electric vehicles have been affected as well, because electric power plants were forced to reduce their emissions. Kilometers per charge have increased from 120 kilometers in the 1990s to 360 today.

Magnetic levitation has also contributed to energy efficiency. Maglev has benefited from the gains in efficiency by electric utilities over the last 35 years. In the United States, for example, electric utilities have been primary movers behind the reduction of energy consumption from 85 quads in 1990 to 69 quads today.

Transportation's share of energy consumption has declined slightly. A near tripling of gasoline prices has caused many people to switch to alternatives such as PRT, mass transit, or bicycles for local trips. The transition first gained momentum in big cities, because these areas have more transportation choices. People began to use their bicycles, or ride the PRT (in Chicago and Boston) and mass transit, to save money.

Bicycling tied in nicely with many people's desire to get more exercise. Increasing ridership of the PRT and mass transit has generated the revenue needed to extend their coverage, within and outside the central city, which has in turn further boosted their use. Rural residents, unfortunately, have fewer choices and in many cases simply are forced to reduce their travel.

Electric vehicles, or gas-electric hybrids, are beginning to compete effectively with those vehicles depending solely on gas. Hydrogen-powered vehicles are just coming into use, gaining 0.01% of the market last year. They got their start in fleet operations in California. They are expected to steadily gain market share over the next decade, with estimates ranging from 5% to 10% by 2035.

WORLD 1—Progress on all fronts

The affluent nations have markedly upgraded their transportation systems over the last 35 years. The European Community, Japan, and Korea (using Japanese technology) were the first to adopt nonautomotive alternatives, such as superconducting maglev trains, helicopters, hydrofoils, and hovercraft on a wide scale. In the case of the EC, maglev prospered early because the air system was chaotic and non-integrated until 2000. Maglev remains strong, forming the backbone of the strongest continental travel system in the world, even as air travel has been upgraded to world-class standards.

World transportation leaders				
Region	**Downtown**	**Polynuclear**	**Continental**	**Overseas**
North America	1	2	3	1
Europe	2	3	1	2
Pacific Rim	3	1	2	3

Rankings: 1 is the best score

Source: *Jane's World Transportation Survey, 2025* (London: UK, 2025).

The annual ranking of the overall performance of the transportation systems of Pacific Rim, the EC, and North America show strengths and weaknesses everywhere. The criteria for the rankings include cost, travel time, energy efficiency, safety, comfort, and ridership satisfaction.

The EC and Japan have been the primary users of hovercraft and hydrofoils. Collision avoidance technologies have increased their safety, and noise reduction advances have made them quieter. These improvements, however, have not satisfied U.S. citizens, who still object to the noise and to the adverse effects on the land that hovercraft travel over. As a result, manufacturers have been reluctant to enter the U.S. market.

Japan has invested heavily in noise reduction technologies for tilt-rotor aircraft and helicopters, which addressed a primary objection to extending their use in the past. The Japanese have a network of over 1,000 high-tech heliports today.

Japan's Great Quake in 2005 wiped out part of its IVHS and other highway systems. Strict enforcement by their civil engineers of building and construction codes since the 1960s prevented complete disaster and enabled to the system to make a strong recovery.

U.S. transportation is on par with the EC and Japanese systems today. This is a remarkable achievement, given the large lead that the EC and Japan had around the turn of the century. The "Selling Sustainability" campaign begun in 2002 boosted the application of innovative transportation technologies that increased energy and travel efficiency. For example, electric vehicle sales surged in 2003, and maglev projects starved for funds found investors. Public-private cooperation on upgrading the transportation infrastructure and integrating it with the information infrastructure, enabled the United States to close the gap quickly and move ahead. The U.S. situation is described next in a detailed case study.

A CASE STUDY—The United States integrates its many systems

People are traveling more miles by a wider variety of systems than ever before. The wide range of transportation options, which once were a logistics nightmare, has been rationalized by computerized distribution systems. Airports coordinate with hovercraft, drones, maglev, and IVHS. The digitization of information, and progress in the arduous standards-setting task, means that computers aboard any transportation systems can communicate with one another. Expert system assistants sort through masses of information to provide the most efficient means of getting goods or people from point A to point B.

Travel has been a primary beneficiary of growing disposable incomes. People want to see more of the country and the world and are increasingly willing and able to pay for it. Routine travel, such as running errands or delivering a package across town, is more pleasant today, as the traffic jams that plagued cities have been reduced by computerized traffic control.

The travel experience today is faster, safer, and more comfortable compared with the last century. Information technology has made all travel at least a neutral, if not rewarding, experience. The lack of information that so often hampered travel in the past has been filled by entrepreneurs providing technologies and services designed to answer the typical questions of the traveler. An interesting side effect, is that people are increasingly fond of "mystery trips" where they do not know their route or destination and rely on an expert system courier to handle all the information. Portable geographic information systems (GISs) access massive databases with the touch of a finger.

Better information, typically provided by the growing field of travel-information services, has enhanced business travel as well. The business traveler has timely access to company information relevant to the trip. One no longer needs to "check back with headquarters and get back to you." Distributed work and the advent of videoconferencing, E-mail, and groupware have reduced the need for travel solely for business. Business and pleasure travel have become increasingly intertwined.

The movement of goods has also prospered over the last 35 years. Goods can be tracked within meters while in transit. More precise logistics have enabled the proliferation of just-in-time systems that have significantly boosted the competitiveness of U.S. businesses. Businesses that need to get their products to market quickly have been willing to invest in upgrading the transportation infrastructure and in new transportation technologies.

Many of the advanced technologies that have improved personal and business travel were pioneered by freight companies. They have long been under strong competitive pressures and have been willing to experiment with the latest technologies to gain an edge. And it is still true today that moving freight is more profitable than moving people. The first maglev line built in the United States, for example, was funded by a consortium of state and local governments in tandem with Federal Express and local business leaders.

The transportation infrastructure privatizes

Federal, state, and local governments are no longer relied on as the sole sponsors of improvements in the transportation infrastructure. The role of government spending for the infrastructure is down from 2% of GDP in the 1970s to 1% today. The cost for upgrading, however, continues to increase.

The private sector has made up the shortfall, for two primary reasons. First, businesses have been willing to invest in the infrastructure where they use it, because it improves their productivity. Second, user fees have been effective in recouping infrastructure investments. Businesses or consortiums that fund an infrastructure, such as IVHS, PRT, or the various maglev systems, charge user fees each time their system is used. Private toll roads are more prevalent today. Automatic toll readers identify vehicles and weigh freight trucks passing through a checkpoint and deduct fees from credit accounts that road users must set up in advance. There is open access for all who are willing to pay the fees.

Governments typically grant privately funded infrastructure tax-free status, contingent on adherence to sustainable practices. Tax policy today encourages sustainable travel practices in metropolitan areas as well. Taxes on auto travel in downtown areas, for example, have helped reduce traffic volumes.

The trend toward increased private funding of infrastructures followed the model of the financing of the information infrastructure. The state-of-the-art national information technology infrastructure, in place since the turn of the century, was constructed and connected largely by private funds. Forty-two thousand kilometers of information highway are operated by private companies, typically in 150-kilometer units.

Restrictions on foreign ownership of the transportation infrastructure, a relic of national security concerns, have been relaxed as well. This trend began in the late 1990s, when the airline industry was desperate for capital investment, and has spread to all transportation sectors. Governments have found ways to use foreign capital without compromising national security.

2025

The rising costs in using the transportation infrastructure have been compensated for by reduced costs elsewhere. The move to leasing rather than purchasing vehicles, longer-lived vehicles, fuel efficiency improvements, and the more durable infrastructures have reduced costs as well.

Personal travel continues to outpace business travel

Growing disposable incomes over the last 35 years have increasingly been spent for pleasure travel both home and abroad. Trips over 160 kilometers within the United States are three times more likely to be for pleasure than business. Similarly, personal errands and pleasure trips account for more than three-fourths of local trips, with commuting and business reasons accounting for the remainder. The number of trips abroad by U.S. citizens is now 27 million annually. About 60% of overseas travel is primarily for tourism, and 40% for business or business combined with pleasure. Greater amounts of vacation time have translated into more frequent, but shorter, vacation trips within the United States

Why people travel		
Annual travel, U.S. residents (# of trips to places more than 160 kilometers from home, in millions)	1990	2025
Visit friends/relatives	231	260
Other pleasure trips	226	251
Business travel (does not include commuting)	169	170
Trips abroad (business and pleasure)	15	27

Air travel progresses steadily

Air travel today is faster, safer, more comfortable, and more energy-efficient. Incremental gains have added up to enhance the air travel experience. The revolutionary advances have been in advanced supersonic transport (AST) and floating airports.

Technologies aboard the aircraft

The added amenities of today's travel have been accomplished even while fuel efficiency has doubled since the 1990s. Lighter-weight, more-durable materials; ceramics; the use of composites; hotter-core engines; an increase in thrust-to-weight ratio; and smart onboard sensors and controls are the reason.

Advances in aircraft center around information technology, expert systems, and sensors coordinating data inside and outside the craft, with real-time air traffic control, constant adjustment to flight conditions to optimize aerodynamics, and efficiency.

Fuel efficiencies have doubled since 1990 with integrated structures, propulsion, and controls and with adaptive wings that change shape to match flight conditions. Ceramic matrix composites for jet engines became standard in the 2000s. They can operate at much higher temperatures than metal engines.

The overseas traveler today takes advantage of chronobiology to neutralize jet lag. A pill or skin patch adjusts the traveler's circadian rhythms by influencing their reaction to light.

The commercial airline industry has stabilized over the last 35 years. Since 2004, there have been the Big Three airlines in the United States. They are supplemented by a growing network of regional airlines that typically serve a dozen or so cities. They compete with overseas carriers as well, as the airline market has truly globalized. ECAir, for example, has a significant chunk of the United States overseas travel market.

A key reason for the success of ECAir in the United States was that it was the first to grasp the trend in air travel from being almost exclusively business to combining business and pleasure. A scaled-down version of the luxury cruise model is adopted for upscale, supersonic air travel. AST planes with average speeds over Mach 2 came on-line around 2015. ASTs equipped with noise cancellation technology reduced noise problems from sonic booms. The cost of building them has come down, making the cost-competitive with subsonic crafts.

Planning trips in flight

I wanted to go fishing on Wednesday in Eastern Germany, preferably for trout. My computer terminal directed me to the fisheries section of Germany's Wildlife Bureau. Bureau staff transmitted a map of a river where I was likely to find trout. I requested a copy of the map, which they faxed to the hotel room I was to stay at because the plane's fax was down. I asked for a weather report and was transferred to the equivalent of our National Weather Service. I was then transferred to a local bureau where I found that I was in for some chilly, but manageable, weather. I caught five delicious trout.

Source: Downloaded from "Travel Today," *CompuServe*, May 24, 2025.

Most airplane travel today is better than the first-class of the past, with plenty of leg room, a wide selection of videos, and good food. Overworked and harried flight attendants are a relic as well. With the addition of robotic assistants, attendants deliver great service today.

The transoceanic airlines today provide computer links to unlimited entertainment and recreation services. Of course, the personal computer function enables one to work and have the results transmitted to the home or office by digital satellite.

Trip times have been reduced by speeding up the loading and unloading of passengers and luggage. Today's supersonic craft carry twice as many passengers at three times the speed of sound with far higher fuel efficiency than the Concorde of fifty years ago.

A hypersonic (five times the speed of sound) plane was developed by the International Space Agency in 2017. The International Hypersonic Craft Consortium (IHCC), formed in 2019, is laying plans to commercialize these craft by the next decade. They can take off and land on commercial runways, exit and reenter the atmosphere en route. It will take under two hours to fly from the United States to Europe.

Twenty-one floating airports worldwide support the increased air travel of the last 35 years. They are a cost-effective solution to noise and land scarcity problems, at least for coastal airports.

The commuting crunch eases

People are living closer together to save travel time and expense. Urban sprawl peaked in 1998. The metropolitan population is up from 77.5% of total population in 1990 to 85% today. The trend to polycentric cities, a core central city with subsidiary central citylike functions distributed throughout the metropolitan area, may finally be tapering off.

U.S. daily commuting times, each way

Minutes	1990	2025
<15	32%	30%
15-29	36%	40%
30-60	25%	25%
>60	7%	5%

Information technology advances increase the attractiveness and capabilities of distributed and off-site work. It is estimated that 39% of the workforce is engaged in some form of distributed work today. The majority of distributed workers appear at a central workplace at least twice each week. Another large block perform their work at home or at local satellite centers, usually within two miles of their residence. The growth of satellite centers surrounding old central cities means that commutes and business calls move all around the peripheries of large cities, as well as to and from downtown. The effect has been more balanced traffic patterns, which has helped reduce congestion.

Flexible work schedules and more vacations

	1990	2025
Time at work, per week	39 hours	35 hours
Flexible schedules*	12%	75%
Distributed work schedules	3%	39%
Vacation time, per year	2 weeks	4 weeks

* 35-hour workweek on staggered hours

The polycentric city phenomena has increased commuting between suburbs rather than suburb to central city. Better integration of transportation systems has cut the 2 billion worker hours per year lost to traffic jams in the 1990s by 75%. There has not been a solution to the traffic problem, but continuous adjustments are steadily reducing bottlenecks.

The tilt-rotor aircraft and helicopter networks that are strong in Japan have not been equaled here. Japan has an impressive infrastructure of over 3,000 heliports, enabling it to deploy a wide network of short takeoffs and landings (STOLs), vertical takeoffs and landings (VTOLs), tilt rotors (combining vertical lift features of helicopters and the high forward speed of airplanes), for 160- to 800-kilometer trips.

Downtown areas revitalize

City planners looking to revitalize downtown areas often concluded that eliminating cars would free up space for better uses. Chicago, for example, banned vehicle traffic in favor of PRT.

Speed is not critical in urban areas. Travelers use portable GIS technology to plan their downtown travel. One can park at the outskirts of cities and use track trolleys or people movers in closed-off shopping areas, the PRT, or even mass transit. Car computers search out and reserve parking spaces in areas that allow vehicle traffic.

Mass-transit planners finally came to terms with the need for either highly reliable timetables or high trip frequency. Mass transit and PRT systems serve niche markets well, typically in older, densely populated cities. Service is clean, fast, and often scenic.

Conflicts between different transportation concepts, such as IVHS versus light rail or heliports continues. The result is a hodgepodge of transportation systems. But systems integration has relieved the problem of unsynchronized systems.

Automobiles are electrified

The long debate about which vehicle fuel or power configuration should be standardized has been resolved in favor of electricity, in a hybrid system. Vehicles powered solely by gasoline were outlawed 14 years ago in 2011. Gasoline is still used, but only in hybrids. The electric hybrids range from the predominant electric-gas/reformulated gas to electric and ethanol/methanol, compressed natural gas (CNG), hydrogen, and the occasional solar vehicle for specialized purposes. Each hybrid has its advantages and drawbacks. There has been some progress in hydrogen-fueled vehicles, but they are used on too small a scale. Hydrogen fueling stations are few and far between.

Three factors spurred the growth of electric vehicles. First, was the early adoption by companies with vehicle fleets. Then high-occupancy vehicle

(HOV) lanes were converted into electric automatic pilot lanes. Battery improvements have improved the range of electric cars from 120 to about 500 kilometers per charge.

Second, was the proliferation of recharging stations built by utilities eager for new markets. People could rely on being able to charge up just about anywhere. The on-the-road charge takes about 10 to 15 minutes. The vehicles are almost always plugged in at home and charged more slowly overnight.

Third, was the relaxation of antitrust regulation to enable old Big Three automakers to form the Advanced Battery Consortium in the late 1980s and the United States Council for Automotive Research in 1992 to share information and costs in the development of new automotive technologies. Today, two of the 20th century's Big Three are left, but they have nine autonomous divisions. Global alliances have virtually erased all but style distinctions between U.S.- and foreign-made vehicles anyway.

Hybrids, either on the guidebeam or combined with gas, methanol, or liquefied natural gas, are more efficient than their gasoline predecessors. But experiments with the gasoline part of the hybrid engines are still continuing. Two-stroke engines and direct-injection diesels have become conventional. Stirling engines are becoming popular for larger vehicles. Ceramic low-heat rejection engines caught on with the public about 2012 as their prices came down sharply.

Non-hybrid-gasoline vehicles

Improvements in existing transportation energy technologies such as the gasoline-powered combustion engine extended their life and delayed the introduction of alternatives. The fuel economy of gasoline engines crept up and kept competitors at bay years after many experts had predicted their demise. They were able to meet the 15 kilometers per liter requirements mandated by Congress in 1998. California pioneered in these efforts with some success by 1999. Electric vehicles began to catch on there and inspired copycat successes in other states and the country. Gasoline vehicles were still in production, but in 2011, vehicles relying solely on gasoline for fuel were banned nationwide.

The growth in the number of vehicles has outpaced gains in fuel efficiency. However, in the year 2002, the number of automobiles worldwide crossed the half-billion mark. Ten years later in 2012, the overall worldwide number of vehicles, including trucks, buses, and motorcycles, surpassed one billion. The impressive gains in fuel efficiency are being offset by the numbers of vehicles. World vehicle fuel use is up 10% from 2000.

IVHS smartens up vehicle travel

The automobile still dominates personal travel within the United States, thanks to the growth of IVHS. It began as a commercial, off-road system, using digital, cellular, and satellite communications to connect vehicle computers with traffic management centers and other information services. Some form of IVHS now covers the nearly 9 million kilometers of the federal-aid interstate highway system. State and local governments took the initiative in developing regional systems over the last 15 years, frustrated by lack of aid from the federal government. Selling advertising on local IVHS networks helps support these systems.

There are differences between public and private IVHS. Motorola, for example, is the key player in providing access to satellite positioning for both public and private networks. The private ones use advertising for revenue, whereas the public ones use taxes and, increasingly, automatic toll collecting for road use fees.

Learning to drive...anything

Virtual reality is the primary tool for teaching one how to control all forms of transportation—from driving a car to piloting a ship or plane. Licensing agencies credit virtual reality training for helping to reduce accidents across the board.

Vehicles no longer sit at traffic lights when there is no cross-traffic. Sensors provide real-time adjustment to prevailing conditions in a half-kilometer radius. On some highways, the driver can play videos or catch up on work. Roads equipped with automatic vehicle chauffeuring capabilities allow the driver to lock on to the chauffeuring function with the onboard computer. Vehicle speed, as well as safety, is maximized by computerized traffic management. Today's speed limit is twice the 88 kilometers per hour of 1990 on long runs of highway in the west. It is difficult to have an accident, thanks to collision avoidance systems. The Santa Monica Freeway, for example, enables totally automated travel. Vehicles switch on to the electric guide beam on automatic pilot to ensure safety. System failures have been few, so far.

Theft prevention: DNA locks

"Genetic fingerprints somehow got mixed up, and the scan system wouldn't let us in. Genetics and DNA has done a lot for us, but sometimes I think they have raised more problems than they have solved. Simple fingerprint scans have turned out to be not so simple. Criminals soon figure out how to beat each new system. It's like the old arms race. A new system, a way to beat it, another new system, etc. Anyway, we used TravelPlan to call up the rental office. They used an override to unlock the door, but not before a half-hour of questions."

Instantaneous access to the nearest traffic control center allows an operator to report an accident or erratic driving behavior. Advanced systems automatically detect lane crossings and either monitor or phone the driver to see whether there is a problem.

Accidents are uncommon if not rare. Drunk driving, for example, has been practically eliminated by standard-equipment sobriety analyzers in every vehicle. IVHS also keep a constant lookout for wildlife. Road kills are uncommon today.

Traffic law enforcement is less necessary and practically invisible. So much of transportation is automatic, it's hard to break the law. Much of the police force that used to handle traffic is now free to cope with more serious crimes.

PRTs are succeeding in older, compact urban areas

PRT systems have operated in Chicago and Boston since 2017 and 2021 respectively, with plans being drawn up for a half dozen similar urban areas. Chicago's PRT overlayed the old mass-transit subway system with small one-to-three passenger vehicles on a new guideway system. PRTs are similar to the dummy cars that have been in airports for a long time now, or the automatic guided vehicles that have long been used in warehouses. Current PRTs travel exclusively on guideways. A proposed system in Charlotte, North Carolina, however, will experiment with dual purpose cars that will travel on and off the guideway.

The vehicles are designed for safety. They are equipped with collision avoidance technology, and reinforced shock-absorbing bumpers surround the cars. In Chicago, they don't have any traffic to contend with. Automobiles are not allowed into the city. One parks outside the city and takes the PRT, helicopter, or hydrofoil, across the lake into downtown. Pedestrian traffic is either on moving sidewalks or in an underground or elevated walkway.

Bicycling emerges as a healthy personal transportation alternative

Bicycle paths are now a standard feature of all urban roadways. The risks of riding alongside traffic are past. Bicycle paths have been used since 2011 in most metropolises. People with increasingly sedentary work lives have welcomed the opportunity to integrate exercise into their daily routine. Those less inclined to exercise use solar-powered cycles. Boosting the role of bicycles was an important component of reducing energy use in the "Selling Sustainability" campaign begun in 2002. It has also helped reduce metropolitan traffic congestion.

Infotech replaces some business travel

Travel solely for business reasons has declined. Videoconferencing, videowindows, E-mail, and Electronic Data Interchange (EDI) cut into business travel by 30% per person since 1997, but the total volume has been more or less constant. Business travel today typically combines business with pleasure. When information technology did not immediately replace business travel as was forecast decades ago, the possibility of replacing some business travel was dismissed. The completion of a national information network, however, has made this vision of less business travel real.

Rail comes back

The maglev is now a practical alternative to automobile travel, even with the advances of IVHS. It is also challenging air travel for trips of less than 800 kilometers. Conventional high-speed rail still exists, but all new rail systems since 2010 are based on the superior maglev technology, which operates with superconducting magnets that suspend the train a few inches above the rail. It uses electricity, so there are no direct emissions.

The intransigent EMF lobby

In the United States, the surprising strength and intransigence of the electromagnetic field (EMF) lobby has held up maglev deployment. Despite the lack of hard evidence connecting EMFs to any ill health effects, the lobby continued to maintain that there are hazards. There was just enough circumstantial evidence and lack of clarity to keep them credible in the eyes of many. Politicians were afraid take them on. Only in the face of overwhelming scientific evidence to the contrary, were politicians finally able to back maglev projects.

The United States has been following the EC's lead with maglev. EC political leaders decided at the turn of the century that further increases in automobile travel would not be tenable in a sustainable society. They made a strong financial commitment to build maglev systems to connect major metropolitan areas across the continent. In addition, business practices in the EC demand efficient continental travel.

The maglev systems are clean, comfortable, safe, and fast. The average speed is close to 500 kilometers per hour. The Japanese also have an extensive maglev system. But the EC consortium's choice of construction materials, such as superconducting materials and advanced ceramics, and recent experience with macroengineering projects, such as the automated subway system in Paris, give it the edge over the Japanese network.

Just-in-time moves goods more efficiently

Just-in-time principles govern the delivery of raw materials and goods today. Higher transportation speeds attained in preceding decades whet the appetite of consumers and business for even faster transportation. It is now almost unheard of for the delivery of goods anywhere in the United States to take more than three days. In addition, cargo transport is more reliable, as bar codes and taggants (chemical identifiers in or on the item) keep accurate track of goods in transit.

Drones in the fast line

Drone couriers, totally automatic solar or battery-powered runabouts, are commonly seen scurrying around in big cities. They are programmed to pick up a load and deliver it to the customer's coordinates. They use global positioning, with street maps programmed in, to figure out the fastest route given prevailing traffic conditions. They have collision avoidance, and person-driven vehicles get the right-of-way, enforced electronically. Peoples' onboard guidance systems know the drone is coming and signal the drone to slow, stop, or get out of the way. Drone cargo shipments travel during the lowest time demand on roadways.

Computers manage the flow of vehicles and goods in a form analogous to air traffic control three decades ago, with the added capabilities of massive computer power and expert systems. Ten times the cargo volume of 1990 can now be handled due to computerized logistics.

Global positioning and related electronic navigation and logistics monitor IVHS, rail, sea, and air transport. Iridium-based global positioning systems (GPSs) track 85% of cargoes and transport vehicles today. The other 15% are tracked with land-based technologies.

A national system of slurry pipelines and pneumatic and hydraulic capsule cargo pipelines has further improved logistics. Oil and gas pipeline networks have been extended as well.

Sea transport stays above water

The once large cost gap between air and sea cargo transport is closing. Air cargo transport is generally faster for small loads, but four factors are increasing the attractiveness of sea transport. First, magnetohydrodynamic (MHD) drives have sped up ship travel by 30%. Second, today's ships are self-cleaning, with biotechnology creating microorganisms that eat barnacles. Third, submersibles enable remote underwater inspections, which eliminate the costly practice of drydocking ships for inspection. Finally, but perhaps most importantly, is the use of factory ships.

Factory ships process materials or goods in transit. The range of products that these ships serve is expanding from the early fish-processing ships to include lumber and paper ships and food-processing ships. These ships are energy efficient because they use totally integrated production systems that take advantage of waste heat for added power. The ships typically run by nuclear power and are cooled by the ocean. The cooling is dispersed over the whole journey, so as not to harm marine life.

Air cushion vehicles serve a niche market, linking long-range land, sea, and air systems. Hovercraft and hydrofoils provide links between airports, for example. They are used for water-surrounded ones because they easily bridge the land-sea interface. They are also useful as high-speed commuting ferries, running as fast as 200 kilometers per hour in trips over eight kilometers.

Modular cargo transport bridges the air, land, and sea

Modularized container and cargo systems are standard today, accounting for 95% of long-distance cargo. Containers go from ship to train to truck to plane to warehouse, without handling by people. This automated cargo handling and logistics depend on ubiquitous information technology. Bar codes and scannable chips enable customers to keep 24-hour tabs on their purchases. Shipments are tracked remotely, and sometimes by satellite, using these identification systems. Packaging itself is more efficient in materials use and is recyclable. Cargo tagged with scannable bar codes or chips are purported to reduce theft and fraud by 78%.

International trade flows increase between trading blocs

International trade has tripled since the 1990s. The trade flows within and between the three big blocs: North American Free Trade Association (NAFTA), the EC, and eastern Asia. Trade with Mexico and Canada in North America leveled off after a 50% increase in the decade following the signing of the NAFTA agreement. Trade with Europe and Asia is now bigger than within NAFTA. Trade with South and Central America grew the fastest but started from a small base. Japan and China dominate trade in eastern Asia, a pattern that began to develop in the 1980s.

Transportation infrastructure builds in disaster detection

Transportation infrastructures are built today using smart structure technology and includes dynamic structures that respond to environmental stresses, including seismic movement, wind, water, and weather stresses. Natural disasters can still disrupt the flows of people and goods occasionally, but there has been substantial progress in forecasting and in some cases preventing them. Improved weather prediction has enabled traffic managers and freight companies to better prepare for severe weather conditions.

The importance of building in disaster detection was highlighted by Japan's Great Quake of 2005, which wiped out part of its IVHS and other highway systems.

WORLDS 2 AND 3—On the road to affluence

The transportation situation in middle and destitute nations can quickly be grasped by looking at the experience of the affluent nations a generation ago. Restrictions to protect the environment from fuel emissions and the higher relative costs of fuels have narrowed the range of options. For example, these nations are not able to build the massive vehicle-based transportation fleet of the United States. At the same time, the combined increases in their vehicle fleets has boosted the global vehicle fleet (including e.g., autos, trucks, buses, and motorcycles) over the one billion mark today. Their transportation systems closely mirror the mixed systems of the EC and Japan, albeit of a generation ago.

The mix of low and high technologies is striking. Bicycles and people-powered taxis travel alongside maglev systems in many countries. Financial assistance from World 1 is typically required to bring in high-technology systems. Affluent nations requiring access to markets in these nations have been willing to aid the development of modern transportation infrastructures and technologies. Just-in-time systems in Tokyo, for example, require that regions such as Indonesia be brought up to speed. The integration of transportation systems in affluent nations is spreading to middle and destitute nations.

A CASE STUDY—Indonesia combines the old and the new

In Indonesia, transportation options range from Java's "bemo" version of the rickshaw to battered merchant ships of the 1990s to hydrofoils that run 200 or more kilometers per hour.

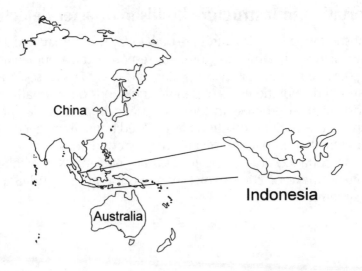

Bicycles are ubiquitous, as the Javanese do not have many private cars. Since gas prices shot up around 2011, more and more people use bicycles in combination with the downtown people mover (aeromovel) and the interurban trains. The transportation options in Indonesia today are outlined in this case study.

There is plenty of direct air service to Java, particularly from Australia, Japan, Singapore, and North America. Flights within Indonesia are still expensive. High fuel costs make them so. Garuda, the national airline as yet has only three of the more efficient turbopropfan airliners. They will probably not soon join the new Hypersonic Craft Consortium for obvious economic reasons. Those planes will certainly fly to Jakarta anyway, perhaps in 7 or 10 years, as JAL and the U.S. carriers will use them on their Pan Asian routes. The table below outlines the options for overseas access to Indonesia today.

Overseas access to Indonesia: a tourist's guide

From	Route
Australia	Hydrofoils take many visitors from Darwin to Bali.
Japan	Cheap supersonic flights to Japan are available from eight cities in the United States. One can also try flying to Narita Airport and then catching one of the four subsonic flights to Jakarta that are scheduled daily. This saves hours.
Singapore	There is fast and efficient superhydrofoil service from Singapore. Five trips are made daily over the 1,000 km at 200 km per hour. This is more than four times the speed of the first hydrofoils to travel this route in the 1990s. Before that, the diesel powered ships took 36 hours.
	The adventurous traveler shouldn't miss a chance to combine the old and the new. One can call at the harbor in Singapore and try to find a yacht or fishing boat that is crossing to Sumatra, ideally to Palembang. From there, see *Sumatra,* below.
Sumatra	The best bet is the hourly hydrofoil to Jakarta from Telekbetung running from 0500 to 2300. Telekbetung is served by the bullet train that crosses Sumatra longways. Or one may take various modes of transport on the Trans-Sumatra Highway, one of the country's best. (Plans to build a maglev on Sumatra are still just plans).
	Two plans seven years ago were rejected that would have made travel between Sumatra and Java much easier. There was talk of a tunnel or bridge to cross the 37 miles. In the days of Suharto Sr., one or the other might have been accomplished. However, seismic instability is still a problem for mass-scale underwater tunnels, and a dynamic bridge crossing would have cost hundreds of billions of Rupiah.
United States	The Supersonic makes it in 4.5 hours from San Francisco or Los Angeles, going Mach 3. But supersonic tickets to Indonesia are nearly twice the cost of subsonic tickets. It's a seller's market, there still aren't enough flights for all the business trade. Other options include conventional jet service.

Source: Excerpted from *A Travel Guide to Indonesia,* Department of Tourism, Jakarta, Indonesia, 2025.

Cargo ships modernize and add just-in-time

Cargo vessels range from the older, conventional-hulled craft to high-speed hovercraft and hydrocraft. The latter are rigidly scheduled elements of the just-in-time manufacturing and raw materials scene in Southeast Asia—their cargos cannot be delayed in any way. They are so closely scheduled and tracked that they make continuous speed adjustments as they travel so as to reach port on a precise schedule.

Searching for environmental offenders

The republic's police are apt to board and search cargo vessels at any time, as environmental laws are strictly enforced today. For example, waste in the Java Sea can be chemically traced to offending ships. The ship may have flushed out its hold during a previous voyage, and the chemical fingerprinting traced the waste to the ship.

Complex logistics computing and satellite communications are used today. Traditional navigation prevailed in the 1980s and 1990s, and then hand-held GPS in the 1990s and turn of the century. Captains have more responsibility for customs manifests and other regulatory compliance work than they do for governing the ships. Ships can be loaded in Jakarta, sail to Bombay or Yokohama, or anywhere else, dock and unload without any human help. Much of today's cargo never or almost never is touched by human hands.

Bemos: the gray economy alternative

Bemos are by now a long-standing tradition in Indonesia. Any vehicle can be a bemo. It is simply a private vehicle, usually a small electric-powered truck, with benches rigged on its flat bed for paying passengers. Despite decades of attempts, the government has not eliminated this form of transportation, though the vehicles must be electric to conform to antiemission laws that came on the books two dozen years ago.

Hitchhiking lives on

Hitchhiking is still safe on Java. Rides are not necessarily frequent and most likely will be small grocers' trucks and other commercial vehicles. Most Javanese, outside the wealthy sectors of Jakarta and its suburbs, do not have their own vehicles.

A more ancient tradition, the use of becaks and dokars are all but gone. Becaks are backwards rickshaws; the driver pedals to drive the vehicle from behind. Dokars are tiny, ornate donkey carts. They might be found on some of the smaller islands or within the sprawling tourist complexes on Bali, but not in Java or Sumatra.

Driving loses ground

Indonesia's roads have deteriorated in the past two decades. Roads are not a priority. The government has worked hard to prevent people from buying and driving cars, and road conditions are getting worse and worse. This is in great contrast to other developing countries where the infatuation with automobiles, mirroring that of the United States last century, led to car-based infrastructure development.

Although Indonesia has made great gains, particularly in rail and ferry service, traditional transport on the ground has stagnated or lost ground. It was worse just two decades ago. Getting from point A to point B by road in the country used to involve hours or days of jolting over roads constructed of trenches big enough to swallow half a truck. Earthquakes and tremors often destroyed roads as well.

Today, Indonesia uses some of the latest Dutch and Japanese road-building technologies. These technologies lay a continuous fabric of pavement with a machine that levels and conditions the ground in front, lays pavement, and paints traffic lines, all in one pass. Indonesian engineers have been able to recycle components of urban waste, such as rubber and glass, into pavements. The whole operation is laser- and GPS-guided.

A positive development that may soon be on the market, to be distributed by the Worldwide Automobile Association in alliance with Tropicars (Java) Ltd., is a device like a collision avoidance module that also monitors lane choice and warns if the vehicle crosses lanes. It senses the lay of the roadway and other vehicles. Most Indonesians with cars have simple collision avoidance systems. This new wireless device can be attached to any vehicle.

Hydrocarbon-fuel-powered vehicles are heavily taxed and the fuel expensive, despite getting 14 to 19 kilometers per liter. Gasoline or gasohol are about two-and-a-half times the cost in the United States, and one-and-a-third higher than in Europe. There are no biofuels yet.

Bicycling: part of a primitive PRT

Indonesian bicycles are not the fine, sleek, strong composite frame ones found in the World 1. The bicycles here are often old, and many have the rust to prove it. Yet they fill a niche in the transportation system. They are allowed on trains, in third- and fourth-class cars, forming the closest local version of a PRT that exists in Indonesia.

Rail as a spur to development

Indonesia's maglev is as good as any in the world. Built by the Japanese in 2017, the maglev line was intended to spur development and create a lucrative new market for Japanese trade goods. Unfortunately, there is just the

Trans-Java line, all other rail service in Java and the rest of Indonesia is nonmaglev. Those other trains are no different from and often worse than train service in the West 40 years ago.

The October 2018 earthquake wrecked part of the Trans-Java maglev line, then brand-new. This led to reconfiguration of the guideway and its supports. The piers and the guideway in the fault zone have been rebuilt as dynamic structures, with continuous adjustments made for vibration, weight stress, and future earthquakes. Prequake monitoring provides a warning a few hours before the quake, enough time to shut down the line if necessary. Of course, Indonesian earthquake prediction is no more accurate, and probably less so, than at San Andreas or Central Honshu. There is no program of prevention in Indonesia as there is in Japan and the western United States.

Airport transport features people movers

The Aeromovel, a completely computer-run people mover, is the best option for the link between Jakarta and Sukarno-Hatta International Airport. It is powered by air currents and runs silently and smoothly at as much as 150 kilometers per hour. The first Aeromovel, still in use in Taman Mini Park in Jakarta, is now 36 years old. The airport line was built in 2004 and connects to the main downtown and suburban loops. There are a total of 22 stations in the system.

The trip from Sukarno-Hatta International Airport to Jakarta used to take as much as two hours. The Aeromovel will get you to any station in Jakarta, including one in the port, in 15 to 35 minutes.

Boats and ferries circumvent crowded roadways

Hydrofoils, including licensed regular-service ones and private ones run by solo entrepreneurs, ply the coast of Java. They can be an efficient way to circumvent bad roads, bicycle traffic, and crowded trains. They run at 150-200 kilometers per hour, a fast alternative to land travel. Ferry travel is used to reach almost any of the islands adjacent to Java.

Critical Developments, 1990-2025

Year	Development	Effect
1990	Clean Air Act Amendments passed.	Tighten vehicles emissions requirements. Improve prospects for electric vehicles.
1998	Motorola's Iridium Satellite Network deployed.	Significant enhancement of global positioning capabilities improving logistics and navigation.
1998	2% policy requiring vehicles with zero emissions in California.	This failed policy spurred production of electric vehicles.
2000	U.S. computer network infrastructure largely in place.	Provides backbone for IVHS and computerized logistics.
2001	Breakthrough in battery technology for electric vehicles.	More than doubles kilometers per charge, from 120 to 250.
2004	Airline consolidation leaves the Big Three Airlines in the United States.	Small airlines serve niche markets. There is also greater competition from overseas airlines in domestic market.
2005	U.S. individual pleasure trips exceed 250 million annually.	Growing disposable income goes increasingly to travel.
2010	The EC and Japan complete . innovative transportation projects, developing premier maglev and tilt-rotor and helicopter systems	Spurs the U.S. construction of a coast to coast magnetic levitation system.
2015	Development of autonomous vehicle technology in the United States	Begins deployment of the technology on major highways.
2017	PRT trial in Chicago successful.	Boost to plans in other cities across the nation.
2019	International Hypersonic Craft Consortium forms.	Plan to commercialize hypersonic craft by 2030.
2020	Metropolitanization of the U.S. population hits 85%.	Trend toward urbanization levels off.
2025	40% of the workforce involved in distributed work.	Commuting volumes continue to reduce.

Unrealized Hopes and Fears

Event	Potential Effects
Gasoline cost of over $4 per liter reduces automobile travel; alternative fuels do not provide a cost-effective alternative.	Severe disruption in goods movement and lifestyles.
Substantial federal government investment in mass transit increases commuter ridership from present 3% to over 50%.	A viable alternative to vehicle-based travel.
A definitive landmark study concludes that EMF from maglev causes damage to human health.	Maglev operations are suspended.
Breakthrough in hydrogen storage aboard vehicles.	Switch away from electric as the preferred alternative.
Social pressure and regulation requiring human piloting of vehicles, and disallowing autonomous operation of machines because of a high-profile disaster involving such machines.	Significant setback to IVHS deployment. Traffic congestion mounts once again.

12

THE WORLD OF PRODUCTION

Custom production has raised product variety, quality, service, and reliability to unprecedented levels. Customers who choose to can design or modify products to their specifications. Today's manufacturers anticipate and quickly adapt to evolving customer needs. They get most products, except heavy goods like tractors or refrigerators, to customers in hours or days, compared with the four to six weeks it took a generation ago. The speed with which an order is filled directly affects pricing. Delivery can be overnight in many cases, if one is willing to pay the premium.

Information technology long ago integrated the factory. In the last decade or so, it has brought the customer closer to the factory floor. Decisions made by the customer, in a showroom or from an office, vehicle, or home information system, are directly transmitted to the machines that will fill the order. The customer sometimes chooses not to choose. As a result, customization specialists or expert systems that make decisions for the customer, are a booming business today.

Today's manufacturing economy accounts for 20% of gross domestic product (GDP), up from 18% in the 1990s. The economy then was only half of today's size. Safety is so ingrained today that it is practically taken for granted. Most of the manufacturing technologies used today were seminal 30 years ago. Although they are not new, strategies and applications for using them are. Until 15 years ago, manufacturing technologies were more advanced than the strategies for using them. This slowed productivity growth. Steady progress over decades, rather than a sudden quantum leap, eventually led to productivity growth rates approaching 5% per year in U.S. manufacturing over the last decade.

Comparing Mass and Custom Production

1990 Mass	2025 Custom
• weak customization: limited choices from a prefabricated list	• strong customization: customer designs
• economies of scale	• economies of scope
• just-in-time delivery: focus on inventory reduction	• just-when-needed delivery: focus on rapid turnaround for customer
• specialized workers	• multiskilled workers
• slow turnover	• rapid changeover
• hierarchies	• networks and teams
• vertical integration	• alliances and virtual organizations
• landfill and incinerate waste	• recycling, reclamation, and remanufacturing
• separate departments and incompatible information systems	• enterprise integration
• focus on technology	• focus on the organization
• price competitiveness	• quality competitiveness
• labor and machines compete for work	• people and machines work together

---

Improved productivity, however attained, is the universal goal. Manufacturers do not overlook opportunities to further automate or to bring in AI, but replacing people with machines is not a general priority. Overzealous attempts to create dark factories to manufacture everything foundered in the 2000s. Although the percentage of manufactured goods from truly dark, unattended factories has never risen above 5%, many analysts still consider them the benchmark for the future.

Disenchantment with the results obtained in trying to automate everything turned the tide toward an approach of people and machines working together. The important lesson from trying to eliminate people from the factory was that people and machines working together are a powerful, flexible combination better than either alone. Of course, people are more highly trained today. They must run CAD/CAM (computer aided design/manufacturing) and CIM (computer integrated manufacturing) systems, and work with robotics, expert systems, and other forms of artificial intelligence. They must also be computer and network literate, have detailed process and product knowledge, and constantly learn and retrain.

The key themes in manufacturing over the past 30 years have been:

- a shift from mass to custom production (the box above compares mass production in the 1990s and custom production today)

- a changing world division of labor for producing universal and culture-specific products

- improved manufacturing productivity through the integration of management and organizational strategies with technologies

The road to custom production

Custom production had its roots in the 1950s. An early variation was weak customization in which companies offered options or services with standard mass-produced products. The big advance, strong customization, allows deeper changes in the product design. Software programs, for example, could be tailored to suit the user. Strong customization, in which the customer is involved in design, was standard procedure for most manufacturers by 2010. Since then, AI has been the primary technological advance. Expert system customization assistance is instrumental in weak and strong customization.

Many goods are still mass-produced. Mass production is highly automated in World 1 nations and more labor intensive in World 2 and 3 nations. For commodity products, customizing does not make sense. Sugar, toilet paper, and computer casings, for example, as highly standardized products compete on price, because customers can assume quality. Customization offers little advantage for commodity products. Closeness to markets is the key to where mass production is done, because it reduces shipping costs.

Economies of scope complement economies of scale. Today's manufacturers strive for the reduced costs of economies of scale and the differentiation of products from economies of scope. Production is to order, not to stock and sell. There are generally fewer machines in today's custom factories than in old mass-production ones, because each can do multiple functions and work seamlessly with each other and with people. The box below shows the percentage breakdown and gives examples of mass and custom production in the United States.

Mass and Custom Production, United States, 2025*

U.S. Manufacturing*

Mass production (economies of scale)		Custom production (economies of scope)	
29%		71%	
Labor intensive 2%	Highly automated 27%	Person-machine partnerships 71%	
		Weak customization 43%	Strong customization 28%
no customization			
primarily craft production	fabrics sawmills	paper	pharmeceuticals
(labor-intensive mass production has migrated to middle and destitute nations)	bulk chemicals leather tanning modular shipping containers meat/dairy products	tires glass computer casings engines carpets	specialty chemicals vehicles scientific instruments garments computer chips

*91% of products are marketed universally, of which 31% are exported. The other 9% of products are culture-specific, that is, they are made strictly for domestic markets, e.g. craft products.

The ancestors of custom production went under many different names, such as "lean," "flexible," and "agile" production or "mass customization." Each built on or complemented the strengths and weaknesses of the other. Custom production today is an amalgam of these strategies. By 2010, it has raised expectations for diverse and tailored products beyond the capabilities of mass production.

High-skill, high-wage labor outcompetes low-skill, low-wage labor

Manufacturing with low-cost labor as its competitive edge consistently lost to automated custom production in direct competition over the last two decades. The case was not so clear at the turn of the century. Many manufacturing companies had come to rely on outsourcing parts of their operations to take advantage of cheap labor.

As far back as the 1990s, labor cost was a decreasing portion of total production cost in the United States—in industries like electronics, it had dropped below 10%. It turned out that greater cost savings and higher profits came from proximity to markets and improved just-in-time performance. Proximity to raw materials is no longer as compelling as in the past, because most materials supplied to manufacturers are today synthetic.

Up to 2010, most companies were inching toward dark, peopleless factories or outsourcing, which resulted in the elimination of jobs. Other manufacturers kept most workers and relied on training them to run flexible, automated operations and did not outsource significantly. The latter were criticized because they were not making the expected productivity gains. The payoffs, however, proved to be a matter of plodding along the learning curve. Once companies and workers became experienced with the flexible automation central to custom production, productivity climbed. Temporaries supplement a professional core of workers for particular tasks or for cyclical or seasonal variations. Today, these firms are far ahead of competitors who did not make the necessary investments in technology and training.

The education level of a country's workforce affects the type of automation it is likely to use. An educated, skilled workforce that can operate complex machines and systems, favors automation that challenges workers to continuous improvement through experimentation and feedback. Workers and AI programs interact to optimize production. They have discretion on how much to produce, how many runs to perform, and other scheduling issues. Workers make decisions on the spot.

Where educated and skilled labor is not readily available, then automating people out of the equation, or dumbing-down jobs, has been more attractive. Then, machines make the important decisions and demand little of the operators.

Universal and culture-specific products

Manufactures fall into two broad categories of products: (1) universal or global and (2) culture-specific or regional and national. Universal products are those exported across the globe, perhaps best illustrated by the personal communicator and flat screens. Others are most pharmaceuticals, virtual reality games and equipment, specialty chemicals, machinery and machine tools, motor vehicles, most biotechnology products, foods, clothing, and furnishings. Culture-specific products are those that are manufactured for home or regional consumption, often in protected domestic markets. They are often variations on universal products that are marketed only to a nation or region, due to limited cultural appeal, or because the manufacturers are protected from foreign competition. Clothing and accessories particular to Islam, for example, are manufactured locally in Iran, the Sudan, and Indonesia. The Brazilian automobile industry makes vehicles for Brazilians and the South American market.

The map below shows the distribution of universal and culture-specific products across the globe. The division between the two is often blurry. For example, in the Bangalore region in India, there is custom production of universal products (flat panel displays) side-by-side with labor-intensive mass production of culture-specific products (saris).

Where Things Are Made

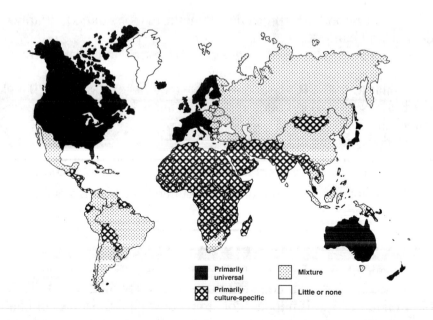

Primarily universal

Primarily culture-specific

Mixture

Little or none

331

Custom production with high-skill, high-wage labor working with advanced automation and AI-based equipment dominates universal product markets. Universal products are marketed globally. Although they are sold in middle and destitute nations, they do not compete directly with domestic manufacturing industries. Zentek, for example, sells its personal communicators in India, but Whirlpool does not sell its refrigerators there, because they are reserved for Indian manufacturers as part of the Bangkok Accords of 2016 that formalized limited forms of trade protection.

World 2 and World 3 nations manufacture over 80% of culture-specific products, which do not compete against universal products of World 1. Culture-specific product manufacturers sell to domestic and regional markets almost exclusively. India, for example, has developed strong agricultural machinery and textile manufacturing selling primarily to countries in southern Asia.

Knowledge is the "value-added"

A key distinction between products is their value-added, or knowledge value, component. Universal products are typically high value-added products. Brands are an example of knowledge-value. Products made of identical material are often priced far differently, if one is made by a firm with an outstanding reputation like Applebaum and another by an unknown. One pays more for the knowledge and reputation of Applebaum. Customers rely on a brand to deliver on its good name.

There was intense competition between affluent and middle nation manufacturers from the 1990s through about 2010. It destroyed or badly damaged many World 2 domestic industries. Political leaders in middle and destitute nations responded by protecting domestic manufacturers primarily with trade and tariff barriers.

Competitors subsequently agreed to an informal truce, later formalized by the Bangkok Accords of 2016. The agreement set up an institutional mechanism, similar to the WTO last century, for trade negotiations. The Bangkok Accords focus on making trade restrictions fair, while GATT's goal was to eliminate them.

Markets are not altogether closed. World 1 sales to World 2 countries like China are still enormous in areas not competing with domestic industries. The outcome is that affluent and upper middle nations concentrate on universal products, and the lower middle and destitute nations concentrate on culture-specific ones.

Improved manufacturing productivity

Advanced manufacturing technology does not improve productivity unless accompanied by a coherent strategy and operated by a skilled, capable workforce. A truism today, confirmed by countless studies of nations' manufacturing competitiveness, is that the best technology does not necessarily win. Some companies, for example, Nummi in the 1980s and 1990s, gain competitive advantage from teamwork, without the latest technology. But there are no cases of companies building a sustainable competitive edge around the best technology alone, although many have tried and failed.

Successful manufactures typically encourage feedback, experimentation, and self-correction in the workforce. Highly skilled employees work in tandem with machines. Hardware and software designers have made great progress in improving the person-machine interface. Generic, off-the-shelf automation software, for example, has usurped more expensive, less compatible, and user-unfriendly custom predecessors. The integration of people and technology in the factory is now accepted as essential to improving productivity.

The figure gives examples of how people and machines work together.

How People and Technology Work Together in Manufacturing

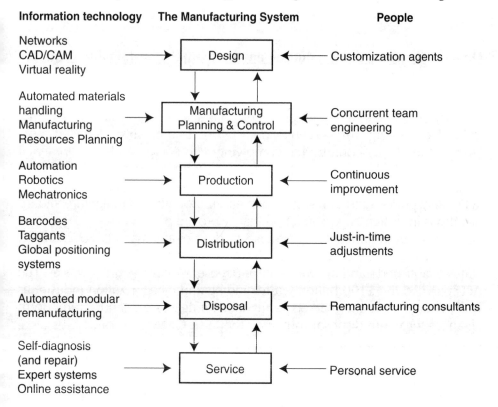

Information technology	The Manufacturing System	People
Networks CAD/CAM Virtual reality	Design	Customization agents
Automated materials handling Manufacturing Resources Planning	Manufacturing Planning & Control	Concurrent team engineering
Automation Robotics Mechatronics	Production	Continuous improvement
Barcodes Taggants Global positioning systems	Distribution	Just-in-time adjustments
Automated modular remanufacturing	Disposal	Remanufacturing consultants
Self-diagnosis (and repair) Expert systems Online assistance	Service	Personal service

The road ahead

Many manufacturing analysts today are projecting a future with all-purpose factories that make many similar products, for example, specializing in wood, automobile, appliance, or machine tool producers. They would draw on stored product specifications and a range of raw materials. Modifications would be simple. Future factories might be designated as large-, medium-, and small-product manufacturing centers, depending on the materials and the facility.

Key forces affecting manufacturing over the last 35 years

Six forces have been primarily responsible for reshaping global manufacturing. They are:

- doing well while doing good, as environmentally sustainable practices improve productivity
- trade blocs slowing progress to a single global market
- cultivating brain power as a valuable capital asset
- miniaturizing products and making them smarter
- biologizing manufacturing
- upgrading the transportation and information infrastructures

Doing well while doing good as environmentally sustainable practices improve productivity

Manufacturing was the first sector of the economy to improve productivity by adopting sustainable practices. Manufacturers found, for example, that reducing energy use—a central objective of sustainability—reduced costs. Studies continually confirm that complying with energy and environmental regulations raises business costs in the short term but reduces them in the longer term. Sustainability-driven changes in national accounts, which more accurately priced natural resource depletion, propelled manufacturers to innovations in design and production that improved productivity.

Innovation sometimes creates new industries and markets. For example, the passage of the Recycling and Reclamation Act of 2013 in the United States established standards for resale and reuse of manufacturing wastes. The Act spawned the $100 billion waste grading and categorization industry as well as burgeoning aftermarkets for manufacturing waste. Manufacturers form alliances for waste disposal routinely. They use GMAN to solicit bids electronically, depicted in the box below.

On-line Auctions for...Waste?

Welcome, Lee Federman, of Zarmex Chemicals.

What can we do for you?
— On-line auction

Sure. What are you offering?
— Waste chemicals

What kind and how much?
— 20,000 kilos of NH_3; 5,000 kilos of HCl

What companies should we alert?
— 3Rs companies within a 300-kilometer radius

What is the closing date and time?
— Noon tomorrow.

Is there anything else we can do for you?
— No, signoff.

The Clean Air Act Amendments in the 1990s and the Sustainability Acts in the 2000s compelled manufacturers to make more durable products. As a result, the average age of people's automobiles has almost tripled from seven years in the 1990s to over 20 years today. Automakers recondition or remanufacture their products every five to seven years. Most people in World 1 do not keep vehicles more than five years. Their remanufactured vehicles are resold in secondary markets, usually in lower middle and destitute nations.

The end of once-through products was forecast in the 1990s. Companies with foresight capabilities were prepared. They reconsidered product disposal, which required rethinking and reconfiguration of design and production. Factoring in disassembly changed assembly.

Today's manufacturing companies develop an in-house remanufacturing capability or form alliances with one of the thousands of 3Rs companies that have sprung up over the last 20 years. Reconditioning and remanufacturing operations are profitable. In-house remanufacturing capabilities help replace lost income brought about by more durable, longer-lived products.

Trade blocs slowing progress to a single global market

The three large trade blocs—North America (led by the United States), the EC (led by Germany), and east Asia (led by Japan), are smaller-scale previews of a single global market. The blocs have always been weak. They have offered trade advantages to member countries, but do not exclude imports from outside the blocs. Trade among blocs has been brisk.

The smaller blocs or regional groupings, such as South America, the Organization of African States, and the India-led south Asia bloc, have been more restrictive. They maintain special trade benefits and protection for manufacturers of culture-specific products. For instance, domestic and regional industries that would otherwise have been battered by foreign competition have been kept in business by import restrictions and domestic content requirements.

There are few differences in manufacturing practices among the three large blocs today. At the turn of the century, global benchmarks were based on Japanese practices, such as quality, KAIZEN®, just-in-time, and mechatronics. In the last decade, U.S. AI-based manufacturing techniques have set the global standard.

The U.S. Retooling Manufacturing Act of 2001 formalized industrial policy and coordinated and increased government assistance to manufacturing. A key provision was the revision of the Sherman Antitrust Act to enable greater industry cooperation on precompetitive research.

The emergence of trade blocs of roughly equal economic strength opened standards setting to negotiation. U.S. companies no longer set de facto product standards as they did last century when they were the dominant economic players. Today, each bloc can set standards within its bloc if pressed. This still happens when trade disputes get ugly. More commonly, however, the blocs agree to international standards, such as the metric system, EDI, and the Machine Tools Protocols of 2009.

Cultivating brain power as a valuable capital asset

The manufacturer's brain power is its most closely cultivated and guarded asset today. Manufacturers in affluent nations are currently spending over 50% more on R&D than equipment, reversing the mass-production era relationship. R&D spending has outpaced capital spending for the last 20 years in most companies—even longer in leading-edge ones. For example, pioneering Japanese firms were spending an average of 30% more on R&D than on capital equipment back in the 1990s.

Flexible manufacturing ultimately reduces capital spending. Existing equipment easily adapts to new processes or products. Workers rapidly reconfigure machines for new products. Modular components enable customization without requiring new machinery. They also make remanufacturing easier. Old or defective modules are easily slipped out and replaced or reconditioned.

The terms "thinking" or "learning" organizations characterize the shift in importance from physical to intellectual capital. Today's leading companies enlist their knowledge assets (roughly the sum of the technology and people assets) to support the company's strategic plan. They assess skill needs and offer incentives to employees for pursuing them.

Skill and knowledge requirements run across the custom production workforce. Management empowers workers to make decisions on the spot. For example, a customer may change an order in midproduction. The worker must instantly decide whether to stop the run, how much it would cost, what alternatives are available, etc. To make decisions like this, the worker must understand the company's culture, its commitment to customers, the manufacturing process, and some basic economics. Underlying these requirements are computer literacy, systems science, information searching and application, statistical quality control skills, teamwork, and decisionmaking skills.

There is no room for unskilled workers. Although they are willing to accept low wages, they cannot work successfully in today's factories. The EC, especially Germany, pioneered vocational education for those not going to college. Rising skill requirements have challenged education systems to pay more attention to the bottom half of the class. The United States, for example, routinely produced top-notch college graduates but failed to adequately reach those who were not college bound. Productivity gains were not made in the United States until the skills of the "bottom half" of manufacturing teams were improved, beginning about 2005.

Miniaturizing products and making them smarter

Materials are the building blocks of manufacturing. The materials revolution has direct impact on what and how things are manufactured. Ceramic engines, for example, are made differently from their steel and aluminum predecessors. Manufacturers today monitor the latest developments in materials science and technology to keep their factories state-of-the-art. For example, new diamond coatings for bearings, shafts, and other mechanical parts have improved machine precision and extended lifetimes, and dimensionally stable materials are improving die, blank, and mold performance. Continuous, real-time, nondestructive testing raises quality while reducing costs.

The unfulfilled promise of nanotechnology....so far

Nanotechnology would involve manufacturing by manipulating atoms or molecules directly or through chemical or biological means. It would use enzymes, proteins, and imitations of biochemical processes for customized materials designed for highly specific functions, such as biological computer chips, molecular switches (which are being used today), and molecular magnets. The goal is to build devices from the bottom up, atom by atom. This would theoretically make products cheaper, stronger, lighter, more efficient and more reliable.

Nanotechnology has not had the success its proponents have been predicting. One breakthrough was designing a protein molecule from scratch and then adding nanomotor assemblies to create protein machines. Manufacturing applications have not been forthcoming—scale-up is so far impossible.

Proponents still hope to build molecule-size assemblers and replicators for thousands of specialized nanomachines, with applications from the microscopic—tiny robots to process waste— to the macroscopic—skyscrapers.

Materials have been moving from natural to synthetic to custom-made. Some materials, like photochromatic glass, have smart properties in their structure. Others have intelligence embedded in them with microprocessors.

Miniaturization has been going strong since the 1980s and 1990s and is culminating in micromachines. Micromachines are etched out of silicon much as computer circuits once were. They are fitted with valves, gears, and motors for fine-tuning robot control, as artificial nervous systems interfacing with digital computers—such as in planes with flexible wings. Their most practical applications have been as sensors. Many experts feel they will eventually make good interfaces between nanomachines and macroscopic devices.

Biologizing manufacturing

Biomimetics and biotechnology are growing influences on manufacturing. In biomimetics, manufacturers apply the lessons of natural structures to new processes and producst. Biotechnology is making inroads in the manufacturing of pharmaceuticals, chemicals, agriculture, lumber and paper, food and beverages, and sensors. Synthetic and biosynthetic molecules carry out chemical reactions, probe cell functions, and manipulate DNA. Biosensors use cell, proteins, and other genetic engineering products for process analysis in food, beverages, and pharmaceutical manufacturing. Biotechnology applications are improving process control, biocatalysis, and recycling and waste treatment in chemical manufacturing.

Links between biotechnology and manufacturing technologies are growing. For example, biotechnicians are directing engineering of most naturally occurring genes, molecules, cells, and organisms, while nanotechnology researchers are working on building molecular-scale manufacturing devices. A collaborative project for biophotonic research was recently announced. Its goal is an alternative to conventional semiconductors.

Upgrading the transportation and information infrastructures

Transportation and information infrastructures are the backbone of manufacturing. Reduced delivery times have come primarily from improved logistics by information technology rather than faster trucks, trains, ships, or airplanes.

Just-when-needed delivery has upgraded the old just-in-time requirements by continually reducing the time from order to delivery. Automobiles, for example, are now routinely delivered within three to four days of the electronic order in most of the United States. Distribution systems are alerted when an order is placed and are continuously updated on its progress. When a product is ready for shipment, the fastest means for delivering it have already been arranged.

Suppliers, customers, regulators, or anyone who needs information about products in transit can have it instantly. Manufacturers use the global network of computer networks for linking factory operations and distribution worldwide. Today's EDI systems not only exchange raw data between machines, but also use artificial intelligence to make decisions based on the data. GMAN, the global manufacturing on-line network of databases, links FAN and similar networks worldwide. Manufacturing companies seeking alliances query GMAN, which identifies suitable partners anywhere in the world.

WORLD 1—Convergence on best practices

Different manufacturing systems and approaches fought it out over the last 30 years and custom production emerged the winner. Its technologies and strategies are benchmarked and are widespread. Gathering, manipulating, storing, and communicating information characterized custom production in affluent nations.

World 1 nations and their bloc partners dominate export markets with universal products. These are typically complex, high value-added products, such as custom autos, appliances, airplanes, or virtual reality equipment. Embedded intelligence is making them more complex and more capable.

Manufacturing continues to make up a significant share of GDP in affluent nations: 24% in Japan, 27% in Germany, and 20% in the United States The figures are little changed from the 1990s, although calculating them is more difficult today. Proliferating alliances and virtual organizations sometimes make it difficult to distinguish a company's or product's national origin. Japanese firms, for example, have been making an average of 31% of the cars made in North America over the last 27 years—including a high of 38% in 2004 and today's low of 27%. Balancing national accounts is tricky in cases where perhaps 29 companies, representing a dozen nations, are involved in a manufacturing project. Figuring out national origins is also important for jobs, taxes, and profit distribution. The tax people have the information, but its significance to the consumer and to marketing is confused. "Buy American" campaigns, for example, are difficult when product origins are unclear.

A CASE STUDY—U.S. manufacturing catches up by linking up

Manufacturing was analyzed incessantly in the United States in the 1980s and 1990s as foreign competitors took over many industries long dominated by domestic firms. Many remedies were proposed and tried: reengineering, value engineering, benchmarking, anthropocentric production, lean production, and agile production all had their day. The common denominator was to reexamine how manufacturing should be done.

The 1989 classic Massachusetts Institute of Technology (MIT) study, *Made in America*, identified six factors at the root of U.S. industrial productivity problems in the 1990s:

- outdated strategies
- short-time horizons

- technological weaknesses in development and production
- neglect of human resources
- failures of cooperation
- government and industry at cross-purposes

Value engineering lives on in practice, not name

Value engineering is a 1990s term describing the search for alternatives to accepted ways of doing things—evaluating systems, equipment, facilities, services, and supplies for lowest cost at required performance, reliability, quality, and safety standards. The term value engineering is no longer used today, but the principle of re-examining operations to reduce costs and maintain quality is firmly incorporated in custom production today.

Other studies reached similar conclusions. Together, they stimulated the formation of a U.S. industrial policy that led to concrete actions, such as the Waste Reduction Act of 1999 and the Retooling Manufacturing Act of 2001. Government and industry cooperation has also led to information- and strategy-sharing mechanisms, revised investment and tax laws to encourage longer-term perspectives, apprenticeship programs for training students who are not college bound, and relaxed antitrust laws to ease R&D cooperation.

Manufacturing in the knowledge economy

A decline in manufacturing's percentage of GDP (from a peak of 30% in 1953 to a low of 18.7% in 1989) stabilized by the turn of the century. It has increased slightly to 20% today. Manufacturing employment, however, continued its free fall— from 30% of nonfarm jobs in 1960, to 17% in 1990, to less than 8% today. Knowledge-based industries have not replaced manufacturing, but have complemented it. The United States makes things as well as producing information and services.

Chemical manufacturing today

The chemical industry has had more problems with its image than its manufacturing. Improved science education, however, has largely overcome the chemophobia and NIMBY attitude that hampered the industry and led to many legal battles. The industry launched a successful public relations and education campaign to demonstrate how the industry had gotten in step with sustainability. Zero discharge processing, for example, has been a public relations boon.

In manufacturing, modular processing equipment and robotic materials-handling systems eliminate many pipelines. Reactor vessels are moved to and from modular processing stations by robots. Pipes are typically the weak link in safety. Replacing them and their propensity to leak eased public fears and reduced maintenance costs as well. The 3Rs have proved cost-effective to chemical manufacturers once initial costs were recouped. What was once waste material is now raw material for another process. GMAN will find someone, somewhere, for just about any material.

Biotechnology brings fermentation, biological feedstock conversion, enzyme immobilization, and cell culture technologies to chemical processing. Biotechnology-based specialty chemicals now account for 25% of the specialty chemicals market. Only the pharmaceutical and food industries have been affected more strongly by biotechnology.

The gap between the value of manufactured goods and knowledge and service industries has closed steadily. Last century, manufacturing produced 20 times the volume of service exports. Today the two are roughly equal. About 15% of the service-sector workforce is involved in manufacturing. Design and engineering services, financial services, insurance, training, waste disposal and remanufacturing, transportation, and communications all have direct ties to manufacturing.

The net job loss from automation in manufacturing has been fewer than many experts forecast. Factory jobs were lost and there are fewer today than 30 years ago. At the same time, jobs have been created in activities associated with manufacturing. Ancillary sectors, such as software design engineers, machine vision experts, and information service providers have prospered in part due to manufacturing.

Integrated fashion packages

The once separate, and often antagonistic, textile and apparel industries are allied in integrated performance systems today. A single organization, typically a large retailer, integrates all aspects of clothing production, distribution, servicing, and disposal. In addition, image and style consulting, wardrobe management, even hair, skin, and fitness services are offered. One-stop shopping is available for the customer who wants it. Virtual organizations are a common mechanism for delivering the integrated performance. The customer deals only with the retailer.

Automation continues to advance in the factory. Composite layup technology enables seamless garments, which reduces the need for hand work. Flexible manufacturing technologies enable choices in form, texture, color, and finish. Tailoring is automated as well. Retailers in affluent nations no longer sell off-the-rack clothing. Shopping can be done on-line or one can visit a showroom and simulate thousands of combinations. When a choice is made, the customer's measurements are scanned into the database, the order is queud up, and delivery is made in days.

Retraining programs did not save the jobs of many less-skilled workers. Retraining implies preparation for a new job. Yet the new jobs available required more skills and abilities than a three- or six-month retraining program could provide. Jobs went to multiskilled candidates with undergraduate or even graduate degrees. Although some displaced workers went back to school for retraining, far more workers simply left the manufacturing sector.

Manufacturing industries still account for 53% of all R&D, down from a high of over 75% last century. The decline reflects the explosion of R&D spending by the information industry. But even half of information industry R&D is for manufacturing-related hardware. There has, however, been a steady increase in R&D funding from services.

Industrial policies and subsequent legislation (such as the U.S. Retooling Manufacturing Act of 2001, which eased antitrust laws) have strengthened cooperative research, especially in areas that policymakers designate as important to the economy, such as software, aerospace, and pharmaceuticals. Financial incentives bring suppliers, customers, and producers together for R&D that would otherwise not get done, to stimulate sharing, and to eliminate redundancy.

New additions to manufacturing

There are new products and new product categories. Information-based products have flourished, such as HDTV, the countless variations of personal communicators, virtual reality systems, and other computer and telecommunications equipment.

Manufactured housing becomes the norm in the United States

Factory-manufactured housing controls over 90% of the new-home construction market in the United States today. Site-built homes are almost exclusively for the very wealthy. Quality gains enabled by manufacturing components of the home in the factory and assembling them on-site are too much for site-built builders to match. Factory homes enable the customer to a high degree of input. Either at the factory, or on-line through the home information system, home buyers can specify what they want—as long as they are willing to pay for the customization. Virtual reality and simulation capabilities enable prospective buyers to try before they buy.

Space manufacturing—manufacturing in microgravity—is a new, albeit tiny category. Five space factory modules are in low Earth orbit. They are almost totally automated. Earth-based teleoperation supplements the automation, such as for docking vehicles to the modules. The modules make products requiring the high precision and quality environment provided by microgravity. Novel mixing and combinations of materials are current areas of prominent activity. For the future, the race is on to build the first factory in space to process minerals mined on the Moon.

SIC categories that are considered part of manufacturing today have changed over the last 35 years. Traditional categories subsume some new manufactures. Energy systems, including photovoltaic cells, fuel cells, batteries, etc., are still grouped within utilities. Manufactured housing, robots and automation, biomanufacturing, reclamation and remanufacturing, and space manufacturing are now separate categories within manufacturing. The graph below shows their growth.

New Additions to Manufacturing

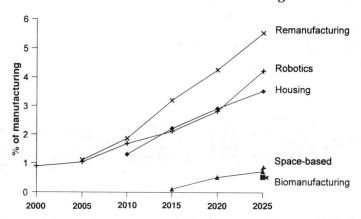

Government statisticians have shifted some industries out of manufacturing. Printing and publishing, for example, are now in the information sector, since so many of the operations are done on-line. Large-scale production runs on printing presses, once standard operating procedure, account for less than 30% of production today. In many cases, printing is done on-site. For example, instead of purchasing a newspaper or magazine as in the past, readers, or their expert system assistants, scan a table of contents on line and print only the articles they want to read. This has reduced large production runs, but not replaced paper use.

The table below compares manufacturing's share of U.S. GDP, along with other sectors of the economy, in 1990 and today.

U.S. Manufacturing's Share of GDP, 1990, 2025

Industry	% of GDP (1990)	% of GDP (2025)
Agriculture, forestry, and fisheries	2.2	1.9
Mining	1.5	0.7
Construction	4.8	3.2
Manufacturing	18.7	20
Durable goods	10.5	12.2
Nondurable goods	8.2	7.8
Transportation and public utilities	8.9	10.3

Wholesale trade	6.6	2.6
Retail trade	9.4	10.3
Finance, insurance, and real estate	17.4	14.7
Services	18.8	19.8
Government and government enterprises	12.0	16.5

The table below compares categories within manufacturing in 1990 and today.

Breakdown of U.S. Manufacturing Industries, 1990, 2025

Industry	% of manufacturing	
	1990	2025
Durable goods	56	70.3
Lumber and wood products	3.3	1.8
Furniture and fixtures	1.7	0.9
Stone, clay, and glass products	2.7	0.7
Primary metal industries	4.6	2.7
Fabricated metal products	7.0	4.2
Machinery, except electrical	10.0	12.3
Electric and electronic equipment	10.0	16.1
Motor vehicles and equipment	5.2	7.2
Instruments and related products	3.2	8.0
Other	8.3	16.4*
Nondurable goods	44	29.7
Food and kindred products	8.4	10.7
Tobacco manufactures	1.7	0.2
Textile mill products	2.2	1.1
Apparel and other textile products	2.6	0.8
Paper and allied products	4.9	2.4
Printing and publishing	6.8	12**
Chemicals and allied products	10.2	8.2
Petroleum and coal products	3.5	1.8

| Rubber and misc. plastic products | 3.2 | 4.2 |
| Leather and leather products | 0.1 | 0.3 |

*"Other" includes new additions to manufacturing

**Printing and publishing is part of the information sector. The figure is included for comparison.

Integration within the factory

A primary objective of custom production is integrating operations and improving the coordination between workers and technologies. Information technology is the principal integrating and coordinating tool: from design, to marketing, to production, to final sale, to maintenance, to disposal. Information technology improves the links of people to people, people to machines, and machines to machines.

Management and workers use decision support systems to quickly form teams for a particular task. Team members may never physically meet. E-mail, desktop vidoeconferencing, and groupware make it possible to work remotely. When the task is completed, new teams are formed for the next task.

Linking technology is often more difficult than teaming workers. Getting multiple generations of technologies—single-purpose automation, numerical control and reprogrammable machine tools, robotics, automated materials handling, group technology, CAD/CAM, CIM, and AI—to work together seamlessly continues to challenge even the best systems integrators.

Teams are up to the tasks

Cross-functional and cross-task teams consist of core and contingent workers. Well-trained and well-paid elite workers are the core of work teams. They are committed to the organization and understand it intimately. They possess the knowledge that distinguishes one organization from another. Contingent employees— part-time, contract, and temporary workers—supplement the core workers to meet seasonal, cyclical, or special project needs. They range from manual laborers to technicians to highly skilled programmers.

The corporate culture of today's team-based factories is far removed from the typical mass-production predecessor. Manufacturers have finally decided that there is no longer any value to be gained from the time and motion scientific management that once characterized mass production. In most factories today, workers are expected to contribute to continuous improvement. Hierarchical organization charts are practically gone. Networks of people with different ranks replace them. Power within the factory is measured more in terms of what people know, rather than their formal position.

Suppliers and subcontractors are also integral to production teams. Many large manufacturers give long-term contracts to key suppliers, so that they can make long-term investments in people and technology. Less important supplier arrangements are typically virtual ones, developed as needed through the GMAN network.

Job losses fewer than anticipated

A gradual social change over the last 30 years has been the rise of person-machine partnerships. Today, the partnership creates new jobs, eliminates others, and continues to reduce the average manufacturing workweek — from 40 hours in 1990 to 32 today.

Robots and automation, properly applied, reduce manual labor and boost productivity. Machines do undesirable, risky jobs. They also do mundane tasks, freeing people to spend more time on higher-level thinking tasks. For instance, CAD programs long ago eliminated the mechanics of drawing and freed designers to test ideas. Artificial intelligence has pushed up the threshold of tasks that machines can do. AI machines challenge people to ever higher levels of thinking. Although there has been a net loss of jobs from automation, societies have created new ones. Manufacturing has followed the agricultural model, in which fewer people are needed to produce a unit of output. Agriculture and manufacturing have maintained their relative shares of GDP over the years, but require fewer workers to do so.

Information technology reshapes the pharmaceutical laboratory

Information technology makes the pharmaceutical laboratory primarily an information processing center that incidentally makes pharmaceuticals. Computer-assisted design, modeling, and simulation, including virtual reality, combine to construct molecules, manipulate sites of biochemical action and track biochemical pathways. These technologies reduce time to market by enabling rapid exploration of alternatives.

Expert system assistants make specialized expertise readily available. Teleconferencing with colleagues and electronic links to databases and networks makes further skills accessible.

Information technology also opens the laboratory, excepting proprietary information, to citizens, physicians, public interest groups, and government regulators. Incorporating feedback from these stakeholders early reduces the number and intensity of problems down the road.

The notion of factory work itself is changing. The 9-to-5 or 8-hour shift, 40-hour workweek, which once seemed sacred to manufacturing, has given way to flexitime, telecommuting, and annual rather than weekly work schedules. There has been a net reduction in the number of hours worked per person in the factory. Many core workers, however, are still working more than 40 hours. Working today meets self-actualization needs by continually providing creative and intellectual challenges.

Design involves the whole team

Design and production are carried out concurrently by AI-assisted, CAD/CAM technologies. Production factors shape design choices by eliminating iterations between design and production. Instead of creating an optimal design and then figuring out how make it, designers test each design with a production and end-use simulation.

Desktop manufacturing in today's state-of-the-art facilities bypasses the factory completely for small product runs. Engineers use virtual reality to test prototypes and make them. More common is rapid prototyping, in which computer peripherals use stereolithography and laser sintering to set plastics and produce small-scale, three-dimensional models rather than finished products.

Virtual reality equipment is succeeding 3-D CAD/CAM as the preferred design tool. Virtual reality is a routine simulation tool for all kinds of physical planning and product design.

Bringing people and machines together

Today the whole factory is linked. People and their machines easily communicate with one another, inside and outside the organization. Compatibility problems have virtually disappeared since around 2010. Productivity gains of 2% to 5% a year since then are common. Even these figures may be understated, because the indirect benefits from reorganizing operations are not always attributed properly.

Mechatronics: linking machines and electronics

An important step in linking the factory was integrating assembly through mechatronics. Mechatronics embeds microprocessors and associated sensors in devices to enable them to respond to their external or internal environment. It enables small-batch production of diverse products in the factory by improving coordination between machines.

Mechatronics also describes the Japanese practice of using integrated teams of product designers, manufacturing, purchasing, and marketing personnel to design the product and the manufacturing system.

Mechatronic products have been around for a long time. Automobiles, for example, have long been arrayed with microprocessors connected to mechanical parts.

Last century, group technology coordinated machines and robots in small groups or cells. Then, flexible manufacturing systems (FMSs) built on group technology by linking cells together. This made smaller production runs economical by reducing changeover times. Computer integrated manufacturing was the next big step toward custom production. It linked FMSs to the rest of the enterprise—marketing, finance, human resources, etc. Connecting the enterprise electronically improves input and feedback. Human resources, for example, can alert production that the people they will need for a proposed project will not be available, enabling production to devise an alternate strategy.

Improving integration and feedback enabled a greater focus on process technologies, which are acknowledged today as crucial to custom production. In the days of mass production, however, it was product technologies that received the most attention. Productivity studies in the 1980s and 1990s found that the United States consistently lagged foreign competitors in adopting new process technologies.

The shift to emphasizing process technologies gained strength around the turn of the century and had profound implications for training and education. Operating complex process technologies requires that even "ordinary" workers be highly computer literate, understand systems science, possess statistical quality control skills, and most important, quickly learn to use new technologies. This dictates an extensive commitment to training and has pushed manufacturers to demand educational excellence—particularly in the area of vocational and technical education for those not college bound. Improving workers' education and skills is integral to manufacturing success.

Robots and AI: The new frontier

Artificial intelligence and robots fine-tune custom production. Intelligent machines or robots, simply defined as those performing functions that require intelligence when done by people, incorporate AI. AI, however, is not restricted to robots. Expert systems enable machines, systems, and devices to mimic or occasionally surpass human learning. They consist of three parts: a database, a knowledge base of rules for drawing inferences, and a high-speed inference engine for applying the rules. Knowledge engineers study human experts in a particular field to develop the experts' usually subconscious rules for drawing inferences.

Dark factories

Dark, lights-out, or totally automated factories are typically parts of the larger manufacturing enterprise rather than the stand-alone entities forecast by enthusiasts last century. They are more prominent in automated mass-production than custom-production facilities.

Japan's labor shortage has pressed it to develop labor-saving robotic and automation devices. The resultant productivity gains drove interest in them across the globe.

Dark factories stimulate a rethinking of how to organize manufacturing. The most obvious advantage of operating without human labor, is that the factory environment need not be adjusted to satisfy people. Hence, lighting, temperature control, bathrooms, etc., are not needed.

Their mean time to failure is measured in years. Reaching this high level of reliability, first obtained in Japan's Fanuc plant (which appropriately is a robot maker), required a great deal of progress in self-diagnostic sensors.

Workers in dark factories program machine tools and robots, and do occasional physical labor, such as moving machines around the workplace.

Today's systems incorporate fuzzy logic, which provides a mathematical basis for making optimal use of uncertain information. Although fuzzy logic was invented in the United States, the Japanese pioneered fuzzy chips in systems and products like home appliances. Intelligent fuzzy products expand customer choices by sorting through mundane choices and leaving important ones for the customer. Fuzzy chips are frequently arranged in neural networks. Bringing the advanced algorithms of neural networks to fuzzy logic has speeded up the "learning" time of the chips. Home energy systems, for example, quickly find out which people are in which rooms and signal components to adjust temperature and lighting accordingly.

AI applications are complex. The person-machine interface issues have proved more demanding than early enthusiasts projected. It has only been in the last 10 or 15 years that AI has become integral to manufacturing. AI software developed incrementally and gradually. Each new generation provided greater subtlety and depth, but uncovered new limitations and issues as well.

Japan, Sweden, and Germany pioneered the adoption of robotics, although it too was invented in the United States. Robotics is particularly successful in countries with strong education systems, high standards of living and labor costs, and a stable or shrinking pool of skilled workers.

Three key advances over the last 30 years boosted robotics: improved mobility, improved vision and sensors, and greater onboard intelligence. Mobility has improved to the point that robots can travel without guide rails or embedded floor coils. Materials science advances enable today's robots to be much smaller, which in turn greatly aids mobility. Improved vision enables robots to compile 3-D constructs of their environment. Three-dimensional vision, combined with improved sensors, enables today's robots to easily solve bin-picking problems that confounded previous generations.

Improvements in mobility and vision are largely due to greater onboard AI in robots. Simple speech recognition and voice synthesis were early features brought by AI. Today's robots learn, even if at a simple level. For example, autonomous guided vehicles (AGVs) for automated materials handling learn the best routes around the factory. Memory stores every experience of the robot for meeting future challenges. Robots can do multiple tasks by voice command without being reprogrammed. Mechatronics coordinate groups of robots.

Closing the loop: integration outside the factory

Custom production completes the integration by reaching out to customers, suppliers, regulators, and competitors. As CIM integrated the factory internally, custom production integrates its externally. Product maintenance, servicing, and disposal are inseparable from planning today, although they were once only afterthoughts. Finding out exactly what the customer wants, complying with regulations, keeping tabs on competitors, and arranging alliances are routine activities in today's manufacturing firms.

Manufacturing enterprises continue to shrink as small size enables a sharper focus on the core business. Manufacturing firms employing under 20 people have grown from 66% in the 1990s to 76% today. It is not unusual for an industry leader to have fewer than 25 employees. The proliferation of virtual corporations allows small firms to tackle large projects calling for expertise outside the core business. Scale and location are less important today—information technology has eliminated most time and space barriers.

Virtual organizations as alliance tools

Virtual organizations are temporary networks of companies—including suppliers, competitors, or customers—that come together to make a particular product or meet a particular need, and then move to a new goal or disband. There are usually electronic arrangements that exist primarily on computer networks. Most participants need not meet face-to-face.

Virtual manufacturers share expertise, costs, and market access. They create temporary, flat organizations without central offices, hierarchies, or vertical integration. Transforming corporate culture to accommodate virtual arrangements has often been difficult. Cooperating with rivals requires levels of interpersonal skills and legal expertise that are still challenging manufacturers today.

Virtual manufacturers spawned virtual products. These products are not made, or readily available, until a customer needs them. They are not catalog items sitting on shelves. Manufacturers devise them when a customer need arises.

GMAN: linking manufacturers

At the turn of the century, manufacturers established FAN as a matchmaking or prequalified partnering service. The size, experience, and expertise of thousands of manufacturing companies, suppliers, and customers were made available on a computer network. FAN led to a proliferation of virtual manufacturing enterprises. GMAN incorporated FAN and extended its scope across the globe.

Companies enter their capabilities onto the network, for a fee, for others to browse. Legal issues, such as patent and copyright protection, were difficult issues largely resolved today—although there was some dissatisfaction along the way. The courts refused to act on intellectual property matters, referring them to Congress. The passage of the comprehensive Intellectual Property Act of 2014 provides a framework still used today to sort out ownership, licensing, and royalties issues.

EDI and groupware, or computer-supported cooperative work, are important capabilities supported by GMAN. EDI links the computers of suppliers and manufacturer's. Groupware enables collaborative work outside and within enterprises. Teams can work on projects simultaneously or in shifts around the clock.

Just-when-needed delivery

Custom production's just-when-needed distribution incorporated and improved on mass production's just-in-time by emphasizing customer needs. Just-in-time improvements concentrated on reducing inventories and costs and improving distribution between suppliers and producers, with reduced delivery times to the customer an added bonus. Just-when-needed modifies the goals to make rapid turnaround to the customer the top priority. The price of goods is directly related to how fast products are delivered.

Factory ships

Factory ships are the ultimate just-when-needed innovation. Raw materials or goods are processed aboard the ship enroute to market. They do fairly simple manufacturing due to the uncertainties of sea travel. The range of products is continually expanding: from early fish-processing vessels to other foods, lumber, paper, and even some petrochemicals. In the next decade, premanufacturing ships are expected to come on-line. These ships will perform simple manufacturing procedures on a wide range of products to prepare them for the factory and cut down the time spent there.

Fiber optics or satellites relay orders and designs in seconds. Computerized air, ship, rail, and vehicle traffic management systems have greatly improved distribution efficiency. Although raw speeds have improved, information management has done more to speed deliveries by reducing delays and bottlenecks. Manufacturers and customers can track goods precisely across the globe with GPSs. Bar codes and taggants provide access to in-depth information on specifications, warranties, and 3Rs data.

Shippers are alerted immediately when an order is received, usually by an EDI system. Included in the order is an estimate of when it will be ready. This gives the shipper time to make arrangements. The shipper, if a trucking company for example, can align rail or air connections if necessary. The shipper can also check in remotely with the factory and monitor the order's progress and make any necessary adjustments. Often today, logistics specialists arrange shipments, without getting involved in the shipping themselves, working much like commercial travel agents.

Closing the waste loop

In 2006, manufacturers in the United States were made legally responsible for disposal of items whose weight in kilos multiplied by their dollar value was greater than 2,500. This requirement, enacted by Germany five years earlier, forced a rethinking of operations. Accounting for disassembly has reshaped assembly. A new industry, remanufacturing, sprang up and complements recycling and reclamation operations.

Closing the waste loop has been a principal contribution of manufacturing to the nation's sustainability campaign. Provisions for disposal are now routinely built into design. Modular components allow easy upgrading. Although sales of most durable goods are down, a significant portion of the shortfall has been made up by remanufacturing capabilities. Customers bring their cars, computers, or refrigerators to manufacturers for servicing or an upgrade, for a fee, after a contracted or legally required time, depending on the law.

Today's competitive edge

Today's manufacturers offer strong customization and rapid response to customer needs. Customers can participate in design by virtual reality or standard simulation. In the weaker forms of customization characteristic of mass production, customers could pick options only from a prefabricated list. Customization creates high value-added products.

Custom production is capable of:

- wide economies of scope

- rapid changeover to meet new process/product demands, including significant changes in volumes

- KAIZEN®, or continuous improvement, in process and product technologies and strategies

- photonic links within the factory and to suppliers, distributors, competitors, and customers

- flexible, scalable production facilities

- nonhierarchical power distribution

Where innovation comes from: technology fusion

A striking strength of successful manufacturers today is their ability to bring together previously unlinked technologies to create new processes and products, dubbed "technology fusion" by Fumio Kodama last century. A classic example is the Japanese fusion of electronic technology to mechanical technology to create mechatronics, which in turn furthered the prospects for numerically-controlled machine tools and industrial robots.

A primary source of innovations over the last 30 years has been through combining the capabilities of different disciplines. It typically begins with company R&D branches investing in a product field outside its primary product line. The two industries engage in reciprocal research that may lead to a new field, such as the new ceramics that resulted from the fusion of the electrical industry and ceramics research.

A key asset of the manufacturing firm today is the ability to think and learn. Intellectual capital is considered more valuable than equipment capital. Far more money is invested in R&D than plant and equipment. Knowledge power, for example, easily reprograms flexible machinery. Changeovers take place orders of magnitude faster than acquiring new equipment.

Competitive advantage based on continuous innovation is characterized by:

- effective communications

- comprehensive organizational intelligence

- quick organizational learning

- rapid technology diffusion

- horizontal information flow systems

- technology fusion (merging capabilities of different technologies)

- concurrent engineering

- translating core competencies into new business development

WORLDS 2 AND 3—The turn inward

Middle and destitute nations primarily manufacture culture-specific products for domestic and regional markets. Most manufacturers in these nations cannot compete head-to-head with highly automated World 1 manufacturers in universal product markets. As a result, governments or trade blocs protect selected domestic industries. Around 2005, a drop in exports from middle and destitute nations between regions and blocs began. Just 10 years later, the drop was a tailspin. World 2 and 3 manufacturers since then have shifted to interregional flows. Affluent nations dominated export markets by 2015.

Middle and destitute nations are not restricted to culture-specific products. Some are strong in resource-based production, such as Indonesia, Chile, and parts of India. There are also fledgling recycling, reclamation, and remanufacturing industries in countries ranging from Panama to Bulgaria to Pakistan. They typically support affluent nation manufacturers, such as Mexico, remanufacturing for the United States. In the last 10 years, however, more manufacturers are doing their own remanufacturing rather than outsourcing. Middle and destitute nations do some assembly for affluent manufacturers. People can do some of the assembly of sandwichlike products designed for robot assembly, although this function is also increasingly automated today.

A fundamental principle implicitly followed by the affluent nations — based on Henry Ford's idea of raising the pay of his workers so they could buy his cars — is that "people without money don't buy our products, people with money do, and people with jobs have money." Occasionally World 1 manufacturers have deliberately not driven out a less-efficient middle or destitute nation manufacturer. To do so would toss thousands of workers, and potential customers, out of work. Manufacturers have shared and divided markets so that there is roughly enough for everyone.

Manufacturers relying on low-wage, unskilled labor have fallen on hard times. China is making the transition to custom production, while others such as Pakistan, Brazil, and the CIS nations have struggled. In many cases the shift away from labor-intensive manufacturing further exacerbated many countries' struggles to get bank loans to pay for imports, especially for getting the capital goods necessary to build new production capacity. As a result, countertrading has flourished over the last 30 years, opening trading opportunities where conventional methods stalled.

A CASE STUDY—Brazil's politics of manufacturing

Brazil typifies manufacturers in middle-income nations, which serve primarily local and regional markets. Their stories have not changed much over the last 30 years. Manufacturing surges and recedes according to the latest government policies. Attempts to compete in universal product markets have failed, and protection barriers went back up. These situations stand in sharp contrast to China, whose manufacturing industries have consistently grown.

Corrupt politics has held back this nation, once hailed as an up and coming star. Brazil has never enjoyed the political stability and government committment characteristic of China's success. The back-and-forth procession between military and civilian rule diverted attention from economic issues like manufacturing. Affluent-nation competitors nearly wiped out Brazilian manufacturers in the 2000s. Company after company closed or sold a majority stake to foreign competition.

The political situation is finally stabilized. The government has turned its attention to rebuilding manufacturing. Brazil restricts multinational corporations (MNCs) but does not exclude them. For example, MNCs share information technology equipment manufacture almost evenly with Brazilian companies.

Brazil and some other World 2 nations participate in export markets in alliances with World 1 nation partners. In some cases, the manufacturer is brought up to speed with advanced technologies and expertise to improve regional trade prospects. In others, the manufacturers concentrate on lower value-added components or goods and the affluent partner makes the higher value-added ones. As a result, some Brazilian manufacturers are becoming skilled at custom production. It makes economic sense for the United States to aid Brazil because it further opens the large Brazilian market for the United States. Chile and Argentina are Brazil's largest trading partners. Discussions about linking North and South America into a "bloc of the Americas" have gone on informally for years. Most experts expect this to happen within a decade.

China's success challenges Brazil

China and Brazil have historically been at opposite ends of middle-nation manufacturing—China has exceeded analysts' expectations, whereas Brazil failed to meet them. China is the world's largest economy. Its economy has quadrupled in size since 1990. Its *per capita* GDP is approaching that of some World 1 nations. Its large domestic market gives its economy flexibility and options much like those the United States had when it was industrializing last century.

The Chinese success story

China began trading raw materials for the electronics and machinery integral to a manufacturing base last century. By the end of the 2000s, it reached a threshold where it was relatively self-sufficient and could focus on its enormous internal market. The move away from central planning was complete by this time. The death of de facto leader Deng Ziaoping halted progress temporarily. Five years later, however, the conservative old guard that had halted further liberalization was ousted once and for all.

Reform surged again. The special economic zones along the coast, such as Guangdong, were the locus of development. Hong Kong was integrated smoothly, as it was granted special economic zone status and left alone. While income disparities between the coast and the countryside rose, there was trickle-down from the coast, as some manufacturing operations were outsourced inland.

The breakdown of huge state-owned industries led to a proliferation of large numbers of small firms. There were many missteps along the way. The firms were often weak in reading market signals. Customer focus has been especially difficult to instill. The years of command economy controls did not disappear without a trace.

Manufacturing's growth has not been without costs. Air pollution, especially from the coal powering manufacturing, has been severe. China is also the world's leading producer of greenhouse gasses.

Manufacturing prospered early this century as the government encouraged foreign investment. Chinese manufacturers moved solidly into midlevel manufacturing by the beginning of the century, selling primarily to other middle and destitute countries. They continued to sell lower value-added products to affluent nations. Rising living standards in the 2010s, however, challenged manufacturers by pushing up wages. Simultaneously, robotic competition was becoming more productive than masses of cheap labor. Chinese manufacturers were reluctant to admit that cheap labor, a linchpin of their previous success, was no longer a competitive edge. The necessary transition to custom production has not been completely smooth, but it has been eased by the fact that China has long been producing a large supply of technically trained skilled workers.

Even successful nonaffluent countries like China, which employs some advanced custom production, are hard-pressed to find markets. Domestic markets are crucial. Productivity gains and increasing *per capita* GDP have made more durable goods, such as automobiles, available to more people.

Infrastructure troubles

Heavy industry in Brazil concentrates in large metropolitan areas. Sâo Paulo is the most industrialized city in Latin America, and one of the most crowded cities in the world. Its infrastructure was overwhelmed by the turn of the century. Today, 25 years later, it is safe to say that the situation is turning around. Housing, sewers, transportation, communications, and power networks are at least adequate and generally improving.

Pollution damage was severe, due to the unfortunate combination of climatic and topographical features of the area. Light winds, air stagnation, and frequent temperature inversions hold polluted air near the ground. Pollution is being mitigated and cleaned up today. Prevention is getting off the ground. Policymakers are planning to directly control pollution, instead of relying on indirect measures of the past like zoning restrictions.

There are some successful manufacturing centers outside urban areas. Manaus, in the Amazon, is second to Sâo Paolo in manufacturing. Built on protectionism last century, it became a free trade zone in 1998, and today has reverted to a modern protectionism.

In the late 1990s, Brazil tried to reenter export markets by adopting stripped-down free trade policies. Import bans were dropped and tariffs reduced. Many manufacturers were driven out of business by trying to become internationally competitive. It became increasingly apparent that little manufacturing capacity would be left if firms were left to fend for themselves. Brazil's Congress passed the Manufacturing Refocus Act in 2013 to save the remaining manufacturers.

Continentwide improvements in the transportation infrastructure have helped manufacturing. Brazil's transportation networks were in bad shape until about 2015, despite an advanced telecommunications infrastructure compared with other Latin American countries. Stories of logistical nightmares were common. For example, it was once cheaper to ship a ton of cargo from Korea to Sâo Paulo than from Manaus to Rio de Janeiro. The 2011 SATI (South America Transportation Initiative) agreement to overhaul continental transportation spurred domestic initiatives in Brazil.

Sources of competitive advantage

Brazil, and other Latin American semiindustrialized countries like Argentina, have a strong natural resource base, a relatively skilled labor force, and a fairly high income *per capita*. Internal markets have been large enough to avoid substantial diseconomies of scale and to provide domestic competition. Brazil is among the main interregional exporters within the developing world.

The difference between middle-income countries like Brazil, Argentina, and India and the Asian Tigers is the former's reliance on culture-specific products and the latter's success in universal export markets. Even the Asian Tigers, however, had to make the transition from low-cost, labor-intensive production of low value-added products to the high-skill, automated production of universal, high value-added products. Each country experienced the phenomenon of being highly productive and successful, followed by demands for higher wages and rewards from the workforce. This led to a drop in competitive standing in export markets, but also increased the attractiveness of domestic ones. It also drove the search for a new edge, i.e., custom production.

Brazilian government trade strategists are using investment policies to get advanced manufacturing technologies for local production. They are looking to China, which is the country most successful in acquiring and applying foreign technology tailored to domestic needs. In particular, state officials are concentrating on the strategic industries of steel, energy, petrochemicals, and communications. The Brazilian ethanol industry is world class. It is the fuel of choice for the domestic automobile industry.

Brazil still has one of the most uneven distributions of wealth in the world. Sixty percent of national wealth is in the hands of 1% of the population—little changed from the 1990s. It is also unevenly distributed geographically, with far more wealth in the south than north.

Countertrade flourishes

Countertrade, which accounted for about 20% of world trade at the turn of the century, accounts for 25% today. It bypasses conventional barriers and circumvents the limitations on what you can trade in impoverished countries. Brazil is a primary user of countertrade, especially in trade with peers like Mexico and less fortunate partners like Bolivia and Nicaragua.

Types of Countertrade

Barter	Direct exchange of goods and services between two parties.
Counter-purchase	Reciprocal buying arrangements designating a set percentage of the total payment in cash, with the remaining percent settled by a transfer of commodities within a specific time frame.
Offset	An arrangement similar to counterpurchase, but the seller is required to use goods and services from the buyer country in the final products.
Buy-back	An arrangement requiring payment in goods and materials in exchange for capital investment or technology. The trader may provide technology, construct a plant, supply a limited raw material or special service, or provide capital equipment, receiving products created by the project in payment.
Switch-trading	Traders establish accounts with specific monetary limits. Each party draws on the account over a period of years. If transactions exceed agreed-upon monetary limits, the deficit parties settle in currency. Switch-traders frequently work through these trading houses to substitute trade credits from clearing accounts for goods in future contracts with other parties or countries.

Economists estimate that as much as 40% of Brazil's economic activity went unrecorded, as recently as 20 years ago, to avoid taxes. For example, illegal mining by garimpeiros surpassed official gold output for decades. Although countertrade is not intended as a means for illegal trading, the lines between legal and illegal were often murky in Brazil.

Japanese trading companies are well established as countertrade clearinghouses. They have created new markets, wealth, and alliances. Europe got involved next, and the United States last.

In barter transactions, goods of equivalent value are exchanged without cash changing hands. Participants fix a cash value for accounting, taxation, and insurance purposes. Exchanges may be a one-time exchange or they may take up to 20 years to complete. They may be bi- or multilateral, public, private, or mixed. Some compensatory arrangement, or reciprocity, beyond the basic exchange of goods and services in cash or kind is part of the deal. Export sales are often conditional on agreements to accept imports.

Critical Developments, 1998-2025

Year	Development	Effect
1998	Iridium satellite network deployed.	Improves just-in-time distribution with its global positioning capabilities.
1999	U.S. Waste Reduction Act.	Mandates 50% reduction in packaging by 2010.
2001	U.S. Retooling Manufacturing Act passed.	Establishes a formal U.S. industrial policy; Sherman Antitrust Act revised to enable greater industry collaboration.
2001	FAN established in the United States.	Creates an on-line prequalified partnering network for manufacturers.
2002	U.S. Energy Transition Act (the sustainability legislation) passed.	Mandates further reductions in energy use; tax incentives for switch to alternative sources.
2005	Drop in exports, aside from those within the region from middle and destitute nations.	Signals advent of affluent nations' high-skill, high-automation manufacturing driving out low-skill, labor-intensive competitors.
2006	A remanufacturing clause is passed as part of the annual environmental bill in the United States.	Manufacturers responsible for disposal of items whose weight in kilos multiplied by its dollar value is greater than 2,500.
2009	International Standards Organization establishes materials characterization standards.	Impetus for an international recycling and reclamation market.
2010	80% of U.S. manufacturers producing 96% of all goods pass environmental audit without penalties.	Manufacturers have firmly committed to sustainability; keep ahead of regulation.
2011	South American Transportation Initiative passes.	Overhaul of South America's transportation network; boost to regional manufacturing.
2016	Bangkok Accords set up an institutional mechanism to formalize limited trade protectionism.	Leads to affluent and upper-middle nations concentrating on universal products, and lower-middle and destitute nations concentrating on culture-specific products.
2019	GMAN established.	Upgrades national prealliance service networks, like FAN in the United States, to include nations across the globe.
2020	Zero Waste Proclamation signed by 61% of medium to large U.S. manufacturers (more than 20 employees).	Sets a target of zero waste by 2040.

Unrealized Hopes and Fears

Event	Potential Effects
Nanotechnology enables manufacturing at the molecular level.	Reshape manufacturing by building thousands of tiny, specialized machines for specific tasks.
Low-wage, labor-intensive manufacturing in middle and destitute nations wipes out competition in affluent nations.	Some redistribution of wealth from rich to poor nations; affluent nations focus on information and service industries.
Dark factories drastically reduce the manufacturing labor force.	Neo-Luddite groups carry out industrial sabotage.

13

A QUEST FOR VARIETY AND SUFFICIENCY

Of every 20 people on Earth today, 5 struggle to get enough food and suffer recurring famine, 12 get enough to eat but are not prosperous, and 3 have a nearly endless menu available to them.

Since the year 2000, science and technology have revolutionized food and agriculture in World 1. Developments have included:

- genetically designed crops and livestock raised on automated farms,

- ecologically sound agriculture practices, drawing on the field of agroecology,

- food grown and synthesized in factories,

- information technology for decisionmaking, process control, and distribution from field to plate,

- robots that manage warehouses, truck loading, unloading, packaging, food grading, processing, and cooking,

- smart appliances that serve restaurant and home kitchens,

- food that replaces or augments medicine for disease prevention and cures.

Destitute World 3 sees little of those benefits. Its people have moved by the millions to cities, where food distribution often fails, and economics reduce access to food. Violence and political forces further aggravate food scarcity and poverty.

Haves and have nots

Despite seven decades of international development aid, the world in 2025 is still divided into haves and have nots.

For 22 or so countries of the world that today are struggling for subsistence, development has failed. Those countries urbanized and modernized, at least superficially, but remain in deep poverty. Their peasantry has moved en masse to the cities. Many have 65%-80% metropolitan populations. Hunger afflicts billions of them. The poorest countries of the world still have the highest population growth.

The figure below shows world population growth by region.

World population climbs with high growth in Africa and Asia and slow growth in the developed world

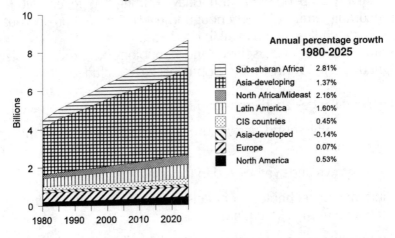

Annual percentage growth 1980-2025

Subsaharan Africa	2.81%
Asia-developing	1.37%
North Africa/Mideast	2.16%
Latin America	1.60%
CIS countries	0.45%
Asia-developed	-0.14%
Europe	0.07%
North America	0.53%

Metropolitan populations create new problems for food sufficiency. In rural areas, despite recurring famine and other problems, traditional practices allowed most people to eat directly from the land. They had options and flexibility. They could fish, hunt, plant, or trade locally for food.

The politically driven breakdowns in agriculture and food supply that became recurrent events in the late 1990s are now commonplace in the poor countries of the world. No longer can people easily migrate to new herding or planting grounds. They have abandoned their agricultural skills. Fortunately, this circumstance does not afflict all of what were once called the world's less developed, countries. Some countries have made it to the bottom rungs of the development ladder and can sustain themselves. Others have valuable resources or tourist attractions and thus earn the money to buy food internationally. Still others have been taken under the wing of benefactors, including prosperous neighbors and, in the case of several, former colonists. The Adopt-a-Country program of the United Nations in the early 2000s led to several enduring relationships between affluent and poor countries for food and fuel. Some adopters have a historic link, others benefit from a strategic resource.

In 23 countries, population has exceeded the ability to produce food	
Africa	9
Southwest Asia	11
Central America	2
Southeast Asia	1

The basic food issue: Enough

Scholars still disagree about how many people are too many for the Earth to sustain. The Earth's population today totals 8.4 billion. Clearly, there are too many people in sub-Saharan Africa and South Asia for the ecologies and societies there to sustain. On the other hand, the food production potentials in Europe, the southern cone of Latin America, Australia, and North America are much greater than what is produced in those areas today.

Theories also vary as to what levels of agricultural production are permanently sustainable. One key development has been the recognition that what appears sustainable at one time in agriculture may have to be reassessed and redefined later, when new knowledge becomes available. Practices of the first Green Revolution (1965-1985), which led to more tilling of land and large-scale irrigation, created unforeseen environmental problems, notably, destruction of aquifers and salination of soils.

Worldwide, the momentum of food production growth slowed in the 1980s through the 2010s. In some regions, especially Africa, food production *per capita* shrank from 1980 to 2010. Its growth has finally returned in much of Africa and parts of southern Asia but remains stagnant or is still falling for some countries on these continents. Among them, some can afford to buy food internationally, and some have experienced waves of famine. The patterns are shown in the figure below.

The World's Food Production Per Capita
Stagnates and Falls, 1980 = 2025

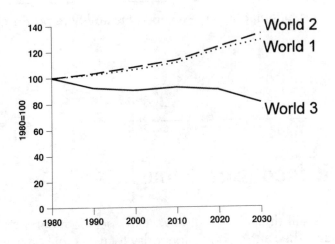

Technology and the Second Green Revolution

The past 30 years have brought the second Green Revolution (1995-present) in agriculture. Agriculture worldwide has been reshaped by four dominant trends:

1. The greening of agriculture—with greater attention to the environment

2. The genetics/biotechnology revolution

3. Specialization in agriculture

4. Automation and information technology

Some of the technologies that have helped drive the Second Green Revolution appear in the box below.

Technology promotes greater food productivity around the world

- *Waste reduction* including recycling more agricultural wastes and animal wastes

- *Genetics* for new plant varieties:
  — symbiotic cocrops
  — nitrogen-fixing varieties
  — crops better suited to rotation farming
  — greater photosynthetic efficiency
  — salt- and drought-tolerant crops (though these have become pest plants in coastal wetlands)
  — engineered-in pest resistance and control

- *Education and extension* for farmers to reduce practices that cause land degradation

- *Innovative irrigation schemes* including drip irrigation, soil moisture retention applications, synthetic soils

Several decades of the greening of agriculture have brought greater attention to the environment and better understanding of agroecology—the place and role of agriculture in nature. This movement has fully saturated international agricultural assistance and technology transfer, and so has penetrated most countries of the world. However, subsistence farming continues in many poor countries that is not in harmony with the environment and continues the degradation of land and other resources.

With genetics, agriculturalists around the world have been able to further fine-tune their crops to local conditions. Early successes in advancing plant hybridization came through greater knowledge of genetics and new tissue-culture methods. New plant species that are self-propagating are beginning to find their way into famine-prone areas and hold the promise of raising productivity while reducing or eliminating environmental damage.

World agriculture today is specialized by region. Efficient global trade and information flows allow regions of the world to produce the crops that they are best at producing for the global market. Twice as many foods are global commodities than were in 1990. Most of the world participates. Market forces usually win out over petty and high-stakes international conflict. The 2011 Nairobi Accord on Food Security provides for continued food commodity trade in the event of international conflict. In subsistence economies, this pattern does not hold. Also, under certain political circumstances, it is common for some countries to go it alone or form blocs that rebel against the world system of food trade. The best current example is the Cooperative Alliance of Oceania, led by Tonga.

Whereas over the last century agriculture became mechanized, during this century agriculture has become automated in World 1. Robotics, information technology, process control, and smart farms characterize agriculture since 2000. Sixty-seven percent of farms in World 1 countries are at least partially automated. Automation brings gains in productivity and resource savings, especially in the use of water.

A double blow to the environment

Ozone damage and greenhouse warming have hit agriculture hard in poor countries, but affluent countries have coped more readily. The 1 degree Celsius rise in mean annual average temperature since 1985 has affected ecologies, sea levels, coastlines, and polar ice. Ecological and climate zones and, therefore, agriculture zones, are shifting as global warming continues and rainfall patterns change.

The 30-centimeter sea level rise since 2000 threatens to destroy wetlands, silt up river channels, and clog river deltas. The world's poor countries cannot effectively build the civil works necessary to hold back the sea if it rises further. Damage so far has been moderate. Flooding happens more often and more readily, and low lying countries, notably Bangladesh, have seen sometimes massive flooding, with considerable loss of life, such as in 2016 and last year. The affluent countries laid plans in the 2010s to build civil works in threatened coastal zones to prepare for additional sea rises over the next few decades. Some are in place today. Though some damage has been done to wetlands and river channels, most developed nations are coping effectively. Greater problems are expected by 2050 with an expected additional rise in sea level of one meter.

Damage to the Earth's ozone has grown worse since it was first diagnosed in the 1980s. Instead of seasonal damage afflicting the south polar region and northern parts of the northern hemisphere, permanent holes have formed, including over western Europe and eastern Canada. Identified in 2014 and tracked since, the full effects of these holes are still unclear. People in World 1 today cover themselves against ultraviolet light with hats, creams, and clothing. Greater rates of human melanoma, as well as some cases of melanoma in farm animals and wildlife, have been a proven outcome. Instances of wild-animal blindness are reported regularly.

Less visible and as yet unproven, is damage to microfauna and microflora from greater ultraviolet light penetration of the atmosphere. These organisms are at the base of ecologies, and integral to higher and lower forms of life worldwide. Massive die-offs of microflora and microfauna would upset ecologies, lead to extinctions of some species, and eventually alter the geography of the Earth. Fisheries already suspect die-offs of some of the microorganisms their catches feed on. Fisheries production is down 17% over the past 15 years, despite better conservation.

World organizations exchange food aid for reforms

Since the Geneva Conference on Hunger of 2007, the trend in food assistance is for the UN to provide assistance only in natural disasters such as floods or extreme drought. Thirteen countries have been warned that they must reform politically and economically and exercise population control to be eligible for food aid. Some private relief agencies continue to aid the starving where the UN will not. Additionally, some countries with traditionally close ties to impoverished nations, or with a security interest in the stability of those countries, provide aid themselves as the United States has done for the Philippines.

The UN and other international development agencies will transfer technology to any country to encourage sustainable agricultural methods.

THE AFFLUENT 1.3 BILLION—CHOOSING FROM A BIG MENU

For the rich of the world, 1.3 billion mostly in Europe, parts of Asia, the United States and Canada, and parts of Latin America, food is entertainment, hobby, medicine, cultural experience, but rarely is it only subsistence. There is plenty of food in tremendous variety for the affluent. Poor nutrition is rare, confined to urban and rural slums and to those ignorant of good nutrition. Proper nutrition is easily ensured by nutritional supplements in food, pharmaceuticals, and nutriceuticals. This frees people to enjoy food as they enjoy music, in endless variety and even frivolously.

For agriculture, this makes gourmet, organic, specialty, and niche crop farming a booming business. Such agriculture is a high-payoff business that operates alongside commodity food production. Commodity foods are often supplied by middle tier countries, where agricultural production has risen steadily since 2000. Agriculture and the food industry work to bring a big menu of choice to the rich of the world. Tastes, ethnicity, medical needs, and packaging formats indulge nearly every desire of the world's wealthy. The following case study of the United States reflects developments in World 1. The descriptions and commentary here have their analogs in several dozen other rich countries around the world.

> **Technology continues to reshape food and agriculture in World 1:**
>
> - Fully integrated, sophisticated agricultural practices and smart farms
> - Genetic engineering of crops for pest resistance, greater productivity, other traits, food science, food synthesis, food design, and food processing
> - Packaging and storage technology
> - Distribution information and logistics technologies, including automatic ordering, distribution, and inventory of food
> - New food preparation and kitchen technologies

A CASE STUDY—The United States typifies affluent World 1

The United States and a few other affluent economies are at the top of the world in terms of food availability. Food is rarely a subsistence issue. Though populations in these countries include poor people, most people have food in unprecedented variety and quality. Safety and healthfulness are paramount. Advanced technologies in U.S. agriculture, food-processing science, and food retailing make the big menu of choice possible. With next to no human effort, everyone at a 10-person dinner party can have his or her own choice of food, fully natural or concocted to suit any medical or nutritional needs. A wide range of choice is available not only at expensive restaurants, but also in the home. Meanwhile, greater quality in fiber production is possible through the same advances in agriculture.

**Food production in the United States grew steadily
from 1980 to 2020
1980 = 100**

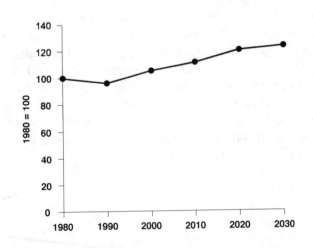

A changing population drives change in the food business. Critical demographic trends include:

- Slowed population growth;
- The aging of the population and retirement of the baby boomers (the median age of the whole population is 41, up from 32 in 1990);
- The baby boom echo reaches prime working age;
- The average American today, compared with the average individual in 1990, is 2 centimeters taller and 6 kilograms lighter;
- New immigrant flows, particularly from Southwest Africa (East and West Angola, Namibia, Zaire), and Amazonia;
- New household structures, such as the rise of cohabitation among old people and more varied sharing of households among adults.

Americans embrace a growing menu

American eating habits have changed through the 2020s along with food products and food technology. Attitudes and values, especially regarding health, quality, and the desire for new experiences, shape food consumption in the United States. Key influences have been:

- Packaging and food-product options
- Attitudes about waste and environmental impacts
- Health and safety concerns
- Ethnicity in the United States, and a revolving cycle of fad foods
- Synthetic fats, sugars, and other food components
- Food preparation technologies, including automated kitchens
- Expectations of choice, quality, and diversity in food

The American menu changes: trends in u.s. consumption of certain foods, 1990-2025

Red Meat	↓↓	Sugar	↑
Fish	↑	- refined sugar	↓↓
Poultry	→	- synthetics	↑↑
Eggs	↓↓	Fresh fruits	↘
- cholesterol free	↑↑	Packaged foods	↑
Dairy	↓	Fresh vegetables	↘
- skim milk	↑	Packaged vegetables	↑
- yogurt	→	Synthetic foods	↑↑
- cheese	→	Coffee	→
- ice cream	↑	Alcohol	↓
Flour/cereals	↘	Nutriceuticals	↑↑
- rice	↓		
- corn	↓		

Key

↑	up
↑↑	up markedly
↓	down
↓↓	down markedly
↘	down slightly
→	unchanged

Production

In 1990, the largest 4% of farms in the United States produced half of the food. As of 2023, nearly two thirds of U.S. food production is done by about 200 of the largest corporate farms. Most of the rest of U.S. farms produce specialized, lower-volume products, such as designer vegetables, nutriceuticals, gourmet foods, and all-natural foods.

Fewer Farms, More Acreage in U.S. Agriculture
Thousands of U.S. Farms

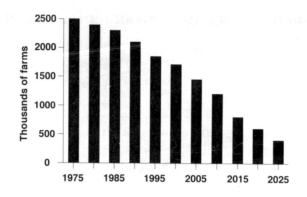

Eight-tenths of a percent of U.S. employment is on farms today, compared with 2.1% in 1990. Additional people live on farms that are operated as hobbies or are used to supplement owners' incomes. An estimated 37,000 of these farms produced food for sale in 2024.

Though organic foods are a lucrative market in the United States, the 1970s to 1980s mania for organic farming has changed. Today the important characteristics of agriculture in the United States and other affluent countries are healthfulness and sustainability. Synthetic foods are judged for their human healthfulness and their impact on the environment. Waste is frowned on; recycling of plant wastes and careful soil husbanding are dominant practices.

Fertilizer use is off by 4/5 per hectare in the United States since 2000. Genetically engineered commodity crops have been the main force in lowering use. For example, in 1990, corn production accounted for 44% of fertilizer used in the United States. The new corn varieties, with new intercropping methods have reduced fertilization of corn to just 17% of fertilizer used in the United States. Chemical fertilizers now are most often used with highly specialized cropping. More common is the use of soil supplements produced from sewage and other wastes. The 1998 Conservation in Agriculture Act provided tax incentives that still stand for farm use of processed sewage or other organic wastes.

Too-high nitrate levels in the Ogallala and Imperial Valley aquifers propelled alternatives to fertilizers forward. The Ogallala reached a crisis in 2016, and strict withdrawal laws were imposed after a bitter political battle.

Inputs to agricultural production are now only 36% of their 1990 levels. Most still used are safe chemicals that have no known negative collateral effects or that improve soils. Chemical runoff is rarely a problem. Agriculture has seen a net energy input drop of 53% since 1995.

In open-field farming, traditional wind and water erosion problems have been reduced. Topsoil loss to runoff and blowoff are down 80% since 1990.

Genetics and biotechnology revolutionize U.S. agriculture

The fundamental trend in genetics and biotechnology has been greater fine-tuning of crop and animal varieties to local conditions and for end-use needs. Since the 1990s, scientists have been able to transfer certain genes directly into a plant genome.

Completed agricultural genomes accumulate:	
Corn (United States)	2006
Cotton (United States)	2006
Rice (Japan)	2009
Wheat (Russia)	2011
Chicken (United States)	2011
Cattle (United States)	2012
Pig (Britain)	2014

Advances in our knowledge of plant and animal genetics and microbiology have been driving advances in the fields of biotechnology and human genetics. Begun with the U.S. Department of Agriculture (USDA) Plant Genome Research Program in the early 1990s, international cooperation has completed genome libraries for the major food crops produced around the world.

Additional gains are expected from stepped-up efforts to collect and catalogue genes from across agricultural plant and animal life. This has included attempts to resurrect genes from ancient crop plants and even plant matter preserved in amber.

Somaclonal variation techniques produce offspring from tissues of plants with desired characteristics. Some outcomes have been:

- Genetic manipulation of fruit/vegetable sizes according to needs (individual or multiple servings)

- Pest and disease resistance

- Pesticide- and antibiotic-free foods

- Altered animals and vegetables with reduced tendency to spoil

- Raised photosynthetic efficiency, higher seed production

Techniques for large-scale cloning and synthesis of biomolecules have enabled farmers to purchase pheromones, hormones, growth promotants, and other compounds for field application. Vaccines are produced cheaply in quantities sufficient to inoculate herds.

Genetics and biotechnology developments

- Beta-carotene-producing sweet corn
- Semicubic tomatoes suited to packing and shipping
- Elongated tomatoes for slicing
- Lower-fat pork
- Cows milk more similar to human milk and better suited to human nutritional needs
- Miniature vegetables such as lettuce in serving sizes
- Short season grain varieties allow multicropping in higher latitudes
- Other nutrient-rich varieties
- POTOTAL™, also known as SuperSpud—a potato-derived variety of tuber that has 45% of the recommended daily amounts of vitamins, minerals, protein, and fiber. A favorite for french frying, it was developed jointly by McDonald's and Asante's Restaurants.

Where difficulties with genetic manipulation of complex traits remain, interbreeding by old-fashioned methods can take up some of the slack, albeit at a slower pace. Successful transgenic animals include cattle, chickens, and pigs. Over the last two decades, there have been numerous experiments to breed new traits into farm animals, but more have failed than have succeeded.

New animal breeds engineered genetically must pass muster with the International Commission on Animal Care in Agriculture (ICACA), established in 2017. The breeds must be viable, healthy animals and must not be subject to birth defects, threatening to other species natural or domestic, or harboring diseases. The breeds must not be "unduly discomforted in life" by congenital conditions or traits.

Genetics brings improvements and variety to livestock

Farmers today raise livestock that are leaner, meatier, faster growing, more fertile, faster reproducing, and less disease-prone than in 2000. Biotechnology, genetics, and innovative management strategies are the reason for this success.

Nearly all livestock are now impregnated through in vitro fertilization and embryo transfer. Usually, ova are induced to produce twins or triplets. Large farm operations use breeder cows of ideal genetic lineage, for example, to produce ova, and recipient cows suited to gestating and birthing, to birth calves. There is now a commodities market for frozen implantable embryos.

Pregnancy monitoring since 1997 has been done by sensing a pregnancy associated peptide in livestock. Each animal is tracked electronically through a microchip tag, in use since the 2000s. Most tests are done with a pinprick blood sample, though experiments are underway using noninvasive sensors through the skin of the inside of the ear (generally hairless).

Changing quality of animal protein

For human consumption, animal meat long offered a good source of protein, but too much fat. Today's cows, chickens, and pigs are leaner because of a combination of genetic manipulation and drugs administered during their development. The drugs have no residual effect on human consumers.

Livestock growth and development involves manipulating endogenous growth hormones (rather than introducing new growth hormones) to promote growth at optimal times. Beef cattle, for example, can be fattened on demand when markets and shipping are ready, a technique approved in 2016.

More for less in animal farming

In 1990, it took 42 days to raise a 2-kilo broiler from a chick. Every 2 kilos of feed produced kilo of meat. Today, it takes only 20 days for a chick to develop into a 2- kilo chicken, and feed is converted to meat at a 1.8 to 1 ratio. One breeder hen can produce double the 140 chicks per year it could in 1990.

At the same time, there is a greater variety of chicken products available and greater flavor diversity.

In 1990, dairy cows averaged 6,300 kilograms of milk per year. Today, smaller cows that consume less feed produce as much as 12,000 kilograms.

Half-sized dairy cattle, genetically altered to produce high volumes of milk, consume only 3/5 the forage as their full-size cousins. They are better suited to containment rearing as well.

Better knowledge and tracking of the ethology of farm animals has reduced disease and behavioral problems and contributed to greater farm productivity.

Dairy milk has been altered through genetics to be better suited to human consumption. In the early 2000s, researchers strove to produce improved cow's milk. Today's milk is fine-tuned to human needs, with variations for children, adults, and those who do not tolerate certain milk components. Pregnant women can now drink Natalact^{tm} to get proper nutrition.

Systems thinking and information govern farm management

Today's farming, be it in contained environments or open fields, uses the principles of integrated farm management (IFM), in which a whole-system approach interrelates and accounts for water, soil, air, supplements, pest management, and temperature. Information technology, sensing, and automated machines allow continuous dynamic adjustment of factors and inputs and bring demand and other economic information to bear on decision making.

More data is available for modeling and decision making

Today's farms use information from sensors installed in farm fields, greenhouses, and hydroponic complexes and from remote sensors on satellites to track farm production and measure the variables involved.

Most farming today is indoor work. The farmer can sit with a computer and view fields for moisture content, groundwater levels, key soil nutrients, plant health, growth progress, and in some cases, fruit ripeness. He or she can take action to fill a need or correct a problem without leaving the console. The USDA-coordinated private satellite services provide data for such image study as often as daily, on a subscription or fee-for-service basis.

Remote and proximately sensed data are also automatically fed to irrigation and nutrient delivery systems in fields and to farm equipment. Using GPS-2 positioning (the global positioning system available since 2012), equipment can deliver moisture, nutrients, or do other processes according to needs, down to the square meter. Finer resolutions are, of course, possible but not normally cost-effective.

Pilotless tractors automatically tend fields and harvest produce on today's farms. Guided by buried wires, field-placed markers, GPS-2, or onboard sensing, these drone tractors can work around the clock and suffer no fatigue. Farmers are freed up for knowledge work, such as modeling markets and planning crops.

Factory farming gains an ever larger share of production

Perhaps the ultimate form of IFM is factory farming. Greenhouse farming and hydroponics have grown over the past two decades to 37% of agricultural production last year. Their technologies have evolved most notably in the fine-tuning of plant light, temperature, and nutrient conditions. Computers and real-time sensors give such containment farming systems their precision. Energy technologies, specifically photovoltaics and storage systems for extended light growing, are integral to containment-plant agriculture. Productivity for a typical system has quintupled since 2006. Most are 95% robotically operated.

Fisheries and aquaculture

Developments include:

- Algae and sea plants harvested for animal and human feed

- Fully contained fish farms using wastes for feed—may be anywhere, including under cities

- Use of fish homing instincts to have fish return to the hatchery for capture

- Sexual attractants for open sea fishing

New crop varieties improve soils and cope with climate change

Among the new plant varieties developed through biotechnology are several corn, barley, wheat, rice, cotton, and canola varieties that improve soils and reverse some environmental degradation. Nitrogen-fixation was introduced from legumes several decades ago. Nitrogen-fixing varieties, however, are less productive because they use energy for nitrogen fixing at the expense of seed or fiber production. Today the varieties are rotated with non-nitrogen-fixing varieties for periodic soil enrichment. More recently (2017), grain varieties were developed that scavenge some pollutants and salts from soils. The current favorite is the corn variety Calgene Y-6. New varieties have also been introduced to deal with reduced rainfall in the midwestern United States.

Fiber joins the genetics revolution

Genetics has kept fiber competitive in the face of revolutionary developments in synthetic fibers. Although synthetics dominate the market, superior quality natural fibers, particularly for clothing, are preferred by about 12% of the consumer market. Engineered species of sheep and cotton are more productive and produce more uniform, higher-quality fiber than ever before.

New technologies conserve water

Technology has solved many traditional water problems, but problems remain. Experiments continue in cleansing soils and recharging aquifers, especially in California's Imperial Valley. Most success in conservation has come from fine-tuning irrigation, low-moisture requirements in hybrid crops, and careful use and reuse. Aquifer depletion continues, and salination of soils remains a problem. One effective technology is the soil adjuvant AGRIGEL™, a colloid patented in 2022 that captures and retains water in subsoils. Tractor-pulled equipment injects the material into subsoils.

Managing pests and soils with few or no chemicals

Integrated pest management (IPM) has been used since the 1980s to fine-tune solutions to pest problems and reduce chemical use. Its capabilities have grown in the past 20 years, such that most temperate climate agricultural pests are controlled without chemicals. Pest and virus resistance are bred into or genetically engineered into every major crop grown in the United States. Other approaches used include use of predator insects, intercropping, hormone and pheromone releases, sound, and light.

The environment shapes agricultural practice

The science of climate and environmental manipulation has grown over the years since the 2011 International Agreement on the Troposphere. Large-scale climate manipulation is done under conditions of drought or excess precipitation when those natural forces are deemed extreme hardships for the regions affected. The Council on the Troposphere set thresholds for precipitation, temperature extremes, and air currents, and sits as a decision-making and appeals body for matters related to weather manipulation and longer-term climate manipulation. The council's goals are to prevent activities that are not sustainable, that adversely affect other regions, or that harm another country.

The principal technology for weather manipulation is cloud seeding, using organic and inorganic seeding agents. Silver iodide, used last century and revived in the 2010s in the developing world, produced unacceptably high concentrations of silver iodide in river beds and soils. Public protest lead to the development of the new agents.

A key to weather modification technology has been advances in the theory of ice nucleation, which was fully developed at the beginning of this century. Hail suppression has improved with better cloud seeding materials. Earlier, more accurate, weather forecasting and sensitive radar allow quick intervention before hail conditions arise. Hail suppression is recognized internationally as an acceptable intervention, without requiring a ruling by the Troposphere Council.

Technologies used for climate manipulation include artificial wind barriers, orchestrated over many square kilometers to alter wind currents, and tens of square kilometers warming or cooling zones on the surface of the Earth, based on groundcover plantings. These create thermal zones that increase rainfall.

Agriculture and restorative ecology have converged further in the 2010s to today. With the development of new crop varieties and cropping practices, ruined ecologies are increasingly salvageable without costly excavation and soil replacement. Coupled with waste recycling and applications of sewage, remedial plant varieties, many of which produce salable crops, can be used to reverse ecological decline and restore health.

When necessary, synthetic soils or synthetic soil supplements (such as moisture- loss retardants) are applied to ruined ecologies first. Then, replanting of one or more varieties of chemical scavenging grasses is done. Several crops in rotation remove trace minerals and salts from soils. They may be burned as fuel, and in some cases used as energy biomass, before toxins are reconcentrated and disposed of safely. Subsequent crops are fit for animal and human consumption.

Unprecedented diversity in the U.S. diet

There is more diversity in the U.S. diet today than at any time since the 16th century in Europe. More of the more than 3,000 known edible plants found in nature plus new ones created in the laboratory are farmed today for human food. Some new foods are:

- Tubers, in wide varieties, including the oakra that tastes like a potato with sour cream already added, and the arrachacha, that tastes like a cross between celery and carrots

- New seafoods such as krill paste and kelp

- New food sources such as petroleum, biomass, wastes, weeds, insects, and vermin

In addition to all new products in agriculture, several plants important elsewhere in the world have become commodity products in the United States. Bulgar wheat and yams, engineered for efficient production in the United States, are widely produced today.

Processing

Industrial food processing, a third of which takes place on farms, is a highly automated, robotized process. On-line controls, instrumentation, and sensors, manage the grading, sorting, and sizing of produce. Some grains enter processing streams directly from field-harvesting gantries through dry slurry pipelines.

Farm- and factory-based processing complexes rely on computer modeling and decision making to coordinate available inputs and market demand (e.g., finding and buying the cheapest starch at a given moment). As with automated farms, the typical operator works from a control booth and rarely needs to leave it.

Additional process controls and sensing assure quality throughout the process. Examples of these include on-line, real-time testing for microorganisms, viruses, trace toxins, and chemicals using DNA probes and monoclonal antibodies (MABs).

Totally synthetic foods grow in the marketplace

Several dozen synthetic foods are in commerce today. Hundreds that failed are patent history, some because they tasted bad, and others because they cost too much to make. With Nuvelle[tm], the fat substitute, users can no longer tell the difference between synthetic and real. The technology behind Nuvelle[tm], patented in 2019, is in part based on an indigestible molecular form of fat. Unlike its predecessors, the molecule is designed to cook like natural fats as well.

Packaging

Food packaging is at the center of the food revolution in the United States. Packaging brings flexibility, storability, portability, and preparation ease to the user. Reflecting a long-term social trend toward greener practices, food packaging produces 75% less waste since once-through plastics and other packaging materials came under the Waste Reduction Act of 1999.

Food packaging is central to the development of choice and flexibility in food consumption in the United States. Walkaway foods (see box below) instant meals, and smart packaging bring new food options to the user.

Walkaway foods suit flexible lifestyles

With hectic, varied lifestyles, fewer people eat together in couples and families, and fewer at traditional mealtimes. A widespread trend is five or six nutribreaks per day. This has made walkaway foods boom. Examples include a host of exotic breads and the ubiquitous nutribagel, containing one-third of daily requirements for minerals, fat, calories, fiber, and vitamins.

Edible packaging and packaging that heats a product when the user pulls a cord have spread the popularity of walkaway foods.

Walkaway foods inspired by ethnic cultures come and sometimes go based on migration and popular culture from other countries. Today the two most popular walkaway foods are fruit sticks and Andean spiced dumplings.

• Biosensors for freshness and quality include tags built into packages that turn colors to show whether food has spoiled or the number of days of freshness left.

Freshness labeling: Freshness dot is integral to plastic packaging

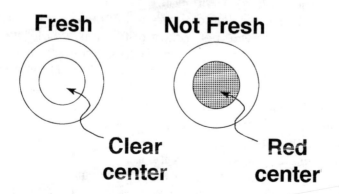

Fresh   Not Fresh

Clear center   Red center

- Built-in food preservation. Edible antispoiling or antiripening agents are coated onto or bred into fruits and vegetables. These allow fruits and vegetables after picking to stay fresh until the user washes or cooks them.

- Ingestible packaging.

- Reusable packaging.

- Bar coding and other labeling, including microchips.

Automation dominates commercial food service

Smaller workplaces have fractionated the market for business lunches and corporate cafeterias. People working in suburban work centers are more often served by lunch caterers and by automated food dispensing and preparing machines.

A stroll down the grocery store aisle

Today the shopper visiting a typical food store electronically or live sees:

- A full range of prepared gourmet meals for microwave reheating.

- A pharmacy console at which the shopper can read his or her smart card and consult computers about foods available that meet the shopper's nutrition needs.

- Self-announcing packaging that beckon the shopper or describe package contents.

- Food packaged with informational microchips on labels that can be read on home computers for menu, recipe, and nutrition information.

- Microwaveables that gang together for exact serving counts.

- Packages that communicate through microchips with refrigerators and microwaves.

Distribution

The logistics of food distribution is tied directly to the transportation system of the United States. Part of the food distribution system in the United States involves a nationwide network of distributors, wholesalers, retailers, and commercial customers and was turned into an information business. Every institution involved can place orders in or take orders off the network automatically. Large-scale buyers keep standing orders on the system, with sellers sometimes bidding electronically for the sale. Most large sellers and wholesalers coordinate goods buying, selling, and handling electronically as well, with robotic warehousing and loading done with little or no human labor. In perhaps 15 years, robotic vehicles will carry cargo from source to user, removing more human labor from the distribution chain.

Large commercial and retail users likewise coordinate their kitchen or plant automation and warehousing electronically through the network. Information is tracked all the way through to point of sale, with orders, reorders, and market planning based on the data collected.

Information technology transforms the retail food business

Information technology today is a tool for management, logistics, advertising, selling, tracking, and quality control. Every store knows what foods people buy or order. Retailers serve customers with standing orders for delivery through fully automated accounts and order placing. They give and receive information directly to and from home and commercial automated pantries and kitchen management computers.

Stores are information centers for the walk-in or televideo customer. Information on nutrition, recipes, new products, and preparation tips is accessed by customers throughout a store or from a home information station. That information can be coordinated with a home or in-store health scan. Smart packaging can be interrogated by handheld devices for the same kinds of information.

End use: The food service industry emphasizes variety

Despite the availability of top-quality gourmet meals for home microwaving, the restaurant business is thriving. The retiring and retired baby boomers have embraced the widespread availability of take-home gourmet meals from restaurants and food services but enjoy the atmosphere, decor, old-fashioned service, and experience of dining out. The habits they developed in the 1980s and 1990s as working adults are now deeply ingrained. Likewise, their children have the habit of eating out for fun and variety several times a week.

To compete with eating at home, restaurants have expanded their offerings of recorded and live entertainment. Some also allow patrons to prepare food using the manual methods of the last century.

The success of McDonald's continues in its seventh decade

McDonald's restaurants have survived decades of shifting food habits and changing food technologies in the United States and around the world. Still a favorite of Americans and others, the McDonald's restaurant of today is a fully-automated version of the original concept. McDonald's used to employ about 50 people per restaurant. Today, most of its outlets are 100% automated modular food service systems. Model H-13, the most common configuration, is a unit of 2.5 meters on a side tha can be set up in office buildings, apartments, on street corners, in parks, and as a drive-through service. A typical H-13 unit is illustrated and explained below.

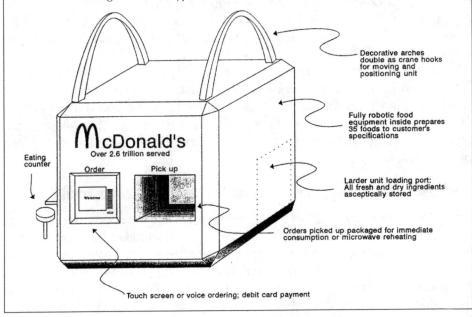

Decorative arches double as crane hooks for moving and positioning unit

Fully robotic food equipment inside prepares 35 foods to customer's specifications

Larder unit loading port: All fresh and dry ingredients asceptically stored

Orders picked up packaged for immediate consumption or microwave reheating

Eating counter

Touch screen or voice ordering; debit card payment

Machines do most of the work of preparing food

Food preparation in the home can involve any amount of human effort from next to none to complete hand preparation. The hobbyist or gourmet purist still prefers to prepare food by hand. Most people, however, rely in varying degrees on automated home appliances and prepared, packaged foods. Some of the home appliances used are:

- *Tunable microwaves,* patented in 2016.

Tunable microwaves cook food with precision

Most commercial and domestic microwaves now employ wave-bending technology for precision cooking. The units have sensors that can assess the shape, moisture content, and density of food as cooking progresses, or, relying on food packaging microchips, the units can reshape and redistribute microwaves so that heating is tuned precisely to the food. Frozen, aseptic, and dried microwave meals are produced for optimal flavor and cooking using advanced knowledge of food chemistry. It is now possible to keep ice cream frozen in a tray next to a hot entree.

- *The automated pantry.* The unit assesses its stock and tells the user or kitchen computer whether ingredients are on hand. Alternatives and substitutions may be suggested. Assuming certain ingredients are missing, the system asks the user whether to order them. Automatic weekly orders can easily be generated based on the automated pantry's self-assessment.

- *Robot chefs.* The robot chef, still a high-priced item, is really an appliance. With a fully automated pantry and robotic cooking appliances, it is now possible to have meals prepared automatically, or with only as much work as a person cares to do. Most major brands, for example, have fully programmable menus (for individual tastes and nutritional and medical needs). The gourmand simply keys in or calls in the desired menu half an hour to an hour before eating time. Current models have a repertoire of about 30 dishes, based on pantry size and options.

- *Zoned cooling in refrigerators.* Refrigerators have either predefined cooling zones or rely on food packaging to signal which foods need what temperatures.

Variety characterizes food in the United States

Today we have unprecedented choice in what we eat. The variety may stem from a consumer's ethnic background, medical needs, lifestyle choices, or, simply, tastes. Intensified flavoring for the elderly is one example. Ethnic foods come and go as fads, often according to new waves of immigration to the United States. In the past five years, we have seen the popularity of rainforest fruits, vegetable stews, Hausa shakes (milk and papaya or mango), and, for the adventurous, spit-grilled iguana imported from Costa Rica and Mexico. A typical family's food day is described below.

A typical food day for a U.S. family

Mom, 37, ate a nutribagel on the way to a 9:00AM supersonic flight to London. Business class served her a California-style lunch, which she chose when boarding. The fruit juice with her meal incorporated a drug to offset jet lag. Then, food was prepared from processed components at the airline's central catering facility and was microwaved in flight. It was well-cooked and tasty. Landing three hours later, she stopped for a snack at a street vendor, buying a hydroponic banana pear. That evening, Mom had a Ploughman's platter catered in from the pub down the street.

Dad, 39, as he begins an early work shift, gets a multivitamin fruitshake from a machine at the hospital. At midmorning, Dad stopped for yam chips and rye cola. Dad skipped lunch because he was in surgery for three hours; instead, he chewed a quick-energy gum. In late afternoon, Dad left for home, stopping along the way for a chicken sandwich at a McDonald's Automat.

The kids, a boy, 11, and a girl, 8, go to primary school. They ate breakfast with Grandmother at home before racing off to school. The boy ate Fruit Rockets™, a cereal for prepubescent boys. The girl had waffles with passion fruit syrup. At school, in mid-morning, the kids drank fortified milk and ate a cookie (state law requires it). School lunch was pizza sticks.

Grandmother, 70, ate a blueberry muffin for breakfast, which was prepared in her Cuisintech in 45 seconds. She could have touched the screen for cranberry, bran, or kiwi walnut. When she orders the batter refills, she always gets the "Traditional Pack." Grandmother lunched after exercising at the Seniors' Club with friends, on angelfish over salad greens.

Dad, Grandmother, Boy, and Girl dine together

Grandmother and the kids picked Dad up at the health club, and they headed for Taste of America. Dad and Grandmother enjoy the variety of food, and the kids can play virtual games in the restaurant's arcade.

Grandmother likes organically grown food. She had open-ocean tuna, Inuit style, with tundra greens. Dad had Polynesian food. His parrotfish were farmed in underground tanks in Milwaukee, fed on brewery wastes. He ate roasted taro and drank coconut milk with his meal. The boy, currently learning about the Amazon in school, wanted rainforest food, and ordered a tapir cutlet with cassava. The girl who had just played a virtual game in which she explored the Austrian alps by parasail decided on Austrian food. She had the zero-calorie strudel. Robotic assistance and frozen and vacuum- packaged prepared entrees make it possible for Taste of America chain restaurants to serve 312 different entrees with a staff of 6.

At bedtime, Grandmother took a snack to prevent Alzheimer's disease.

Part of the choice available to most people is a range of alternative foods, including the increasingly popular nutritionally complete vegetable varieties. Some people round out their nutrition through daily or weekly nutritional capsules or nutritional implants that release nutrients into the body at regular rates or according to need, based on continuous monitoring of nutritional status.

Molecular synthesis of foods has often been written about. The practice is still too expensive to compete with other synthetic as well as natural foods. Agricultural food production has grown so dramatically that the notion of producing foods molecularly, using basic chemical feedstocks, has not gotten established. There are, however, several common feedstocks for other kinds of food synthesis, such as soybeans, corn, and milk. Algae and several other aquaculture products also provide raw materials for synthetics the way corn once did.

Today's fat, sugar, and salt substitutes fool the body and the taste buds. In addition to Nuvelle[tm], noted earlier, several competing salt and sugar substitutes have proven themselves in taste tests and in a dozen or more years on the market.

Vintners now sell more synthetic wines than real ones. Using grain alcohol and chemicals, wine makers produce what they claim is wine of superior quality, matching the great vintages of the past 100 years. White wines have had much greater success than reds.

A final alternative food is no food at all. Recent experiments have renewed interest in stimulating the hypothalamus to fool the brain into thinking it has eaten. There will no doubt be a big market for this technology as a diet aid if the technique proves itself. Forecasters who are advocates anticipate a marketable device by 2032.

Minimizing wastes

Wastes from food processing and preparation are minimized with genetic engineering and better preservation and packaging techniques. The least waste is produced by industrial food processing, in which nearly every waste material is recycled into some use at the facility or sold for the preparation of foods and chemicals elsewhere. In the home, packaging and spoilage-resistant foods have paid off in far less waste. Wastes from home and industry are effectively captured for reuse from sewage systems, most often finding their way into energy-producing units that feed on waste biomass.

Technology Reduces Food Waste

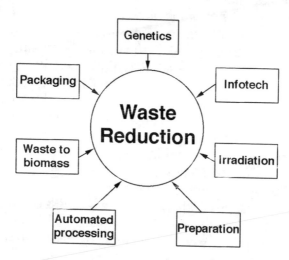

The medicalization of food

Modern technology, particularly genetics and information technology, combine to make food an integral part of human medicine. Since the end of the last century, we have seen the medicalization of food. For people with special problems or particular needs, nutritional assays are done periodically to set up diet plans. Smart cards, which people carry to restaurants and stores and use with their home appliances capture and track nutritional information specific to the person.

Those with less acute food needs (or interests) generally choose to get their nutrition willy nilly as they eat healthful and not so healthful foods. They know medicines and food supplements are available to correct any deficiency.

Pharmaceuticals and food have converged as nutriceuticals—foods custom fit to individual needs and designed to supplement nutrition. These products have replaced many dietary supplements and some medicines.

WORLD 2—5.1 BILLION WHO ARE GETTING ALONG

The middle 5.1 billion live in South and Central America, parts of eastern Europe, the eastern Mediterranean, Middle East, Asian republics of the CIS, Southeast Asia, and Oceania. These countries have well-developed agriculture and industries and a base of resources with which to trade internationally. Their people are educated, and their governments are usually stable.

It is easy for a country in the middle to fall back into poverty, especially under conditions of ecological disaster, long-term political instability, or the loss of an important resource. It is much harder for a World 2 country to rise to the top.

Food and agriculture for the 5.1 billion involves moderate amounts of advanced technology. Genetics and biotechnology, through technology transfer, international investment, and local initiatives have raised agricultural productivity in these countries. It is still growing. Robotics and automated and containment agriculture are unusual, typically undertaken by multinationals and government farms, whereas the majority of farming is done with equipment and processes that would not be out of place in the 1990s.

Food processing, distribution, packaging, and end uses span the spectrum of low to high technologies. Some enterprises use the advanced technologies of food processing found in World 1, whereas others make good business out of traditional methods of food processing.

Distribution typically uses a combination of hydrocarbon-powered vehicles and rail for transport. National or large business distribution networks are fully computerized, with ordering, payment, and inventory done electronically. Small operators may do everything by hand and on paper.

Home and restaurant cooking are largely traditional activities, though the microwave oven reigns supreme as it does in the affluent world. Video instruction is widespread. The 30-year trend toward more eating out and prepared foods in World 1 countries began to penetrate World 2 countries in the 2000s and 2010s.

WORLD 3'S 2.0 BILLION—THE WORLD STRIKES A HARD LINE

Unrestrained population growth in the poor countries of the world and increasing episodes of famine since the 1990s have afflicted 2.0 billion of the world's population. Excessive population growth has eaten up any gains in food productivity from the first and second Green Revolutions, and lowered *per capita* food production in most of Africa and southern Asia. The same countries are prone to violence, warfare, refugee problems, poverty, and unemployment.

The hunger situation has given the affluent world three choices:

- Attempt to feed everyone, perhaps creating soup kitchens on the million-person scale,

- Introduce technologies and practices that enable people to learn to feed themselves, or

- Allow famine and large-scale starvation.

World powers and international bodies have decided over the past two decades on a combination of the second and third options. They have stopped distributing food every time a country falls into famine. The rationale, established at the 2009 Conference on the Planet in Rome, is that every region and country of the world must strive to sustain itself through its own resource development and participation in world trade. The conclusion of the conference, still adhered to today, is that the world cannot feed everyone. Countries are responsible for their own political and economic stability, so long as what they do does not impinge on other countries or adversely affect the global commons.

Famine today is most often caused by political instability, the destruction of the environment, and unemployment. Natural disasters are considered unavoidable, and international bodies are willing to relieve their effects. More pernicious is the inability of millions of people to buy the food they need. Even where food is available, many starve. Unemployment rates estimated as high as 67% in metropolitan slums contribute to much of the world's hunger.

Technology transfer is the international philanthropy of choice today. Despite limited success over the past 40 years, it is considered an economical and ultimately effective way to spur sustainable development. It also creates synergy between nations with the relationships it builds between businesses, governments, and research institutions.

Changing demography of the poorest

The populations of some poor countries have finally begun to stabilize, for example, in Egypt and China. Sub-Saharan Africa is the last holdout and is under considerable pressure from the World Population Centre. Family planning and rising educational levels for women have made the difference. This could allow food production finally to catch up with the needs of the population. But migration and political instability continue to disrupt development. Acquired immunodeficiency syndrome (AIDS), especially at its epidemic peak in 2009, raised the health-care burden and decimated the working populations of a dozen countries in Sub-Saharan Africa.

Forces promoting hunger

Among the forces promoting continued hunger and famine in the poor countries are:
- Unsustainable population growth
- Unemployment
- Stagnant agricultural practices and unused new technologies
- Erosion, especially driven by fuelwood taking, overgrazing, and overuse of water
- Land lost to urbanization
- Land lost to desertification
- Market disincentives
- Political and social upheaval
- Refugees

A number of countries, particularly in Africa and southern Asia, have seen recurrent episodes of transborder migration of hundreds of thousands to several million people. The refugees are increasingly violent in their attempts to reach safety and food sources. Many refugees die in transit.

Metropolitanization has strongly affected the poor countries of the world and was exacerbated in part by international aid in the 1980s to 2010s that made capital cities and port cities the centers of food distribution. Perceived economic opportunity, of course, is the main driver of urbanization.

The 60% to 80% levels of urbanization that have occurred are unsustainable in many cities of the world. Lagos, Nigeria, and Dacca, Bangladesh saw widespread famine and municipal breakdown as they were overwhelmed with population growth and in-migration. Often the first systems to break down are food distribution and clean water.

A CASE STUDY—Nigeria typifies World 3

Nigeria today, with 246 million people, is the world's sixth most populous country. It is also one of the world's poorest. It is worse off today than it was in the 1990s.

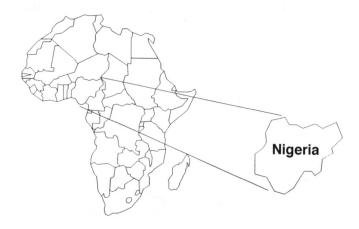

Nigeria perhaps best illustrates the plight of the destitute around the world. Though the country is atypically large in population and blighted with more demographically produced food problems than many of its neighbors, it shows what a chronic demographic crisis will do to many of the world's poorest countries. There are rich Nigerians, but the mass of the population lives in poverty, with about a third subject to hunger and famine. The successes of Nigerian economic and technological development have been insufficient to pull its population out of destitution.

Food and agriculture science and technology have raised Nigeria's overall food production, but that growth has been eaten up by the growth of the country's population between 1990 and 2025. *Per capita* food production is at 82% of its 1980 levels.

Nigerian agriculture, except for three dozen commercial or government large scale farms, is peasant agriculture. Family-run farms produce grains, meat, milk, and vegetables, with considerable hand and animal labor and gasoline-powered tractors. Produce is sold to agents who carry it in trucks to city markets. Value-added processing is done by farmers, agents, and street vendors, and includes meat and fruit drying, preparation of cheese, yogurt, fermented milk, and grain. The processing is low technology, using traditional materials.

Surging population and mass-scale metropolitanization

Nigeria is perhaps the quintessential highly metropolitan, increasingly destitute country. The country is highly industrialized, with a dozen exporting industries, including petroleum, natural gas, petrochemicals, light machinery, fish, cocoa, and rubber. With two-thirds of its 246 million people in cities, it suffers all of the expected problems of urban poverty: overcrowding, lack of housing, unemployment, disease, hunger, pollution, and violence. Millions of Nigerians move to the cities each year with the glimmer of hope that they will find jobs and build modern lifestyles for themselves. Few succeed, but the gamble is worth it to many. Today, at least 49% of Nigerians are unemployed or underemployed.

Nigeria's Population May be Levelling Off
Total Population, 1995 to 2025

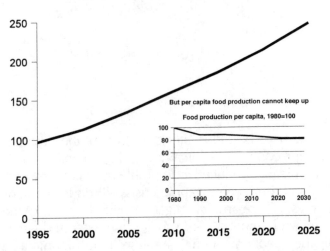

But per capita food production cannot keep up

Food production per capita, 1980=100

With its oil resources, Nigeria should have been an African success story. Two things happened that scuttled that ambition. First, Nigeria, in spite of new finds, overproduced its oil reserves, and effectively ran out of oil in 2015. Driven by raging demand in developing Africa, it could not resist the cash earnings. The money earned over four decades of production was invested poorly, and often expatriated. Much of the money was used to purchase food internationally, reducing pressures to advance and modernize Nigerian agriculture. The oil bonanza long kept government officials from acknowledging and tackling the excessive population growth that Nigeria was experiencing. Serious efforts to slow growth came too late and were only marginally effective.

Second, Nigeria was unable to reverse the runaway metropolitanization of its population. The population changed from less than a third metropolitan in the 1980s to two-thirds metropolitan today.

The Nigerian oil boom also slackened efforts to develop other industries for export. Nigeria has never met the potential for surplus food production that it has, and little food is exported.

Cholera returned to Nigeria to a significant degree in the 2000s and 2010s. The epidemic has yet to run its course despite peaking in 2011. More than a quarter of the disease's victims go untreated.

Nigerians eat more nontraditional foods: trends in Nigerian consumption of certain foods, 1990 - 2025

Traditional		**Nontraditional**	
Fresh fruits	↘	Red meat	↑
Fresh vegetables	↘	Sugar	↑ ↑
Fish	↑	- refined sugar	↑
Poultry	↑	- synthetics	↑ ↑
Eggs	↗	Packaged vegetables	↑
Dairy	↗	Packaged fruits	↑
- fresh milk	↑	Synthetic foods	↑ ↑
- fermented milk	↓	Coffee	↑
- yogurt	↑	Alcohol	↑ ↑
- cheese	→		
Flour/cereals	↑		
- rice	↗		
- corn	↑		
- wheat	↑		
Bread	↑ ↑		
Cassava	↓		

Key
↑	up
↑ ↑	up markedly
↓	down
↓ ↓	down markedly
↗	up slightly
↘	down slightly
→	unchanged

Nigeria has been the site, and even showcase, of many food-production technologies over the past 30 years. Multinational food companies have been at the forefront of these. Despite isolated successes, the country suffers recurrent famine. In the period 2018 through the present, 3.7 million people starved to death in Nigeria, almost 2% of the population.

Nigeria's sporadic successes in development and in food production have often made the country the destination of refugees from neighboring strife- and hunger-torn countries. Thus its 246 million population is swelled in its poorest ranks by all-too-frequent incursions of new mouths to feed. Though its population growth is slowing, new people to feed will contribute to lower *per capita* food production at least until 2040.

Nigerians adopt nontraditional foods

Nigerians, as has been the pattern around World 3, consume more animal protein than they did three decades ago and consume more now than people in the affluent world do, *per capita*. The protein comes from more sources, including more fish, more poultry, and exotic local species such as large rodents and lizards.

Agriculture in Nigeria is painfully slow in changing what it produces. Traditional crops, cassava, wheat, corn, cacao, and milk still dominate production. Other foods are imported. Nigerians now eat more imported foods, including synthetic foods and packaged, preserved foods.

Production

Agriculture and food industries employ 24% of Nigerians, down from 54% in 1990. Production is still labor-intensive. A fifth of Nigeria's food production is basic subsistence farming of a hectare or so plots by poor families. For rural people, cassava and grains are still the dietary staples. However, spurred by government initiatives for food self sufficiency, Nigerian agriculture modernized from the 1960s to 2010s relying on the extensive use of fertilizers, irrigation, hybrid crop varieties, and genetic engineering of cassava. The second Green Revolution (1995-present) expansion took its toll on cropland and on the ecology. No longer were farmers able to or likely to shift cultivation, giving fallow lands the time to be restored to fertility between crops. The strategies used were not sustainable in the long term and contributed to increased instances of famine as soils were depleted, and water tables lowered. New desertification affected 21% of Nigeria's land mass from 1990 to 2020.

Subsistence farming is no longer an option for the majority of the population. Nigeria is 65% urban, and its urban residents rely on truck farming and large scale commercial and government farms to produce their food. Commercial farms make up 30% of Nigeria's land in agricultural production, up from only 2% in 1990. One big actor is the multinational Nestle, which produces processed food products in Nigeria, partly relying on its own farms for inputs.

Spirulina farming could help solve protein shortages

Since pilot programs in the 1980s, spirulina production has grown in recognition as a way to boost protein in poor, warm-climate villages, while recycling human and animal wastes. The nutritious red-green algae grows fast and is more productive per hectare than every commonly produced grain crop and mammal protein source. The technologies at play are simple and appropriate to small-scale production. Energy input is minimal, a small photovoltaic panel or windmill is sufficient to power the paddlewheel stirrers in spirulina pools. The practice, though growing, is still not common in Nigeria. A survey last year found only 87 projects.

Successful production crops include corn, wheat, rice, cacao, cassava, milk, and beef. Poultry and fish are gaining favor as more economical foods that Nigeria can produce abundantly. There is hope for export businesses in these foods, especially corn, over the next decade. Cocoa remains a major export.

Problems and prospects for Nigerian agriculture

When everything goes right in Nigerian agriculture, which it does about 7 years out of every 10, the average Nigerian gets 1,875 calories a day, down from 2,116 in 2005. Government and private programs make nationwide efforts to teach Nigerians good nutritional practices and safe food handling. Agricultural extension work emphasizes sustainable, productive farming practices.

Attempts to build a fisheries base

Nigeria has 853 kilometers of coastline. It appears to be a natural place for a productive fisheries industry. Attempts to develop Nigerian fisheries have been plagued by technical, ecological, and cultural problems. Although it is possible to turn to the urban labor force for fisheries workers, cultural problems have plagued pilot programs. Red tides in 2017 and 2022 killed much of the fish in Nigerian waters and disrupted the marine food chain for several years in each case. Siltation steadily destroys more breeding grounds for fish in the Niger Delta.

Innovations have been proposed and tried for Nigerian agriculture over the past three decades. A notable example is microlivestock.

Nigeria has intermittently been able to feed its population adequately over the past 25 years. Famine has been a recurring phenomenon, most notably in 1998, 2011, and 2016. Food-production expansion has been at the expense of the environment, and unsustainable practices have continued. Eleven thousand square kilometers of agricultural land are lost annually to overcropping and erosion. Some aquifers are depleted and saline.

The ecology of Nigeria has been greatly damaged by agriculture and other activities. Little of the original rainforest is left in the country. Soils are depleted and eroded, the Niger Delta is heavily silted, and forests have been stripped of their trees.

Microlivestock promise greater production from less

Pygmy livestock, including half-sized cows, pigs, goats, sheep, as well as exotic species such as iguanas and large rodents, have joined chickens and other foul in providing protein in the poor nations of the world.

Pygmy dairy cattle are today being introduced to Fulani herdsmen in Nigeria. Advantages include ease of management, even by children, less cost for feed, and less space requirements.

Nigerians now consume an extra-meaty large rodent that is easily reared in a small pen in a yard. Its principal advantage is that it reproduces quickly.

Difficulties remain in preventing microlivestock from turning into pests and harboring diseases. New livestock means new disease vectors and new questions about human/animal interaction.

Farming practices and new technologies are slow to arrive

Biotechnology and genetics have made only marginal improvements for peasant farming in Nigeria. Great success came with improving the genetic strain of Fulani dairy cattle. Coupled with techniques for better care of the animals, introduced through television soap operas, Fulani cattle milk production is at 157% of its 2010 level.

Processing

Food processing is a cottage and small-business industry in Nigeria. Except for some state-run and development-project food factories, most people get their processed food from peasant entrepreneurs and street vendors. Imported food, such as powdered milk, flour, rice, and canned fish, makes up the balance of the Nigerian diet.

Home enterprises have taken over from some food processing, including drying fish and meats, fermenting milk, grinding grains, and baking breads. Government mills using decades old technology make flour from domestically grown grains. The construction of a plastic bag factory in Lagos in 2006 meant home enterprisers and street vendors could package their produce much more effectively for distribution and sale.

Solar meat and fruit drying is done on roof tops and in solar-drying apparatus. The technology involves simple sheet metal boxes that concentrate the heat of the sun.

Distribution

Food distribution is the point of frequent failure for the Nigerian food supply. Conflict, theft, and infrastructure problems inhibit the flow of goods from ports inland, from farms to cities, and from cities to cities.

The greatest imbalance in food distribution for the country is the difficulty peasant farmers and small-enterprise food producers have in transporting their produce to and into the sprawling metropolitan zones. The Lagos Region Light Rail Network, built between 2001 and 2014 meant many more people could reach the central markets with their inventories. Planners did not expect and did not allow for the use of the lines for goods transport. Train cars were altered with alternate seats removed in the 2010s to accommodate people with goods.

Storage problems and appropriate low-tech solutions

The country is finally beginning to emphasize food storage at the local level, instead of insisting on national grain reserves stored centrally. Losses to theft and vermin are common, with little in the way of technology to prevent either. But local storage removes the need for repeated transport and can mean less loss to theft and vermin overall, as long as storage silos are vermin proof and secure.

Packaging is coordinated poorly with logistics in Nigeria. Foods requiring refrigeration or dry conditions in transport and storage are frequently damaged and sometimes lost.

End use

Cooking and eating food in Nigeria combines the traditional methods—and customs of the Fulani, Hausa, Ibo, and other groups—with the realities of urban life and imported western practices. The influences of immigrants and refugees is also great. Chinese food is a tremendous fad today, with shops, restaurants, and street vendors selling it throughout Lagos and Zaria.

Cooking-fuel shortages are the worst ongoing ecological threat

A dozen or more solutions to fuelwood shortages have been proposed and experimented with. They usually try to replace wood as a home heating and cooking fuel. What little woody vegetation still exists is rapidly being taken by people foraging for fuelwood. Fuelwood provides 16% of the country's energy needs, down from 60% in 1990. Still, that is too high a share to sustain. Nature preserves have taken to dumping tons of imported coal outside their gates to discourage wood rustlers from taking preserve trees. Alternatives have included biomass cookers that burn wastes.

Natural gas is being turned to to solve the fuelwood crisis. For decades, most of Nigeria's natural gas was flared off as a byproduct of petroleum production. Natural gas also serves as a raw material for fertilizer production, a role it has held since the 1990s. Increasingly, it is exported for use in neighboring African countries. It is unlikely, however, that a natural gas infrastructure will soon reach many or most Nigerian homes to provide cooking fuel. Electric stoves, with energy produced centrally using natural gas, have a better chance.

Genetically engineered varieties of grains and other plants can be made to require less cooking. Nigeria, in consortium with other African countries, is planning to identify and introduce such varieties in the next five years.

Home food storage technology

Almost no rural and few urban Nigerians, other than the upper classes, have adequate food storage facilities in the home. Many people know little about safe food-storage practices. Imported and some local aseptically packaged foods help considerably, but vermin attack nearly anything stored in the home. One positive development has been the widespread sale of food safes, extremely tough polymer cabinets that latch closed and seal tight. These are produced cheaply in the region with recycled plastic wastes. Nine hundred thousand have been sold in Nigeria since 2017.

Fermenting milk and milk/grain mashes, as well as cheeses are preferred ways to use milk in the absence of other storage technology. The practice is still dominant. However, portable food irradiators are available in larger villages to create milk and milk products that need no refrigeration. About 4,000 are in use. Fresh milk is sometimes made lactose free (for the lactose intolerant), using filtering membranes.

Other potential agricultural strategies

Nigeria has not progressed far in developing its agricultural capabilities. The pace of technology transfer has been slowed by economic and political impediments. With added efforts, Nigeria could look to other potential agriculture and food production technologies for sustainable development that have been tested and often proven for use elsewhere in the less-developed world. Among these are:

Mariculture

- Plant varieties genetically programmed to collect water when it is more plentiful, e.g., at a set time in the growing season, and conserve it for drier times
- Sustainable agriculture approaches
- New advances from biotechnology
- Irrigation technologies, including drip irrigation
- Macroengineering schemes to replenish aquifers, reroute river courses, and raise rainfall levels.
- Farming the cities (see the box below)
- Technology to restore heavily leached tropical soils, including artificial soils, nutrient supplements, and restorative ground cover plantings

Farming the cities makes urban life more sustainable

Urban farming makes it possible for the urban poor to provide some of their own food supplies. It incidently helps clean the air and recycle organic wastes. Urban farming has taken hold in several dozen of the impoverished world's megacities. Once certain techniques are taught, regulations changed, and materials provided, it takes little pressure to get people to sustain an interest in working plots on roof tops, alongside roadways, and in city parks. The remaining problem for most gardeners is theft of produce. Technologies available include:

• Lightweight, moisture-retaining artificial soils

• Systems that combine pigeon or chicken husbanding with vegetable farming

• Plant varieties, through biotechnology, that tolerate urban polluted air and even scavenge pollutants from soil and water

Food

•   Innovative storage possibilities on the mass and household scales

•   Transport and distribution technologies and strategies including modular containers and vermin proofing

•   Information flow to raise distribution efficiency with monitoring of supplies, demand, and logistics planning

•   Delivery on a mass scale of nutritional capsules to the hungry

Critical Developments: 1990-2025

Year	Development	Effect
1965-1985	First Green Revolution.	Widespread use of high-yield varieties.
1991	USDA Plant Genome Research Program.	Spurred completion of genomes for commodity crops and livestock.
1995-2025	Second Green Revolution	Widespread adoption of engineered crops and automation in farming.
1998	U.S. Conservation in Agriculture Act.	Provided tax incentives for waste recycling and water conservation.
1999	U.S. Waste Reduction Act.	Banned once-through plastics in packaging.
2003	Fully automated technology for containment agriculture introduced.	Advent of fully automated factory farms.
2007	Geneva Conference on Hunger.	Established the principles for food assistance only in natural disasters.
2010	Greenhouse warming proven significant.	First projects for mitigating greenhouse warming effects begun.
2010	Direct transfer of genes into genomes practical.	Flowering of genetic engineering in agriculture.
2011	International Agreement on the Troposphere.	Established council to oversee climate manipulation.
2012	Introduction of fully automated equipment for open-field agriculture.	Led to establishment of highly automated open-field farms in affluent countries.
2016	Tunable microwave patented.	Flexible microwaving opens up the prepared-food field to much greater variety and higher quality.
2016	Ogallala aquifer depletion crisis in the United States.	Establishment of tough water withdrawal restrictions for all U.S. aquifers.
2017	International Commission on Animal Care in Agriculture established.	Approval mechanisms and regulations for transgenic animals in agriculture.
2025	World population reaches 8.4 billion.	Uncertain.

Unrealized Hopes and Fears

Event	Potential effects
Population growth in the poor countries slows by 2010 to 1.5% annually.	Food production *per capita* reverses its downward trend and grows to adequate levels.
Successful economic development in the poor countries of the world.	Elimination of hunger in all but isolated situations.
Political instability or warfare in a large developed country.	Threatens international food supplies; exacerbates famine and poverty.
Higher levels of greenhouse warming, temperature rises 2.3 degrees.	Disrupts world agriculture, causes mass desertification in midwestern North America, Amazonia, Western Europe, Georgia, and Ukraine.
Nuclear accident blights millions of hectares of agricultural land.	Disrupts food supplies and causes political destabilization, potentially with retaliation for economic damage done against the perpetrator.
Runaway transgenic animal and plant populations.	Infestation destroys agricultural lands, parks, wild areas; international agreement to outlaw transgenic plants and animals.

14

STRIVING FOR GOOD HEALTH

The three worlds split on measures of health. Affluent World 1's people are healthier and longer-lived than they were 30 years ago. They have gone beyond disease prevention to improving health and enhancing the human organism. World 2 join the affluent, with steadily improving health and success in conquering infectious disease, though they lack universal access to advanced medical and health technologies. World 3's people live in destitution and struggle against disease and early death. This chapter looks at the state of health in the world today, contrasting the health situations of the most affluent and the most destitute.

Putting the Genome to Work

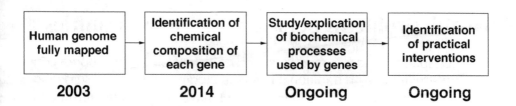

In World 1, life expectancies are in the ninth and sometimes tenth decades. In the poorest countries, life extends to the sixth or seventh decades. The world's affluent have a heightened interest in maximizing human quality of life, productivity, and achievement, along with their continuing interest in curing and preventing disease. In World 3, the emphasis remains on preventing infectious disease and other early killers. Violence, however, continues to claim lives in World 1 and especially World 3 countries. The contrast is shown below:

2025

Affluent	Destitute
Prosperity and good health	Poverty and disease
Genetics dominate	Inoculations and public health central
Drugs based on biochemical individuality	Mass inoculations
Personal responsibility	Governments promote health top-down
Fitness and lifestyles are central	Adequate nutrition is central
More people live to older ages	More people live to late middle age
Mental and physical enhancement	Physical health and survival
Cultural centrality of good health	Cultural fatalism about health

Some details of these stories follow below and in case studies for the United States, representing the affluent countries, and Pakistan, representing the poor countries.

Most people lead healthy lives in World 1

The extent of the health movement can be seen in a fundamental shift in the affluent world. World 1 societies no longer stress preventing and curing disease. Their increasing emphasis is on achieving the maximum quality and health of life, and contribution to society from each person. Human enhancement has taken hold.

Health problems persist in the poor countries

Despite advances in medicine and health in World 1, the poorest countries are struggling with traditional killers, including infectious disease, poor sanitation, and malnutrition. Some diseases have been conquered, but the world's destitute are gaining little from the human genetics revolution.

In much of World 3, health is little better than in the 1990s, besides reductions in illness and death from infectious diseases. Pandemics and recurrences of killers such as cholera still happen. Several new viral diseases have appeared, such as localized diphosphate disorder (Eldred's disease), which disrupts cell energetics. This virus emerged in 2017, and took four years to conquer. Sanitation has not kept pace with runaway population growth and especially urbanization. Some countries once rated middle-income and successful in development have lost ground in health because of massive urban growth. Sewage that fouls potable and garden water is most often the culprit.

Fighting disease around the world

International public health initiatives, such as those of the World Health Organization (WHO), have for decades fought to eradicate disease and were successful in stamping out several infectious diseases. Today, the emphasis is on helping the poorest countries reverse the health consequences of dirty water and poor sanitation.

Over the past 30 years in the affluent world, the public health focus has shifted from infectious disease to the diseases of aging, mental illness, and mental health. From the late 1980s to 2005, an overlapping emphasis was on children's health.

In World 1, most people survive or avoid the diseases that 40 years ago killed them before old age. In the poor societies of the world, more people are surviving the diseases that used to kill people in childhood and adulthood, but they succumb to the diseases of old age.

The table below shows the top killers 35 years ago and their trends in growth or shrinkage to today. The story is not good. Despite preventive treatments, palliatives, and cures for nearly every disease on the list, some have grown in deadliness. The basic reason is the failure of the poor countries to use available treatments.

Causes of death worldwide shift: trends in the leading causes, 1990-2025*

Cardiovascular diseases	↑↑		Meningitis, bacterial	→
Acute respiratory infections	↑↑		Schistosomiasis	→
Cancer	↑↑		Pertussis (whooping cough)	↓
Diarrheal diseases	↗		Amoebiasis	↗
Tuberculosis	↗		Hookworm	↗
Perinatal causes	↓		Rabies	↓
Chronic obstructive pulmonary disease	↑		Yellow fever	→
Malaria	→		African sleeping sickness	→
Hepatitis B	↓		Other diseases	↑
Death in childbirth	↓		Nondisease deaths	↑↑
Measles	↓			

*The list falls in the order of top causes of death in 1990

Key

↑	up	↗	up slightly
↑↑	up markedly	↘	down slightly
↓	down	→	unchanged
↓↓	down markedly		

For the top few killers, hopes for prevention and cure have not been realized for several reasons:

- More people living to old age (cardiovascular diseases, acute respiratory infections, and cancer)

- Massive urbanization, with more people living in close proximity (tuberculosis, diarrheal diseases)

- Failures of sanitation systems (diarrheal diseases, amoebiasis, hookworm)

- Degraded environmental conditions (acute respiratory infections, cancer)

- Microbial and viral resistance and mutation (diarrheal diseases, tuberculosis, other diseases)

- Cultural resistance and ignorance (all diseases)

WORLD 1—HEALTH TRANSCENDS MEDICINE

Genetics revolutionized health over the past three decades. It still dominates health care and medicine in the affluent countries, but its capabilities are less widely available in the poorer countries of the world.

The completion of the first phase of the Human Genome Project in 2003 made it possible to design genetic therapies for conditions and for the predisposition to disease sometimes with as high as 98% certainty. People get therapies from conception onwards, depending on the disease and the therapy. Genetic testing and counseling also play central roles in managing lifestyle and occupational health, to prevent problems caused by a particular environment or situation.

After the genome was mapped, researchers stepped up work in understanding the intermediate biochemical processes involved in the expression of disease and possibilities for intervention. The diagram below shows the many successes of those efforts.

The Genetics Revolution

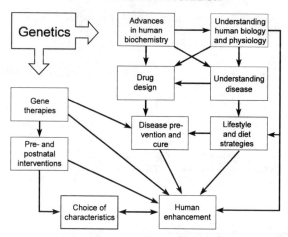

The primary outcome of genetics has been the rise of drugs, foods, diagnoses, and lifestyle health strategies based on biochemical individuality among humans. Historically, the drug industry and medical professionals skirted the issue and designed treatments for the widest possible application. A successful drug needed at least 75% efficacy. Drug manufacturers usually ignored the richness of alternative choices, despite their potential success in some individuals.

Medicine, the hero of the story, steps aside

Medical researchers and geneticists played the central roles in the health revolution. Their knowledge comprehends and spans the prenatal and perinatal effects of environment on health and longevity. They understand the effects of nutrition, fitness, mental health, and the environment on the organism. Most important, they are armed with detailed genetic and biochemical understanding of the human organism. Researchers completed mapping the location and purpose of each gene in 2003. Now, they are working to derive the chemistry of each gene and understand the biochemical pathways involved.

But increasingly, medicine has stepped aside. Information has empowered people to make many of their own choices about health. Lifestyle choices and self-care mean health is decentralized. Individuals control their own health. Genetic testing and counseling allow people to make realistic lifestyle choices according to their predispositions to disease if they choose.

People increasingly give themselves treatments for disease prevention and cure, symptom treatment, and enhancement. They can choose to take or ignore responsibility for making the most of their innate physical and mental potentials. The companion role of information technology in the health revolution is described in the box below.

The electronic and photonic world of health

The proliferation of photonic and electronic technologies in health continues. Notable current and future applications are:

Current

- Health smart cards for combined use for health histories, interactive health diagnosis, shopping for food, ergonomic adjustments, etc.
- Global database of human genes for research consultation
- Photonic medical and health journals and newsletters
- Multi-image sensors and scanners, with interpretive diagnostic software
- Databook/photonic adviser medical consultation for professionals and individuals
- Artificial intelligence triage systems for health-care institutions
- Self-contained, portable sensing and diagnostic equipment also used in communication with remote experts and information resources
- Automated health kiosks in food stores and restaurants for nutrition consultation
- Smart prostheses and sensor implants that may be queried, adjusted, and instructed from the outside
- Interactive home health diagnostic systems
- Automated reporting of health status and problems to doctors and updating of health histories
- Telesurgery for simple procedures with paraprofessionals in attendance

Future

- Home telesurgery, requiring no professionals in attendance
- Global human gene database linked to synthesis equipment for gene insertion therapies
- Worn or implanted health monitors that automatically administer drugs or neural impulses according to need

How health transcends medicine

If this section were written in 1990, it might have been titled "Medicine 1990." In those days, health for nearly everyone meant doctors, hospitals, and mysterious treatments. Health meant the absence of disease, and little else. By the early 1990s, however, affluent societies were beginning to emphasize more than just the absence of disease. They began to recognize the value of advancing human quality of life and productivity, first through achieving more of the innate potential of humans, and later through physical and mental enhancement.

The enhancement movement picks up steam

Enhancement goes beyond achieving good health and the absence of disease. The entering wedge of enhancement were cosmetic surgery, orthodonture, and administering growth hormones to short children, in addition to temporary enhancement such as from steroids—all examples from the last century.

Since then, we have seen muscle-mass stimulation, bone lengthening, and trials in gene-therapy-based intelligence boosting. Traits from tone deafness to short attention span are candidates for adjustment. There are plans to raise the national average cognitive performance quotient (CPQ) by 4 points between 2027 and 2037. The role of enhancement interventions in this is not clear.

Health is decentralized

Health is decentralized in World 1. This is true for four reasons:

- Information technologies and communications give people unprecedented access to information about health.

- Society has come to recognize that a big part of health is the responsibility of the individual.

- Regulators in the late 1990s and the 2000s changed strategies, deregulating much of health care and letting people decide, based on information, what technologies to use or avoid.

- Implants, sensors, and home/workplace/ recreation place diagnostic technologies have taken health care out of hospitals and away from administration exclusively by professionals.

Life span extension strategies are commonplace

People use parallel strategies to extend their life spans to natural limits of 80 or 90 years. We know what those limits are, based on research done over the past 40 years. The human life span is circumscribed by the degradation of tissues, limits to cellular regeneration, and general metabolic breakdown. Geneticists identified the three longevity genes in 2021. Researchers are now working to unlock the secrets to intervening in the biochemical pathways used by those genes to lengthen lifespans for more people.

Increasingly, geneticists administer gene therapies prenatally and in young children to manage the genes that promote and inhibit aging processes. Simultaneously and later, people use lifestyle strategies, including diet and fitness, to promote their longevity. Older people get immune system booster therapies for reversing the onset of autoimmune responses.

The longevity curve has squared off at average ages in the late 80s and early 90s. Such a curve for the United States is shown below.

The American Longevity Curve Squares Off
Survivors from Birth to Successive Ages
Total U.S. Population, 1950, 1986, 2025

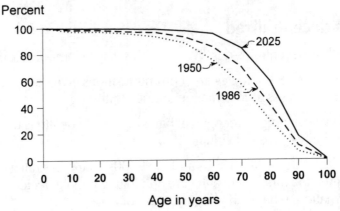

Life expectancies are similar in the affluent nations and are as much as 30 years shorter in the poor countries. Strategies to limit the effects of aging are listed below.

Limiting the effects of aging...
Medical strategies are used on people to check the effects of aging. Some of them are:

... in children
- Precise diet design, involving the principle of reduced caloric intake
- Gene insertion in the blastocyst
- Prenatal and perinatal gene therapies

... through adult life
- Lifestyle strategies, including avoidance of environmental threats and oxidizers
- Lowering metabolic rates through implanted drug pumps
- Tailored antioxidant free radical blockers
- Calorie restriction

... in old age
- Lifestyle strategies such as mental health and exercise regimens
- Testosterone or estrogen supplements, HGH
- Bionics, e.g. heart pacemakers, artificial pancreases, heart valves, artificial kidneys
- Cloning and matching of bone marrow, thymic, and spleen tissues, e.g.
- Drug-induced resumption of cell division or stem cell transplants
- Immune therapy and treatments to stop autoimmune responses
- Lysosome membrane stabilizers to prevent cell breakdown from free radicals
- Cross-link inhibitors to slow collagen aging
- Organ transplants and artificial organs

Old people are healthy and vigorous

To their 80s, most people are healthy and vigorous; however, people in their 80s experience system breakdowns and mental lapses.

Emphasis on quality of life has overtaken an obsession with lengthening life spans for most people. Innovations in health care, self-care, and medical progress mean fewer old people are physically or mentally debilitated. Compared with 20 years ago, elder sports and elder hostel activities have grown in participation eightfold. Arbitrary retirement ages in the 60s are seen more often as a reward than as a requirement.

More people take their lives

Life extension means more people make a choice about when to die. The old, gravely ill, or impaired may find it convenient to die on their own terms and not rely on a medical decision to extend life for a few months or years. Thirty years ago people saw suicide as a social and mental health problem. Today, they accept it as part of life. More often, those who take their lives are people who are physically or mentally debilitated, mentally ill, lonely, or lack sustaining hobbies and other interests.

A CASE STUDY—The United States typifies World 1

In the United States, health is a national pastime. More people spend more effort on their health than ever before. Longevity and the eradication of disease mean the emphasis is on quality of life, into a person's 80s and 90s, until death.

Genetic testing is nearly, but not completely, universal in the United States. Often it is a requirement for health-care programs. The limits of mandated testing are still debated, and 9% of people in the United States, according to a poll taken last year, eschew all genetic testing and counseling, preferring to "let nature tell me." The wealthy, of course, can make the greatest use of genetics, going beyond covered health care.

U.S. society exemplifies the affluent world's approach to health and sometimes has led the way among these countries. Technologically, the developments and forces described earlier exactly fit what is going on in the United States. Socially, there are several fundamental shifts:

From	To
Doctors and institutional decision making	Individual decision making
Regulators and institutions responsible	Individual responsibility
Generalized approaches	Finely tailored approaches
Universally indicated treatments	Patient-specific treatments
Treatment	Prevention

Importantly, medicine recognizes the particular health needs of people who were less often acknowledged by the system in the last century: women, the aging, and people of different ethnic groups. With this, medicine has moved its focus from the end of life to the beginning, with interventions, especially genetic ones, for lifelong outcomes. This shift in focus necessarily places special emphasis on children.

The following is a comprehensive look at how health is promoted and how people use medical science and technology throughout their lives. A central feature in the section is a look at six health histories.

What is health today?

Health in 2025 is a lifelong process, starting before conception and ending with death. Health includes much that is outside of formal medical institutions. It involves lifestyle, nutrition, fitness, and decisions about quality of life and when death is desirable. Medicine's role is most important in educating people about health and in early intervention to prevent, not just to cure or to treat, the symptoms of diseases. People are responsible for their health and make their decisions with the consultative participation of doctors and other experts.

Society emphasizes performance

An emerging trend in health has its roots in the last century, in sports. Athletes in the 1980s and 1990s tried to achieve their maximum physical and mental potentials. They turned to innovative mental, medical, and training strategies. Coaches and athletes perfected techniques including biofeedback, intensive physical training, and training processes. Sometimes they used drugs, steroids, and blood transfusions to go beyond their innate potentials.

Promoting good health—smart cards

The smart card, used since the late 1990s to carry health information, has meant there are more places people go that can become part of their self-care and health promotion. Thirty years ago, stores, restaurants, and recreation places began to install health kiosks for checking blood pressure, cholesterol levels, estimating kilos overweight and other simple health measures.

Today, the smart card allows people to coordinate their food purchases in grocery stores and restaurants, their exercise activities, their clothing purchases, and the ergonomics of home, vehicle, and workplace. The cards start smart and get smarter as they monitor the health history of the individual and collect information.

Society began to generalize this practice late last century with movements in business, self-help, education, and health to meet and exceed traditional levels of performance.

Today we see people identifying and targeting thresholds of performance that they want to achieve or cross. This can be the hobbyist athlete competing against others, much the way professionals have. It can be the aging person trying to maintain or increase physical vigor or mental acuity against his or her own past records or an age group standard.

This trend is likely to see its full flowering in the next few decades as more people examine their genetic potentials, decide what they want to do, and turn to self-driven and medically supported practices to reach them. Advances in genetic interventions will also serve them.

Health care and health-care institutions

As noted, people take a much greater part in maintaining their health than they did historically in the United States. However, the United States has strong professional and institutional commitments to health care. Beyond doctors, we today have much more widely developed practices in nonphysician medical consulting, health advisers, nutrition counseling, fitness, home hospice care, managing death, midwifery, etc.

Radical change has reshaped hospitals

The hospital as a place for sick people was the 20th century notion. Today, we think of hospitals as places for getting information, consulting experts, learning how to use home diagnostics, having a custom lifestyle and nutrition plan done, getting a full genetic assay, etc.

The hospital building itself looks more like a combination research laboratory/office complex. Gone are the hallways lined with patient rooms. Few people are put in beds for multihour or multiday care.

Medical technology is genetics and information technology centered

The revolution in health flowered with the rise of information technologies. Genetics is the basis of U.S. health care, but information technology empowers individuals to make decisions and provides easy access for remote consultation, diagnosis, treatment, self-maintenance, and home treatment. From the earliest diagnosing software and home medical testing in the 1980s, to today's televideo consultations, expert systems, and remotely controlled medical instruments, information technology has enabled more people in World 1 to take their health into their own hands. Today, health information technology centers on smart card record keeping—the same cards used at home, at medical facilities, gyms, stores, etc.

The box below describes a typical encounter with a home health system.

Donny Ambaye gets a clean bill of health

Donny Ambaye is 12. Donny is in excellent health, and plays speed tag. Because of hereditary diseases in both parents' families, Donny had a full genetic assay when he was a second-month fetus. Of particular concern to Donny's parents was sickle-cell anemia, which is present on his mom's, Connie's, side. Had their fetus carried sickle cell, the Ambaye's might have decided to abort it in the fifth week.

If Donny were born today, his parents might have had the sickle-cell gene replaced in Connie's fertilized ovum in vitro at the blastocyst stage. Twelve years ago, those procedures were only experimental.

Donny and his dad go into their techroom to use the HealthXOne system. Donny is old enough to do his own checkup and faces the wall screen to begin. His dad, Robert, stands nearby, and watches the monitor as Donny places a HealthXOne sensor on his earlobe. Robert chuckles as Donny rolls his eyes at the appearance on the screen of Doctor Mastodon, who says "Hello, Billy." Donny's brother Billy is five and used the system earlier. Donny says "health check" and the system recognizes him. The image of crossball star Curtis Pangabe greets Donny: "Hello, Champ!"

Reading Donny's blood, the sensor picks up nothing outside the normal ranges. Then Curtis asks Donny whether he is still playing speed tag and whether he's going out for any new sports. Curtis says, "Donny, since you're going out for soccer, let me go over some food tips with you," and suggests how to use foods currently popular with kids.

Then the HealthXOne does a few other tests with Donny. The grip test, lung test, and eye check all show Donny right on track for a kid his age. Curtis shows Donny a key move the pros use in speed tag.

Finally, Curtis comes back on screen and reminds Donny to get his yearly injection of anticancer cocktail. At the same time the system makes an appointment at the clinic, and puts a reminder in Connie's video mailbox.

Curtis says, "Hey Champ, don't forget to put your Health Card in the slot for updating."

Donny's health stats are stored in the system, and on his medical card. Had he had any problems, the system would have reported them to Connie, and to the family's health plan. There some problems are evaluated and treatments prescribed by the plan's computers, while special cases are reviewed by doctors.

Allocating health care in the United States

U.S. health care now more sensibly allocates its resources. We no longer follow a save-the-patient-at-all-costs strategy or buy a few more days or weeks of survival with massive medical care. Prognoses dictate care. Decisions are in part arbited by triage software that makes objective judgments based on widely accepted standards. People who refuse to be rationed thus may still find care if they can pay for it, or may try home care or alternative medicine. The box below includes a famous example.

People who refuse to die

Medicine has established that human lives are finite. Still, it is medically possible for doctors to extend lives indefinitely with complex treatments on old people. Only a handful of people, typically the super rich, have had the battery of treatments needed, possible since the 2010s. Japan's national treasure, the poet Shinagawa, is an example.

Another is Costas Papadopoulos, the transportation trillionaire. He is now 73. His procedures were begun in 2017. Though it is unlikely, he plans and expects to live to be at least 145. Among the treatments he began in 2017 were:

- Fetal brain tissue therapy for maintaining mental acuity

- Marrow infusions and tissue therapies for maintaining blood cell production

- Heart, kidney, and liver transplants

- An artificial lung paired with his natural lung

- Anticancer therapies (ongoing)

- Immune system boosters (ongoing)

"Papadollar," as he is called, eats only a laboratory-concocted diet, designed for maximum nutrition and antiaging. He is looking for experimental treatments to rejuvenate his lymphatic system and to further forestall metabolic system breakdown.

More people live to their natural life expectancies

For females in the United States, average life expectancy today is 87.3; for males, it is 81.4. Though some scientists used to think that people could live to as much as 140, no one has.

Calorie restriction is proven to lengthen life. From childhood onward, perhaps 47% of people in the United States, according to a recent survey, reduce their caloric intakes. More people need to adopt the caloric and nutrient guidelines from the FDA that are fine-tuned to lifestyle, ethnic, and age differences among people. Though the guidelines have been published since 2011 and updated regularly, only about 17% of people in the United States follow them, according to the same survey.

On the following two pages are six health histories that illustrate the evolution of health and medical technologies over the past 75 years.

Seven health histories

A female fetus (due in 2025)

- Parents chose the date of conception, gender, longevity, and hair and eye color of the child. The genes of the blastocyst were screened for genetic disorders possible from either parent. A several-step genetic manipulation of the preembryo, shortly after conception, replaced the essential genes for characteristics selected.
- At the end of the first trimester, a physician corrected the fetus position in the womb and used a multi-imager to check for any growth and development problems.
- At eight months gestation, the fetus had an immunization complex for five infectious diseases.
- Ready for birth on October 27, 2025. Parents will schedule a convenient delivery time.
- Will get a blood chemistry assay and tissue sampling to catch any problems not screened at conception or detected in utero. This follow-up is not covered by insurance, but the parents want to make doubly sure the child is in and stays in perfect health.

A five-year-old boy (born 2020)

- The child's parents used a hormone treatment prior to conception to select a boy.
- Gene therapy in utero to "correct for" congenital short stature that is common in both parent's families.
- Will be 4 centimeters taller than both parents by age 13.
- Additional inoculation cocktail for other infectious diseases at five months.
- A broken radius at age four reset, with healing promoted electrically. The bone was fully healed in six days.
- Parents use the Nutripedia™ system to optimize growth and development. The system also anticipates the problems of senescence, beginning the child on what is meant to be life-long caloric tailoring for longevity. The system helps parents select foods according to blood chemistry and other measures of nutrition and health that they can do at home.
- Since birth, parents monitor the child's health using home testing and diagnostic equipment. Blood testing is done without penetrating the skin, with a blood chemistry assay.
- Using a Televideo center, parents can make routine diagnoses with computer assistance, or consult a pediatrician on line. The system updates the child's health card for use at the doctor's office, pharmacy, day care, or shopping center.

A fifteen-year-old girl (born 2010)

- Genetic testing was well established, though genetic therapies were new. The parents spent several months with technicians testing fertilized ova to avoid several genetic predispositions to disease. Tay-Sachs and compulsive disorders are present in both families.
- Regular blood tests screen for any abnormalities.
- Cancer prevention, liver, kidney, heart, and lung diseases to be watched for in annual tests through life.

A thirty-year-old man (born 1995)

- Conceived through in vitro fertilization, already becoming common at the time.
- In utero surgery to correct a misshapen left arm.
- Saw a psychologist in preteen years for shyness, stutter, and school refusal and began biochemical therapy for all three conditions.
- Accident caused deafness in one ear at age 15—corrected a year later with an auditory nerve growth stimulator
- Sees a counselor for anxiety and depression and takes the over-the-counter drug Serenity™; part of a photonic affinity group for the clinically depressed meeting via televideo.

A fifty-year-old woman (born 1975)

- Parents used fertility drugs when they had difficulty conceiving.
- Fraternal twins born. The boy died of sudden infant death syndrome; monitors and preventative technologies were not yet available.
- Jaw bone reinforced and induced to grow new bone mass at age 45.
- Diagnosed with ovarian cancer at age 33, the cancer was reversed without surgery; today the disease could be caught and quickly stopped through annual blood tests.
- Vehicle accident at age 37, required microsurgery in brain to prevent permanent damage.

A seventy-five-year-old man (born 1950)

- At age 10, had braces to straighten his teeth.
- An AIDS survivor, the disease was reversed prodromally with immune therapy in 1997.
- Knee surgery on damaged cruciform ligaments—right knee in 1974 done with open joint surgery. Left knee done in 1994 through arthroscopy. Scar tissue recently removed from right knees with catheter laser.
- Gum disease, which developed around a dental implant in 2012, improved but not cured with a cell-growth therapy.
- Radial keratomy in 1994 to correct extreme nearsightedness in his left eye.
- Predisposed to Alzheimer's and is taking drug therapy and may have neural tissue grafts this fall. His health insurer will not pay for the grafts unless he undergoes the drug therapy first.

A person near death (born 1935)

- Will shortly enter home hospice, with attendants around the clock. Life will be extended a few days with a drug to counter cardiac arrest and cessation of breathing, so that family members will be assembled. Anticipates a three-week stay.
- Will control his own pain through neural stimulus, as needed.

Genetics and lifestyle choices

Smoking, all but banned in the early 1990s, has seen a resurgence as more people find out whether they are genetically predisposed to lung cancer. By 2006, genetic tests established with 98% certainty whether a person was predisposed to the disease.

Other examples abound in which people follow or reject lifestyle advice based on genetic testing. Because a person may now know, with nearer certainty, the health consequences that will result from a behavior, more people can choose behaviors that were once thought to be universally deadly.

Still fighting cancer

The variety of cancers and variety of processes and tissues affected by cancers have kept the disease at the top of the health agenda in the United States. Despite impressive gains in treatment, prevention, and cure, cancer kills 7.3% of the U.S. population. The central problem is that cancer is elusive, evasive, and persistent in looking for inroads into the human organism. The vigilance required to prevent cancers is too much for health care and people to manage, and the disease continues to make people sick. According to cancer specialists, too much of our approach to the disease is on stopping it after it has started. To fight the disease better, they say, everyone should be genetically screened at birth for predispositions to cancer. Researchers are still identifying interventions for people predisposed to cancer.

The still common cold

Although we cannot prevent colds, treatments available today typically stop symptoms and beat the viruses within a few hours of the onset of cold symptoms. Because there is no preventive remedy, the common cold adds up to tens of billions of dollars in health care costs in the United States.

It would also be effective though expensive to monitor every person's health for the onset of cancer. Although all school children and 78% of adults are estimated to be screened regularly for blood indicators of tumor growth, the disease still strikes.

Replacing parts—often the cheaper solution

Synthetic blood revolutionized medicine in the late 1990s. Since then, we have seen advances in other areas of implants, prostheses, and donor organ replacements. The age of medical prostheses and mechanical implants is ending. However, the use of medical implants is an important adjunct to preventive therapies, genetics, tissue culturing, and cloning. It is often cheaper and less stressful on a patient to implant a heart valve, artificial pancreas, or spleen, pacemaker, lung blood oxygenator, or stomach membrane lining replacement than to use a donor organ or other expensive and complex therapies. Because the basic technologies are well-established, devices are virtually failure-free.

The 2010s—the era of tissue transplants

Tissue transplantation was a stopgap method of treating disorders, especially prenatally and perinatally. When researchers identified the tissues responsible for a particular disease, they quickly were able to find healthy tissue to substitute for the unhealthy. Sources for tissues included accident victims and aborted fetuses. More recently, the full force of genome outcomes has come into play. Tissue transplantation has given way to direct intervention in the biochemical pathways of the body, and to genetic manipulation of the sperm, ovum, and fetal tissues in utero. The latter strategy was illegal until 2007 but quickly became not only accepted but increasingly central to pediatric medicine.

Organoids

Organoids have also been successful. Their success suggests that we are not far from growing perfect replacement organs either in the body or in vitro. First demonstrated in animals in the late 1980s, these artificially grown organs became effective replacements for the spleen and pancreas by 2013. The potential for more complex systems like the heart have so far been beyond our reach.

Organoids are organs induced to grow in the body on an implanted matrix. Particular cells, such as insulin producing cells for pancreas replacements, are inserted in the matrix, along with growth hormones. The body adapts to the organoid with a blood supply, and it develops the ability to secrete substances and govern biochemical processes. Once triggered, the organ-forming process goes on by itself in the body. The organoids do not only simulate or replace existing human organs. Some, such as the anticancer organoid TNF32, become a biologically compatible implant for introducing an enzyme, hormone, or other compound into the body. Another valuable use is in lubricating degraded joints. Tiny hyaluronic-acid-producing organoids placed, for example, in the knee's synovial sac, can reduce or remove the pain of arthritis. The acid was once obtained from the combs of roosters, among other sources. Today, an organoid can make the body produce it.

Organ donation continues, cloning costs remain too high

Donor organs are transplanted in record numbers, despite advances in preventing and curing diseases and advances in artificial organs. Organ banks and preservation technologies have transformed this practice since the 1990s, when most organs had to be obtained fresh only hours from use.

Examples of the scope and extent of medical interventions: a schematic

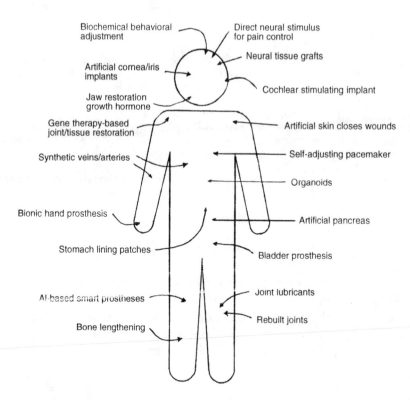

The artificial kidney

Membrane technology allowed the development of today's artificial kidneys. This highly successful technology has ended all but the occasional donor kidney replacement.

The most important changes over the past two decades have come in donor organ handling. Immunological responses to donated organs is a thing of the past. Tissue matching is no longer necessary, because we now control the body's response to foreign tissues.

Donor organs are cheaper than artificial organs. Sometimes, they work better. However, fewer organs are available as more trauma and other accidents are made survivable by medical advances. More donors today come from outside World 1.

Reproductive control grows

Reproductive technology has coalesced in recent years to the point where people have complete control over the conception and delivery of babies. The experience of a typical couple planning to have a baby is described below.

Emily and Carl plan to have a baby

Emily and Carl have decided to have a baby. They want a girl, and they want her to be born in October. They are interested in choosing characteristics such as height and hair and eye color, but those techniques add too much to the cost. They look forward to being surprised. Their genetics counseling assured them there were no diseases or other problems to worry about.

Emily and Carl have elected to use a more cost-effective gender-selecting enzyme, available from the pharmacy, to prevent conception of a boy. To time things just so, they will use ovulation-enhancing drugs to extend their window of opportunity to four days.

Before conceiving in mid-January next year, Emily will have a full examination to see whether "all systems are go." They have not had a child together before. She will get a blood scan for hormonal levels (which she can do at home). At the clinic, a clinician will use a noninvasive probe to give her a pelvic exam, checking her entire reproductive tract.

Emily and Carl plan to have a birthing specialist on hand at their home for delivery, if Emily's pregnancy goes well. They will use a drug to induce labor when everything is ready. They do not expect any problems.

Technology has promoted reproductive planning. By 2023, 94% of births in the United States were planned. In 1995, about 52% of births were planned. Family planning initiatives seek to reward planning with health and other benefits, and provide counseling and access to technology.

Geneticists can manipulate more of a fetus's DNA to select traits and prevent disease. Gene replacement, once illegal, will become a central element in reproduction technology in the next few decades.

Remaining on the reproduction agenda is the need to get 100% of pregnant women to use the pre- and perinatal care they need for healthy pregnancies and healthy babies. This has been a decades-long battle. The remaining problems are social, not medical.

Contraception has benefited from decades of research in human hormone biochemistry and immunity. The every-three-month male injection, for example, relies on an extremely subtle adjustment of hormonal levels. Women can have an induced immune response to sperm, get a five-year contraceptive implant, or take a regular injection. Progestin receptor-blocker based drugs originating in France with RU-486 (Mifepristone) became legal in the United States in 2000. Other postcoital drugs to induce menses, anticipated in the 1990s, were finally perfected in 2011.

In vitro gestation and potential for the artificial womb

We cannot expect artificial wombs for another dozen years. Some parent-candidates have recently become part of experiments in having the first 40 days of gestation in vitro. These experiments follow successful monkey experiments done at the Chicago Institute of Fertility. This procedure will require the mother to get drugs to induce her body to respond as though pregnant, so a fetus may be implanted later.

The laser, still shining bright

The decades-old technique of using lasers in surgery and other procedures is still the safest and most efficient technique for dozens of routine medical procedures. Our lasers are more finely tuned than yesterday's, and many are sent into the body through catheters, and may sometimes be implanted for ongoing work.

To cut or not to cut

Most people today, when faced with traditional surgical procedures wonder why doctors sometimes have to cut, with all the noninvasive techniques at their disposal. Does open thorax surgery still make sense? Sometimes, of course, it does. A doctor facing a patient with multiple internal injuries usually finds it more efficient to open the patient's thorax, the way it was done last century, to make multiple repairs. Each procedure could be done with catheters and other less-invasive techniques, but in combination, those techniques would take too much time. Also, the effects of one internal injury can obscure others, making the doctor's exploration and surgery with catheters and scopes more difficult.

Pain control possibilities

Pain control has come a long way since the knock- 'em-out drugs used last century. Greater refinement of drug design, advances in brain science and neural stimulus mean that pain control rarely has unpleasant side effects. The latest drug implants are implanted tissues that produce natural narcotics for local or general pain control.

Still, the long-heralded direct brain anesthesiology has not yet arrived. Today's electrical simulators are a stopgap technology. Look for greater advances by the end of the decade. It is likely that neurotransmitters will be blocked non-invasively to prevent sensations of pain for the duration of a medical procedure or for routine control of headache, etc.

Nevertheless, most of today's surgery not tied to trauma is done from the outside, with probes, radiation, catheter lasers, gamma knives, etc.

Getting surgery is easier for those in remote areas

Greatly simplified and cheaper telesurgery equipment is now available for clinics in outlying areas. Like the existing teleoperated surgeries, these devices let surgeons work remotely on most basic procedures, usually with a nurse assistant at the patient site. The systems may be soon adapted for mobile paramedical crews as well.

Working with the brain

Electrical brain stimulators for pain control are just one application of chip power to the human brain. Paralytics regain the use of limbs with implanted stimulators and a software package called "New Step," which helps them walk by recreating natural motion. People can sometimes treat mood and depression and some headaches with stimulus. Victims of recurrent headaches can have an implant that they can trigger for pain relief. Other successes include:

- Restoring hearing
- Restoring and improving taste and smell in old people

Direct computer/brain interfacing requires more control over neurological processes than we have today. Nevertheless, the experiments are well underway in interfacing the brain stem with computers to correct neurological problems. In as little as 15 years, some researchers believe, we will connect computers to the cerebral cortex for successful direct interfacing. The popular sitcom from the late 1990s, "Brainoid," anticipated that development.

Neural grafting cures Alzheimer's and Parkinson's

Particularly as more of us live to old ages, Alzheimer's and Parkinson's diseases, though preventable, turn up in disturbingly high numbers in the population. Each year, about 500,000 people in the United States are diagnosed with Alzheimer's. Today's elderly, who were born in the 1940s and 1950s, did not have today's genetic and lifestyle therapies available to prevent the disease. Many also lived most of their lives before researchers discovered the lifestyle and environmental factors behind Alzheimer's, Parkinson's disease, which is genetic, also afflicts a substantial portion of the population. For both Alzheimer's and Parkinson's, there are cures with 60% to 80% success rates. The treatments involve brain tissue grafting and tissue regeneration. Most often, doctors patch the diseased tissues with healthy tissues from donor brains.

Imaging makes steady improvements

The photographic quality of a multi-imager is a long way from the murky, grainy images of magnetic resonance imaging (MRI) and positron-emission tomography (PET) 30 years ago. The key is in how the systems combine the sensing abilities of positrons, photons, magnetic resonance, and molecular probes into the same full video image.

Color is useful to the clinician to accent the properties of tissues analyzed and to highlight diseased tissues and other medically relevant conditions. Artificial intelligence can interpret most images and give the clinic professional an immediate diagnosis. Most systems are fully integrated with molecular probes that take a closer look at the tissues of interest and report precise biochemical findings. The ability to do all this in full motion and real time allows computers or clinicians to make live adjustments to a condition and see how the tissues respond chemically.

Most clinics are networked to the global Med-Imagebase that allows a quick match of the image at hand to archetype images indicating diagnoses. At the same time, unique or rare readings are reported with the patient's consent to research organizations that have logged standing requests for certain classes of things on the system. You could be scanned on a Tuesday and be in a Photonic NewsFlash that afternoon, or an online medical journal two weeks later.

More sensing and imaging available to people

Health professionals no longer have a lock on sensing and imaging technology. Rentable pregnancy ultrasound equipment for home use has been available for 20 years. Technologically, it has also long been possible for people to use multi-imagers and other medical sensing instruments on themselves at home, in health clubs, etc. Until recently, the cost has been too high. Now, coupled with better diagnostic systems and the cooperation of doctors, full-fledged multi-imaging systems are to be marketed for home use.

Molecular probes for gentler sensing and testing

Molecular probes are the favored technique for diagnosing disease, particularly diseases discovered and tracked through biochemical processes. Similarly, geneticists use gene probes on DNA, even for fetuses in utero.

Implanted sensors: making sure

Implanted sensors make sensing even more foolproof. About 17 million people in the United States have implanted sensors. Many monitor for and correct biochemical imbalances. Others monitor body processes for diagnosis. Doctors can read them remotely. Still others give biofeedback signals to the muscles or brain to correct a musculoskeletal or system problem.

Most intriguing is carrying further the possibilities for muscle-computer chip interfacing, for enhancing human physical performance. Today this is done in limited ways in some athletics. An intriguing possibility is having a human enhancement Olympics. The box below describes the idea.

A human enhancement Olympics

The rise of physical enhancement in athletics has led to recent discussions of holding a special Olympics. Instead of shutting out athletes using artificial means to improve their performance, this Olympics would celebrate it. The boundaries would be defined, of course, but contestants with implants to enhance oxygen capacity, muscle performance, and other factors would compete against others in using the same enhancements. It remains to be seen how enhancement researchers will be acknowledged in awards ceremonies.

Mental and physical fitness go hand in hand

We now recognize the central role of mental fitness and attitude in health. Dealing with the psychological effects on physiology are now integral parts of medical treatment and important in the self-health movement. This fact makes the role of ethnic healers important.

Also integral to maintaining health is physical fitness. People approach physical fitness in diverse ways. Older people and naturists often prefer some outdoors activity. Others may choose automated fitness equipment and indoor exercise technologies. Still others use artificial muscle, heart, and lung stimulation.

Orchestrating healing with sound, light, and touch

Music's power in comforting patients and accelerating healing has long been understood. Its use with migraines and during and after surgery is widely practiced. Therapies for depression and anxiety often use music as well. Other sound has been used similarly. Light and touch can also improve a patient's response during and after a medical procedure. All these therapies may be self-administered.

For the mentally ill, and for neurological disorders, there are similar therapies involving music, sound, light, and touch.

Holistic approaches will remain the norm

Since the U.S. NIH set up its Office of Alternative Medicine (now the Institute of Holistic Research) in 1992, formal medical practice has made growing use of practices once thought flaky. In particular, holistic medicine, in which mental stimulus and psychology are coupled with physical therapies, is common. Similarly, the rich heritage of U.S. ethnic groups in folk medical treatments has seen resurgent interest. If nothing else, ethnic approaches are comfortable and reassuring to people of those ethnic groups. Many are based on good science, and many have spilled over into general medical and home health practice. Some classes of medicine considered alternative when NIH first began are mainstream today, including some homeopathy, biofeedback, chiropractics, and acupuncture.

WORLD 3—GRAPPLING WITH THE HEALTH EFFECTS OF POVERTY

Control of infectious diseases and birth control have dominated national and international health programs in the destitute countries. Innovative practices and technologies are used in those areas throughout the world, and with success. Only recently have more money and effort been put into sanitation, especially in urban areas. Sewage and water continue to create health crises and are responsible for killing millions of people annually in the poor countries.

Besides those emphases, the revolution in health care seen in the affluent world has not arrived in the poorer countries of the world, except among the top 5% tier of affluent people in each country.

Public health dominates

Maintaining good health in the poor countries involves substantial public sector health spending and international aid. This contrasts sharply with World 1 where the individual takes control of health care and maintenance.

Infectious diseases kill 23 million per year

A big part of health care in the poor countries is managing infectious disease. Continued high rates of population growth and urbanization make for unhealthy living conditions. Water and waste systems do not keep pace with growing populations. Although some diseases are well-controlled, others, such as cholera, resist control. Africa is still afflicted with river blindness, malaria, etc. Famine exacerbates the problem.

Violence kills 11 million per year

A parallel struggle in World 3 is in preventing violence. Driven by overpopulation, sectarian and ethnic conflicts, and poverty, more people are injured or killed every year. Gun violence kills many in the poor countries who could have been saved by gunshot trauma medicine, available in the rich countries.

Contraceptive technologies are widely used

The biggest public health efforts in the poor countries outside the control of infectious diseases are in birth control. Inoculations are quickly becoming the preferred method of birth control, despite resistance in some cultures and the need to reinoculate people periodically.

Simple, small-scale technologies work best

Successful medical technologies for the poor countries are usually ones that are small, portable, mass-producible, and can be used by people with minimal training. On the other hand, public health officials do not like to count on the public to follow a schedule for taking a treatment or visiting an office.

A CASE STUDY—Pakistan reflects patterns typical of the world's poor countries

With 252 million people, Pakistan is the fifth most populous country in the world, after China, India, the United States, and Indonesia. At 42% urban, Pakistan is less urban than most poor countries. Despite communications and the flow of goods and people, rural areas maintain traditions that affect health status and the delivery of health care. In those rural areas, the age of marriage and the age of first birth are still young. Fertility remains unacceptably high.

The top tier of Pakistan's population, about 20 million people, lives in affluence equivalent to that of the United States and other affluent countries. Another 111 million live comfortable, though modest, lives. The bottom 121 million live in deep poverty and destitution. There are threats today, from overpopulation and system breakdowns, that more of Pakistan's population will be in destitution and poor health, despite strong efforts by the country's government to improve quality of life and health.

Health-related factors in Pakistan

	Urban	Rural
Percent of population with access to safe water	92	66
Percent of population with access to adequate sanitation	76	37
Percent of population with access to health services	100	91
Percent of population fully immunized	95	87

Figures exclude recurring episodes of system breakdown

Medicine in a traditional culture

Pakistan is 94% Muslim, a fifth of whom are fundamentalist. At the end of the last century, Pakistan had no significant fundamentalist population. The heyday of fundamentalism in Pakistan was the first decade of this century. After its war with India in 1999 over the Kashmir, Pakistan briefly installed a fundamentalist government

Instituting birth control practices 20 years ago meant working closely with the mullahs to interpret Islam and guide their communities. Traditionally, people frowned on contraception because they viewed children as the "will of Allah." In fact, Islam itself does not oppose contraception but traditional Islamic cultures often do. Contraceptive use is low in Muslim countries, including Pakistan. Today the percent of married women using birth control is 54%. Low educational attainment of Pakistani women has slowed the growth of contraceptive use and the decline in fertility. Government initiatives have faced widespread cultural and economic barriers but are making impressive gains.

As is typical of Islamic countries, Pakistan's birth rate at 33 per 1,000, is high. It was 44 per 1,000 in 1992. Thus the population has climbed to a difficult-to-sustain size, given the country's underdevelopment, political, and social problems.

Life expectancy at birth in Pakistan is 66 for women and 61 for men. In 1992, it was 56 for men and 57 for women. The population grows 2.3% per year. The total fertility rate is 3.9, down from 6.1 in 1992.

Despite government and international initiatives, poverty and underdevelopment continue to shorten lives in Pakistan and lead to high death rates. Likewise, infant and maternal mortality rates remain high.

Technology seeps into Pakistani health care

Health care in Pakistan is dominated by centrally directed public health programs. The identity card that every Pakistani is supposed to carry doubles as a health service card. It tracks inoculations, infectious diseases, visits to clinics, etc. Broken-down card readers and clinic workers who forget to read the cards plague the system. Still, the government has been able to raise over-all immunization levels to 93%. At the same time, it is beginning to be able to gather data on the health of the country, and to plan its programs better.

Going to the clinic in Sahiwal, Pakistan

Sahiwal is a town of 9,000 southwest of Lahore. The clinic there is well equipped for routine treatment and has an educational facility adjacent to it. For the past 12 years, the clinic has promoted healthy lifestyles, especially among the women of the community. The work is funded and overseen by the government as part of Operation 2040, a 20-year plan to reduce infant mortality, raise literacy among women, and generally promote good health.

Most difficult in the region has been improving water quality and managing sewage and other wastes. Overpopulation, even in Sahiwal, has led to breakdowns in basic sanitation. Cholera hit the community in early 2010s, killing 1,700.

There are only two doctors working at the Sahiwal clinic. Nearly all the treatment there is given by nurses and clinicians.

Most of the clinic's work, outside of educating people in self-care, is in inoculation, contraception, and treatment of people with infectious diseases.

The Sahiwal clinic and others like it around the country have had some success in fighting ignorance and disease, but much remains to be done. Modern treatment, such as microsurgery, eye surgery, and hormone therapy are most often administered in Lahore. The people of Sahiwal are not often likely to get such treatment.

The technology is allowing more health care to be done by nurses, midwives, and paramedics. Doctors are scarce, especially in small communities. Video instruction is making inroads in educating people and health workers in new treatments and healthy practices.

Pakistan's children suffer

Pakistan's children die in large numbers from fever, diarrhea, vomiting, flu, and pneumonia. Their deaths are directly attributable to problems with sanitation, water, and the delivery of health care. Inoculations have successfully conquered malaria, as well as measles, and other infectious diseases. But cholera, chronic diarrhea, and other poverty-driven diseases persist.

Like most of the other poor countries of the world, Pakistan loses children to diarrhea, despite effective oral rehydration therapies. People, especially in smaller rural communities and urban slums, do not properly rehydrate their children, because they do not know how. The government's health-extension programs are fighting the effects of ignorance by face to face, television, and video programs.

Rural Pakistanis suffer most

The percentage of rural Pakistanis without access to safe drinking water rose to 66% in 2023. Generally, access to basic sanitation services in Pakistan was only 53% in 2023. New systems and services have not kept pace with population.

Critical Developments 1990-2025

Year	Development	Effect
1992	Office of Alternative Medicine established by NIH.	Earnest inquiry into new approaches to medicine, many of which have since entered the mainstream.
1998	AIDS vaccine perfected.	Stepped-up public health efforts to stem further infection.
Late 1990s	Synthetic blood marketed.	Reductions in the costs and risks of blood transfusions in surgery.
2003	Human genes and their functions fully matched (Human Genome Project Phase 1).	Basis for being able to test people with near certainty for susceptibility to genetic-based diseases. Far more diseases and traits were found to be genetically based than anticipated.
2006	Genetic testing achieves 98% certainties.	New roles for public health in intervening in environmental triggers of diseases with genetic predispositions. Quick rise of genetic counseling in the following five years.
2009	AIDS complex cured with 82% success.	End to death and illness in those areas where the treatments were made available.
2013	First effective organoids.	Alternatives to donor organs, more efficacious treatments for diseases.
2014	Chemical compositions for all known human genes fully identified.	New powers for intervening in the onset of diseases and in the symptoms and side effects of genetically based disorders.
2022	Monkey fetus successful gestation outside the womb to four months.	Potential for ex utero gestation in humans, probably for early pregnancy only.

Unrealized Hopes and Fears

Event	Potential Effects
Cheap human organ cloning, including commercial organ farming.	Widespread treatment of damaged and diseased organs and tissues, potential for life extension.
Donor organ "farming" in which people are the livestock, selling their organs for transplanting.	Extreme values conflict in society.
Human cloning.	Extension of life through cloned organ replacements or transfer of consciousness to replacement body
Fast mutating viral and bacterial disease pandemics.	Tens or hundreds of millions lose their lives; extreme medical costs of chasing down vaccines and cures.
Complete control of genetics at or before conception.	A eugenics-based society.
Life extension to the 15th decade.	Uncertain.
The end of infectious disease.	Greater health standards worldwide, but concomitant growth of populations possibly exacerbating current problems.
Elimination of all cancers.	Cost savings and quality of life enhancement across society.

15

OUR DAYS AND LIFETIMES.

From 1990 to 2025, technologies transformed daily activities, from sleep and hygiene to work and play. At least in World 1, a growing emphasis on the quality of people's lives accompanied the transformation and raised new personal and public issues. This chapter explores how science and technology have helped people satisfy their social and cultural needs.

What are social and cultural needs?

Social and cultural needs are things large and small that serve to maintain people in their daily lives, hold the fabric of communities and nations together, and fulfill human striving for intellectual and cultural enhancement. Needs include:

Personal life

- Child rearing
- Preparation for adulthood
- Career choice
- Behavior enhancement
- Hygiene, clothing, and personal adornment
- Care of the aged
- Socializing

Intellectual life

- Self-fulfillment
- Spirituality
- Learning
- Language

Community life

- Participation
- Governance
- Finance and exchange
- Crime and punishment

To describe the social transformations that have occurred over the past 30 years, the following sections lay out the development of social institutions and practices and the technologies supporting them. A case study on the United States represents World 1. Briefer discussions look at Worlds 2 and 3. The discussion in each section is organized in the following three spheres:

- **The personal sphere**—daily personal life and needs through a lifetime

- **The intellectual sphere**—the needs of the mind, including lifelong intellectual development, spirituality, mental health, and mental enhancement

- **The community sphere**—needs associated with living in communities and nations, including governance, managing social problems, promoting sustainability, and other public interaction with and without government participation

FULFILLING SOCIAL AND CULTURAL NEEDS IN WORLD 1—The refinement of everyday life

In affluent societies people live in peace, security, and freedom. They rarely suffer from hunger, epidemics, or civil strife. Their lifeways are increasingly environmentally sustainable. World 1 societies depend on stable and competent governments at the local, regional, provincial, and national levels to create the context in which we take care of routine needs.

Having basic social and animal needs met means people in the affluent societies can explore new forms of intellectual and physical enhancement and the improvement of their personal and community lives. Art, leisure, entertainment, and cultural activities flower in a climate of possibility. In the United States, these goals are embodied in the quality of life (QOL) movement. This generally happy story has characterized most peoples' lives in World 1 for the past 30 years.

Affluent societies are also buffeted by change. Technological and social change creates new or newly recognized needs just as old ones are resolved. World 1 societies work at resolving continuing social problems, at making government more effective, and at improving their community and home lives.

A U.S. CASE STUDY—The QOL Movement goes mainstream

U.S. society is obsessed with quality of life. The QOL movement, which gained steam in the 2010s, has its roots in the last century. QOL adherents were viewed as hedonists or fanatics at the turn of the century. Today, they are the mainstream gurus of lifestyle enhancement and promoting community aesthetics.

The QOL Quotient: Tyranny or key to advancement?

The Quality of life Quotient (QOLQ) arose from Rand McNally's *Places Rated* almanacs, first published in the 1980s. Rand McNally's ratings evolved into QOLQs in 2006, and the company extended the ratings to all U.S. Census Places jurisdictions, and 147 foreign cities.

Communities covet a high QOL quotient and work hard to get it. QOL consultants and community makeovers are gaining importance to municipal leaders. Some people think that inter-community competition for high ratings is destructive. Others think it is the key to advancement across the board. It is clear that QOLQ chasing has led some municipalities to do foolish things such as building downtown arts centers they cannot afford, or creating artificial lakes and ornamental canals at high energy and water costs.

By some measures, QOL ratings have raised American geographic mobility since their first publication in 2006. Particularly among affluent retirees, moves as often as yearly are not unusual in the drive to live in the best places.

Most people in the United States strive to improve their personal lives. Social institutions, from learning centers to affinity groups, promote such enhancement. Many people use personal-change facilitators and QOL consultants to work toward personal lifestyle enhancement. Accompanying this trend is the move to more interventions aimed at enhancing human capabilities.

The personal sphere

The following discussion of the personal sphere uses a typical daily round and a typical lifecycle to paint a picture of the successes and shortfalls in meeting human needs.

The daily round

The person "starring" in this daily round is a 33-year-old male, Ashton. He is married, with one child. He lives in a downtown condominium building. Ashton and his family have an annual household income of $124,000.

6:55 A.M.

Ashton wakes up at 6:55 A.M. to a clock video that is tone modulated to bring him gently out of his sleep. Without disturbing his sleep, a system built into the bed has already checked his blood pressure, blood chemistry, and temperature. Ashton has a predisposition to colon cancer and high blood pressure, and so he takes extra care to have his health monitored for early signs of problems. Unless the system finds a problem, it simply logs the data electronically and stores it in a HealthXOne home health-assessment databank. The system gives Ashton's wife Mariah a monthly check of her health.

7:03 A.M.

Ashton gets up and looks at his clock video to remind him of what's on his daily schedule. He will visit his employer today, rather than spending the full day in his home office.

7:08 A.M.

The shower stall recognizes Ashton as he enters. It adjusts five water streams for his 2-meter stature. The shower quickly reaches the water temperature Ashton prefers. The shower video flips to his preferred channel, MTV-5. The shower system emits a microbiocidal skin conditioning soap to gently cleanse Ashton's body. Ashton prefers to be clean shaven, so he applies a hair remover to his face. In two minutes, under the relaxing water streams, he is fully beard-free.

Ashton's hair style is computer designed and needs little care. His hair is washed automatically in the shower and dries momentarily under built-in air currents in his bathroom.

7:18 A.M.

Ashton chooses lightweight clothing, because he will be visiting headquarters today. His Weatherwindow™ shows that temperatures will reach 30.5°C. His clothing is of a thermally adjusting fabric and designed to keep him comfortable in temperatures ranging from 7.5°C to 38°C.

Clothing care

In the 1990s, clothes had to be either washed and dried using two large home appliances or taken to a commercial cleaner for laundering or dry cleaning. Those establishments still do well in many smaller cities and less affluent areas. However, since the 2010s, people have been able to buy home clothing-maintenance machines that automatically identify fabrics and use the appropriate cleaning process, wet or dry, to care for them. Most new homes and at least half of U.S. households have the machines. With the widespread sales of the smart cleaning systems, energy use for clothes cleaning has fallen by 70%.

Ashton uses an appearance consultant to coordinate his clothes purchases with each other, with his eye color, and with his hairstyle. This morning's clothing choice is an ensemble that he knows goes together, because each item has an icon on its label to help him coordinate.

7:38 A.M.

Ashton's wardrobe today extends to several accessories: jewelry, a wrist health computer, and a designer personal communicator handset. He uses a facial cream that subtly alters his skin color and gives his skin a smoother texture. Ashton had eyeliner tattoos done three years before to highlight his eyes. He also had his eyes surgically-corrected to 20-10 to improve his edge in tennis.

He chooses noncorrecting contact lenses according to his clothing, and sometimes, according to the degree of sunniness in the sky. For special occasions, Ashton has evening wear contact lenses suited to night lighting. He expects to try the new temporary cornea dyes when they become more generally available.

Ashton tries hard to keep up his appearance. He had always been a sloppy dresser, but he found that his appearance was important to his job. Now he regularly visits the Bodini Day Spa for full personal makeovers and appearance consulting. He also sees a personality consultant to fine-tune his disposition and demeanor. Ashton attributes his growing business success to this careful attention to the details of his appearance and demeanor.

7:45 A.M.

In the morning, Ashton is interested in news that might shape his day. Local news, news affecting his job, and the weather dominate his interests. After hours, he can roam more freely through the network to get other kinds of news and information. Ashton uses his breakfast room flatscreen to view the news he needs. Voice commands let him flip from weather to local news to world market news with ease.

8:05 A.M.

Ashton is a three-star gourmet cook, certified in 2019. As a hobby, and by the demand of his family and friends, he often cooks exotic cuisine. This morning before going to work, he decides to order some special ingredients for a small Angolan feast he will give tonight. He uses a kitchen video console to search the local fresh-ingredient retailers to find some of the more exotic produce he wants. He also updates his standing order at Meijer's Megamarket through the console.

2025

While he is in a shopping mood, Ashton decides to look for some new furniture for the home's media center and also some art for the walls. He calls his wife who has already left for the day and images her on a few items for the room that catch his fancy. They choose to buy two laser etchings but decide to order a walk-through simulation of their media center from the furniture sellers that they can view that night at home. Ashton instructs his computer to send the room data for preparing the video they will view.

Advances in shopping technology

Shopping has long been an activity relying on advanced information technologies. By the late 1990s, home shopping through cable television had begun to face fierce competition from interactive video shopping. Most analysts think that the original telephone-based cable TV shopping laid the groundwork for home shopping using interactive video.

The food shopper began to encounter information technology by the mid-1990s with nutritional information kiosks in food retailing stores. He or she could also purchase computer software and data to help design menus and recipes combining food preferences and nutritional needs.

By 2011, supplemental devices allowed people to use virtual reality to try out products, see how they looked in new clothing, and explore design options for their homes, among other possibilities. The home-based virtual reality hardware relied on retailer-provided data and, sometimes, computing power.

Since the late 1990s, another powerful tool of shopping is the shopper's smart card. The first cards were spinoffs from credit cards. Often they were the same cards from organizations such as Discover™ and MasterCard™. In the late 1990s, those cards used their magnetic data storage for storing basic consumer preference information such as clothing sizes.

The 12-point clothing sizing system for men and women, spearheaded by Levi Strauss and other clothing retailers, made closer clothing fits possible as it was adopted in the '00s and '10s. Smart cards stored the 12 fitting measurements and coordinated the matching of data-encoded clothing tags with consumer needs.

Incremental change marks advances since these developments in shopping technology. Today's food retailing relies increasingly on standing electronic ordering from automated food pantries. The increasing automation of the kitchen is refocussing the food retailing business on packaging and product design for automation. At the same time, gourmet food retailing has blossomed based on consumer interest in hand-made, non-automated fancy cuisine.

It seems nobody ever sits down to breakfast in Ashton's family. His wife, Mariah, and son, Bertie, always race out the door to go somewhere, or go straight to the entertainment room or the home-work-study center with just a nutrishake. This morning, Ashton decides to have an old-fashioned breakfast. He is not due at his headquarters until 10:15AM

Ashton orders breakfast from the robotic chef in his kitchen. Forty-five seconds later, his coffee, muffin, fruit, and fritters are ready. To his annoyance, the fritters are cooked unevenly, even though the company had remotely re-programmed the unit for the second time.

8:45 A.M.

A little while after breakfast, Ashton prepares to leave for the day, but first, he visits his grandmother. She lives four states away in a cooperative retirement village. She and Ashton chat for a few minutes, and he shows her images of Bertie playing in a speed tag tournament.

Securing the home

An ordinary home security system knows who is home and whether they belong there. Such systems usually arm themselves automatically when occupants depart. Should anyone tamper with the home, try to enter, or vandalize it, the systems alert police or a security service and begin to record every sound made in the structure. Security cameras with motion-sensing directional tuning record who or whatever is moving about the premises.

9:45 A.M.

When Ashton leaves, the home security command module automatically re-sets itself, arming all security devices on the premises.

3:39 P.M.

Ashton returns in the early afternoon to work at home for the rest of the day. His key card gains him access to the main entrance and the entrance to his unit. Each security point senses and reads his card without Ashton having to take out his wallet.

His front door automatically unlocks as he approaches, confirming his identity by voice. Also when he approaches, his home interior climate system quickly brings the unit's temperature, humidity, and ambient sound to the desired levels.

3:42 P.M.

Mail and messages are waiting for Ashton when he returns. Most are on the network, but there are several pieces of paper mail waiting, including a sing-ing birthday card from his mother. He sits down in his home office to deal with the information, queries, and other messages he has received, ignoring the insistent rendition of "Baby Blue Eyes" from the card.

Editors then and now

By the middle of the 2010s, the use of information technology in filtering and selecting information had led to the near demise of old-fashioned, edited paper journals, newspapers, and television and radio information broadcasts. People could use their own software and hardware to monitor and collect information of interest to them.

But by 2020, it was becoming clear that there were advantages to edited information sources, so long as the editor's interests and the user's were in harmony. Electronic information sources that are selective, maintained by informed editors, amounted to a high value-added information source. The editor's name for many is a brand of quality.

For any electronic message or information source, Ashton can rely on his knowbot assistant to winnow the information down to a manageable level. The knowbot he has named "CyberJean" is programmed to anticipate Ashton's information needs based on his interests, profession, place of residence, background, and intellectual type. Ashton believes that over-edited information, as is possible with advanced knowbotics, can sap his creativity, so he has programmed CyberJean to surprise him with interesting stories from news sources around the world, translated as needed. He particularly enjoys video news stories from Asia and North Africa, and added anything on Burkina Faso and Borneo to his regular monitoring.

Bertie arrives home while Ashton is in the media center, and the two of them sit down to write another episode of the story they have been working on. They collect images off the network and create another episode for the robot twins Jarvis and Jasper.

3:58 P.M.

Later, on his desk video console, Ashton views a ballot issue from the city council and registers his vote. The balloting on the fly that is now possible is being overused, in Ashton's view. This ballot issue was about whether or not to change the icon and logo used by the city's environmental management service. He does not bother to respond.

4-5:20 P.M.

Ashton spends part of the late afternoon catching up on some household chores. He spends a few minutes dictating his preferences into the videoconsole and updating his food order. He gets on the network to choose and download to optical disk several recent concerts he will play at his dinner party that evening.

5:20 P.M.

Then, Ashton gets down to work on household finances. His home financial analysis program alerted him the previous day that his credit had passed a threshold level and that he should consider reducing his debt or lowering the costs of servicing it. He uses the program to further examine the situation and decides that he will refinance his mortgage and channel some of the money into paying interest and principal on other debt. He decides to shop for a refinancing deal so he can show his wife what their options will be. The home finance system searches through the network for deals and lines up for comparison five possibilities appropriate to Ashton's financial status. Its expert system quickly winnows through the possibilities. He notes that only two of the five financial institutions are in the United States.

Computer-assisted socializing

Commercially run and nonprofit services that help people find new friends have become a $230 million industry. These services use computerized systems that not only search their databases of clientele, but also use the global network to look for people linked into affinity groups in a client's areas of interest. The programs understand how to mix similarities and complementary characteristics for maximum compatibility.

7:00ish

Ashton's personal-lifestyle consultant tells him he needs to socialize more than he does now. His friends Marcus and Jerzy have regularly tried to get him out of the house too, so he takes the advice seriously.

Based on this advice, Ashton decides to have a small dinner party that evening. Jerzy and the other guests eat the Angolan feast Ashton cooks, for which he consulted an African chef on the Compuserve Video Forum™. Ashton's finesse in the kitchen provides his guests entertainment and edification—most of them are rarely in the kitchen. They are used to either automated cooking, prepared foods ordered in, or relying on an electronic coach to talk them through preparing a dish. Ashton's friend Marcus lives too far away to come to dinner, but he visits by VideoWindow on the condominium's large dining room flat screen.

Managing sleep

People today manage sleep as never before. Thirty years ago, most people established habitual sleep patterns and relied on nature to provide them sufficient sleepiness and wakefulness through their daily rounds. Coffee and other stimulants, plus sleep-inducing drugs for some, fine-tuned the system. Today, international travelers, nighttime workers, and those with sleep problems use hormone and light therapies to adjust their internal clocks and manage their circadian rhythms. First identified and synthesized in the 1980s, melatonin led the field as the first substance used to manipulate the pineal gland.

Sleep is also no longer down-time for people intellectually. As brain science began to recognize the complexity of brain activity during sleep, and the purpose of dreaming, scientists devised sleep-time interventions to adjust mood, ensure the cognitive recharge people need from sleep, and, at least experimentally, to promote subliminal learning during sleep. By 2030, scientists believe, we may be able to use auditory and other sensory stimuli to teach people while they sleep.

11:42 P.M.

After his guests leave, Ashton prepares for bed. He and his wife pile the dinner dishes in the automated dishwasher, which washes, sorts, and stacks the dishes for easy unloading. Ashton and his wife then can catch some late evening entertainment before retiring to unwind from their socializing.

Ashton has frequent insomnia, so he nearly always relies on a sleep inducing hormone. But since he and his guests indulged in a mild narcotic after dinner, Ashton is drowsy enough to fall to sleep easily.

The life cycle

Like the typical daily round described above, the typical life cycle lays out much of the social and cultural needs people encounter and how they solve them. The life cycle presented on the next few pages is a composite, looking in on people at each life stage today.

Prenatality—growing interest in *in utero* intervention

Thirty years ago, the interest in prenatality was focused on the nutrition and health of childbearing women. Today, that need is well-served by instant information and expertise access for pregnant women. Since the 1990s, doctors have designed more and more prenatal interventions to correct problems and ensure healthy deliveries and healthy babies.

Cognitive development *in utero*

A simple device—the fetal stimulator—is available to stimulate babies *in utero*. The device uses human speech, music, and mechanical stimuli designed by prenatal psychologists. Its proponents believe such stimuli give babies a head start on cognitive development. The devices are typically rented by mothers through their obstetricians. On neonatal IQ tests, babies who had the device score 8% higher at 2 weeks, 12% higher at 10 weeks, but only 2% higher at 4 years. Some researchers suspect that this practice creates stress on the fetus and newborn and may do more harm than good.

Besides medical interventions, other conditioning and educational interventions *in utero* became available. People have tried to teach and otherwise influence unborn babies for decades. By 2010, prenatal psychology grew into a fledgling but recognized discipline and began to provide scientifically confirmed strategies for prenatal conditioning and education.

Infancy and childhood—technology supports child care and child development

Infant care has never been easier. The institutions and technologies at play today give parents an unprecedented sense of security. And parents know much more than their parents did about what really matters. In the 2000s and 2010s, technological solutions to child care, by many accounts, were overdone. Electronic monitors and other safety devices made live parental contact with children less necessary. Today, technology is used in better balance with touch and nurturing. Technology helps parents cope with chores and monitoring baby safety, but pediatric psychologists and pediatricians have established the minimum and optimum ranges of attention, touch, and human nurturing for each stage of childhood.

Robositter™: the babysitter's best friend

The Robositter™ is a simple robotic device that makes it possible for daycare workers or parents to more easily and with greater security watch children. It cannot replace nurturing and so remains out of favor with the International Pediatrics Association. Nevertheless, since their introduction in 2019, almost a million of the systems have been sold annually.

The device relies on an electronic badge worn by the child, usually pinned to the child's collar. The Robositter™ is programmed to follow the child and monitor his or her activities. If the child leaves a predetermined perimeter play area, the machine alerts a human sitter. Upgraded versions can monitor the child's breathing, and observe what it puts in its mouth.

The worries of several decades ago—breathing, gastrointestinal problems, genetic disorders, infections, and other illnesses—are largely taken care of by today's medical technology. Today, parents worry most about a child's psychological and social development, and about certain environmental threats such as new materials, excessive ultraviolet exposure, and electromagnetic fields. But the socialization and education of children, starting with newborns, is the biggest focus of most parents' attention today.

Infant sleep patterns are managed and shaped by the use of sound therapies and approaches to daily feeding and sleeping schedules that optimize sleeping and waking time.

Child care goes big business

The big story in child care is the growth of franchised, commercial daycare services and cooperative daycare centers set up by employers or parent groups. Large employers typically offer on-site daycare, or give employees vouchers for buying daycare elsewhere. Daycare, like elder-care, described later, is a routine cost of doing business and an essential benefit.

Artificial breast milk frees the mother and ensures optimal nutrition

Simulated breast milk today is 99.8% chemically the same as natural breast milk. Part of the 0.2% is supplements that improve on nature's formula. The story was far different just 20 years ago, when breast milk advocates could argue for the health benefits of breast feeding against infant formulas. Immunogenic supplements to infant formula make up for the loss of mother-to-child transmitted immunity. They can be designed for cultural, regional, and ethnically specific immunity needs.

Child care often takes place in the home by workers with reduced workday schedules or in cooperative daycare centers at satellite work centers in communities. Flexible work schedules and work at home also enable more parents to participate in their children's education. A problem recently identified with home day care by parents is too little attention given to children. Day care with a full-time caregiver ensures that kids get more attention and probably that they learn more.

The network makes it easy for parents to make video visits from their workplaces or when they travel. They can have intimate time with the child two or three times a day. Children quickly learn to accept that experience as nearly as good as having mommy or daddy visit them in person. The telecommunications network enables interaction with children anywhere in the world, as well as letting the parent who is traveling nationally or internationally to look in. This capability has helped charity groups publicize the needs of children around the world, and adoption agencies use the technology to help prospective parents meet orphans.

Technology shapes and enables learning and play

From prenatality to adolescence, brain science has revolutionized the daily activities of childhood, with the big emphasis on shaping personality and learning. Brain science has led to the development of software that figures out a child's level of ability and interests and steadily ratchets up the challenge. With occasional guidance from a learning coach, and through choices made by the child, the artificial intelligence can seek out information, games, and images from anywhere in the global network. A child who has just learned to say or spell elephant can be treated to images of elephants from a game park or zoo. A child who learns what an optiphonium is can listen to the music made by one, watch it being played, or play it on the learning center's synthesizer.

Adolescence—the challenge of making adults out of children

Adolescents have long posed a challenge to adults in shaping their personalities and accommodating the turbulence of adolescent behavior. Counseling strategies, preadolescent preparation and socialization, and other strategies derived from decades of psychological research are resulting in better methods for smoothing out the experience of adolescence for parents and teenagers alike.

A key task in socializing teenagers, done early in their teens, is to teach them to interact and cooperate with people of their own ages as well as other ages. Rarely has this been done sufficiently. Most middle-class and affluent parents choose for their children to attend experiential programs modeled after Outward Bound or Vision Quest. These programs energize their intellects with intensive new experiences such as exploring the ocean bottom, climbing a mountain, or cooperating in the construction of a bridge or house. The children gain physically and socially as well in learning to use their bodies in constructive tasks and in cooperating with others in doing so. Unfortunately, only about 66% of U.S. kids have access to the programs. Uninterested parents and a lack of available programs are the chief causes of this low rate of participation.

The transitions to young adulthood

By the 1990s it was becoming clear that there was a need for young adults, those 17 to 23 years old, to make a better transition to adulthood. Their passage from high school to college or to vocational training and then to entry-level employment had been, until late in the 1990s, more often than not haphazard and unguided. Employers took up the task of improving the results of young adults identifying professions, experimenting with work and lifestyles, and then planning more focused strategies for entering careers. Employers worked with universities and high schools to design programs to make the experimentation and wandering of preadulthood a positive and shaping experience for the young.

Two keys for preadults are finding out about their educational and work options and having well-rounded experiences in different work and life activities. Information technology serves both needs. On the global network, there are forums for 17- to 23-year-olds to share experiences and recommend things for each other to do. Video and virtual reality experiences can serve those who are dabbling in different leisure and work activities, especially those who do not have the means to travel and try things.

Adulthood and setting up families

Adults, even those in their 40s, 50s, and 60s, have not always resolved how to live their lives, as their needs and interests change. Ad hoc counseling and exploring new professions and lifestyles continue through adult life. Again, information technology, from the global network to images and virtual reality, helps people learn about and try out new things.

Finding a place to live. The simple introduction of advanced imaging technology and virtual presence makes it possible to efficiently market and sell homes to people. They need not travel to another city to preview homes there, and they need not travel the streets of their own city to narrow down their choices for a cross-town move. The video tour, simulating a live walk-through is routinely available on the home flat screen. Real estate sellers and lenders are available, sometimes in real time, to pop up in a screen window and answer questions. More important, perhaps, is the decision-making soft-ware that home buyers use to evaluate practical needs, set financial priorities, and sort out emotional inputs to deciding on a home.

Marriage and divorce. Driven by the divorce rate and family conflict, more people are focusing on ways to make marriage a more rational process. Ideally, people should be equipped with knowledge and advice on how to make a good match and how to build a sound marriage on agreeable terms. Integral to getting things right is preparing people prior to adulthood in early adulthood to make sound choices. In the schools, role playing, the introduction of images, and models of agreeable marriages and other kinds of unions, and the exploration of nonmarriage lifestyles help children get accustomed to adult domestic life, regardless of their own home experiences. For adults, counseling, church and social groups, and other help groups offer advice and support.

Where careful preparation is done, marriage success rates increase to 78% measured by the share reaching 10 years still married. The choice of more kinds of unions is resulting in less family violence and greater ease of terminating a bad union. Although the average length of a marriage is shorter today than it was in 1990, by most accounts better marriages are being made. Unfortunately, many marriages and other forms of union are still done on whims by ill-prepared couples.

Family types for the 2020s

The family types listed below range from least-shared property to most.

- *Cohabitation*—an acknowledged temporary union in which property ownership is kept divided.

- *Trial marriage*—a six-month to two-year agreement to share a household and its finances and care for the explicit purpose of trying before you buy. These may be informal or formally contracted. Physical goods and real estate are separately owned and remain with owner at dissolution.

- *The brokered marriage*—marriages existing under comprehensive contractual terms covering finances, household arrangements, and including some behavior covenants. Property ownership and rights at dissolution are built into the contract.

- *Life partners*—two adults of mixed gender or the same gender who have registered with state authorities. Property is jointly owned and both parties have rights under dissolution.

- *Economic unions*—any combination of adults and children who agree to live together and join their finances for mutual benefit, permanently or for a fixed term. Governed by state and federal separation laws which will equitably divide property if the union ends.

Family unions, including marriage, come in more forms than they did in the last century. From the 1960s onward, people experimented with increasing intensity with new kinds of family and nonfamily domestic relationships, including group homes, communes, and cohabitation. This experimentation led to the variety of family/nonfamily structures seen today.

Mature adulthood—enjoying the fruits of affluence

Thirty years ago, advanced old age meant people were increasingly immobile, and often bored. It was not unusual in the second half of the last century for even younger people of 70 to 80 to have their health deteriorate soon after retirement or after losing a spouse. The causes were in part psychological—depression and boredom can sap the interest of people in taking care of themselves.

Today, old age is a reward for most middle-class and wealthy people in the United States. Through their working lives, they nurture intellectual and hobby interests and social activities that can blossom in their later years. Today's elderly include the baby boomers born 1945-65. They are well off, are in unprecedented good health, and are physically active.

Through their adult lives, mature adults sought adventure, self-fulfillment, and exotic experiences. Now they have the time and often the financial means to go further in pursuing those interests. Unlike their parents and grandparents, most are healthy, active, and mobile into their 80s.

Relative comfort with advanced technology means the mature adult can take advantage of the global communications network to contribute to social and hobby interests, travel planning, and staying in touch with family.

For those with a minor disability, technology can help. For example, robotic aides, dynamic prostheses, and mechanisms in clothing can help people with musculoskeletal problems, keeping them active longer. Health-status monitoring and self-care technologies in the home make it easy for a majority of those in their 70s and 80s to care for themselves.

Coping with extreme old age

Extreme old age brings new family and social issues for some. People in their 80s and 90s who are not in good health, or who have a loss of mental capacity sometimes need daily care. Elder-care insurance came into mainstream use in the first decade of this century to help foot the bill. The insurance provides for five levels of full-time care, depending on what the policyholder can afford:

- Self-care
- Group day care
- Individual day care
- Nursing-home care
- Individual, round-the-clock home care

The dumping ground nursing home is a thing of the past

Although there are still traditional nursing homes in some states, 87% of elderly nursing home residents live in homes inspired by or adhering to the elder-care design principles that emerged in the 2010s. Nursing-home facilities depend on the active participation of their residents. Some are resident-run, including those set up as cooperatives by founding residents. Central to success is the growing practice of handpicking a mix of residents based on personality testing, backgrounds, and health status.

Baby boomers, the dominant demographic cohort when insurers introduced elder-care insurance, had their eyes on their own retirement and had several decades to anticipate and cope with the senescence of their parents. Elder-care insurance was an entirely commercial innovation, but legislation and support at the state and federal levels quickly followed.

The technological tools for elder-care shape the kinds of care available and their cost. Home care by semiskilled workers is supported by instant access to medical and social worker advice provided by elder-care agencies and hospital extension offices. Nursing-home care is made increasingly efficient with the use of sensors and monitoring devices that help elderly people monitor their own health and alert the staff if they are in distress. Similar technologies and video answerlines help older people with self-care. Those in self-care typically have access to insurer-provided health-status monitoring systems and medical-alert systems.

The end of life

The QOL movement has relevance up to and beyond human death. Quality of life is increasingly important to older people and the terminally ill as their expectations for good health, comfort, and successful cures for illness rise. We know with greater surety today whether a disease is terminal or curable, and when. People are empowered to make decisions about continuing or ending their lives. People and their survivors also emphasize the quality of the process of death and ways in which survivors can remember the departed.

Remembering the departed

Ways to memorialize the dead are almost as common as forms of information technology. Holographic images in mausoleums never really caught on. Much more popular memorials are multimedia memorial disks that can be played anywhere, including home flat-screen displays or pocket telecomputers. People seem most comfortable with an edited pastiche of impressions and images of their departed loved one, rather than more traditional video memoirs.

Approaches to death and dying are as varied as are human cultures and lifestyles. They are influenced by the dying or deceased, by the preferences of their families, and by their cultural and ethnic backgrounds and lifestyles. But the bottom line is, people are learning how to die. Dying is a medical and a social process.

Euthanasia is now a choice of 8.2% of those at the end of life, up from about 2.3% in 2000. People choose euthanasia to protect their wealth for their heirs, to end pain and suffering, and to end boredom. Extending lifetimes through medical technology has given more people the chance to plan when to die in their mature years.

Legalization, eventually throughout the United States, helped propel euthanasia into the mainstream; so did gains in medical understanding about illness, pain, and suffering. Today, the medical risk/outcome assessment program Evaluator™ and its competitors help death-care counselors and their clients weigh the medical and personal factors and alternatives in deciding whether to end their lives, and when. The longer lifetimes and better health of aging people in the United States made a clearer distinction between living a quality life and the point at which that quality begins to fall away too much.

The intellectual sphere

The QOL movement and the technologies it spawned figure most prominently in the intellectual affairs of people. The life of the mind is central to the quality of one's life. Never before have there been so many technologies to improve human mental capabilities, mental state, and personality.

Promoting the new spirituality

People put great emphasis on spirituality, not always in conjunction with religion. New forms of secular spirituality are pervading U.S. society. Among them are:

* the recognition, embodied in the sustainability movement, of the importance of living harmoniously with the Earth
* spiritual components of striving to perfect quality of life

How technology can contribute to spirituality in the future is unclear. Today it can be used to open the expanses of people's minds and help them join together with others elsewhere in spiritual dialogue for spiritual learning.

Dropping language barriers

Information technology, particularly in its most sophisticated and transportable forms, enables more people to communicate within and across languages. Machine teaching, interpretation, and translation of languages is integral to education and a standard tool of business. Portable language coaches and interpreters give the nonspeaker of a language greater mobility and competency.

Researchers in the field of machine language translation believe that by 2035 we will have 98% machine comprehension of five major languages. Today, such machines are accurate to about 95% for restricted vocabularies. But those 300- or 500-word capabilities have profound implications for international business dealings, routine correspondence, and the delivery of news and other information. The literary community is still unimpressed. In literature translation, linguistic nuances are still beyond the capability of machines.

English gains 900 new words every year

In 1993, the English language, according to lexicographers, was gaining 400 words every year. The pace of growth of the lexicon is clear. Today, lexicographers recognize and define about 900 words annually.

Drivers of the growth of the English lexicon include evolving technologies, communication with and celebration of more cultural and linguistic variety as ethnic fads and fashions come and go, and the arrival of immigrants from new places.

A sampling of words introduced to English in the past two years includes:

cryoengineering—engineering systems and the management of physical and quantum physical effects in supercold temperatures.
forensic allelography—genetic component matching and lineage tracing through minute genetic samples.
sponge polymers—high-surface-area materials used in catalyzing bioecological regeneration.
wantok—a friend, comrade, a Papua New Guinea pidgin word brought by the more than 140,000 refugees accepted by the United States and Canada after the 2013-5 civil wars in PNG.
socioclastic—adjective denoting a behavior considered destructive of social institutions or the fabric of society. Analogous to *ecodestructive*.

Improving knowledge and learning

The second half of the last century is a useful comparison point for the education system in the United States today. The most distinctive changes are two interrelated ones:

- Education has become almost universally recognized and promoted as a lifelong process.

- Technology empowers individuals to spearhead their own education, by giving them instant access to information and learning resources.

The diagram below emphasizes the outcome of these primary forces, showing the rising emphasis of tertiary and quaternary education, and efforts to push education back in the lifecycle to a new category: preprimary education.

Education Through the Human Lifecycle

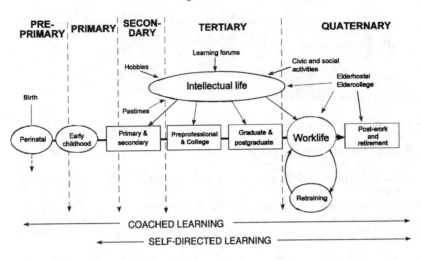

Education's successes and failures

Formal primary and secondary education underwent an extended period of experimentation in the 1990s through about 2017. Schools without walls, distance learning, computer teaching, year-round schooling, and home schooling competed for the attention of educators through that period. A consensus emerged in the late 2010s, embodied in the 2019 National Commission on Education (NCE) final guidelines for primary and secondary education. Significantly, the NCE endorsed the teacher-coach concept as integral to information-technology-driven learning. It reinforced the notion that sending children to schools for live interaction still has advantages to the virtual classroom, at least for 14 to 22 hours per week. In part, this reflects the needs of parents to have time away from their kids for work and leisure. The growing role of nonformal education, including affinity group activities, home learning, and self-teaching, as adjuncts to formal education, round out intellectual development for children.

Learning in the global classroom

Through information technology and virtual presence, the whole world, and, some would argue, the universe, have become the classroom for students. This was first practiced through electronic messaging late last century and progressed through videoconferencing classrooms and then desktop conferencing. Today students of any age can use their own image devices individually or in groups to share lessons and team activities, visit notable places, and learn about culture, climate, and geography. They can even view celestial phenomena and human activity in space.

Learning has evolved a long way since it was done in groups of children sorted by age cohort and instructed by an adult. Today's learning coaches, human and machine, exemplify the individualization of education. With learning software, each student is taken along a customized path to new skills and knowledge.

Another innovation, coming from the mid-2010s, is the student coach. Having children or adults in learning centers teach each other, or having coaching help through the network benefits both learner and coach.

The technology of the brain and moves to human enhancement

There are plans to raise the national average CPQ by 4 points between 2027 and 2037. The National Education Association and the Department of Education are overseeing the effort. The educational system discussed above and the technology of the brain are two primary tools for accomplishing this goal.

Over the last 30 years, brain scientists have gained formidable understanding of the brain including its genetics, biochemistry, cellular and lobal structure, organization, and external links. This understanding brings better understanding of how the brain governs thought, emotions, health, immunity, and action.

By some estimates, scientists have the brain about 75% figured out. The easy work is done. The mechanisms and linkages governing discrete behavior and chemical and physiological disorders are understood. The subtleties of personality, mood, emotion, and thought are less well-understood. Within two decades, according to brain scientists, we will have mastery over those aspects of the brain as well. Important gains already made in brain science include much better understanding of how genetics shapes personality, environmental influences, and psychosocial forces.

With greater understanding of the brain, the precision and efficacy of medical interventions will grow more complete. Interventions will come with few side effects. Public attitudes are likely to remain divided on how much we should strive for perfection of character through such interventions. Some people will not accept tampering with brain functions and personality characteristics.

Behavior and its modification

Behavior modification centers on the physical and mental enhancement of people through medical interventions, training, and other conditioning. The decision to apply therapies to permanently alter behavior may come from authorities through a court order, from the individual involved, or from family members, again through a court order. Before the courts can require such therapies, the interested parties have to prove that the individual's behavior is self-destructive or harmful to others. Though it is possible to improve the personality and behavior of people who are socially unpleasant, over-bearing, mean, or overly nervous, those disorders can be corrected only at the request of the person having them.

Tools of the behavioral engineer

Soft approaches

- Environmental design
- Psychotherapy
- Behavioral therapy

Hard approaches

- Biochemical & genetic intervention
- Surgery
- Drugs, hormones, and pheromones

In the past five years in the United States, 36 million people have had personality therapies of some sort. Only 154,341 have undergone court-ordered therapies to permanently alter their behavior. That number includes treatments done on the brain to mitigate the effects of antisocial predisposition. The practice is still a tumultuous public policy issue, despite scientific evidence of its success.

More common, affecting tens of millions of people in the United States is brain hygiene. These therapies and treatments are designed to allay or reverse memory loss, improve general mental function, and extend human analytical capabilities. Involved are brain stimulating chemicals, acoustic and image stimulation, coaching strategies, facilitation, practice, and, occasionally, surgical probing and lesioning for error-prone sections of the cortex.

The pheromone in behavior shaping

Using pheromones in marketing took off in the late 1990s with their use in direct mail and retail establishments. By 2008, 28 states required that the use of pheromones be announced, with sign postings in shops, and with a printed warning logo on mailed items. Their use expanded to include recreational mood altering and manipulating the moods of crowds at sports and entertainment events.

Today, pheromone research is breaking new ground. Four new human pheromones were discovered between 2017 and 2023 that were unknown at the turn of the century. Research into pheromones led to the discovery of antipheromones — substances which block the effects of human pheromones. Four have been identified so far. It is also possible to excise or microlesion the pheromone receptor sites in the nose and prevent any further effects.

Brain scientists today are armed with better measures of cognitive style, creativity, honesty, ambition, personality, and aptitude. These make the diagnosis of problems, capabilities, and areas for potential improvement routine.

Remaking personalities

The QOL movement has led more people to pursue self-improvement. Partial or complete personality makeovers are as popular today as cosmetic and surgical makeovers were 20 or 30 years ago. As never before, people are considering how they can redesign themselves to recreate their personalities, world views, and mental states to more closely resemble their ideals. Consultants and franchised storefront clinics work with clients to redefine their goals and aspirations.

Sometimes people are under pressure from family, friends, and colleagues to alter some aspect of their behavior. Others do it without any pressure from others. Most people do not have comprehensive makeovers, sticking instead to minor adjustments to their behavior, such as improving mood, ambition, sense of humor, or correcting an addiction. Methods for making such adjustments include drugs, hormone therapies, biofeedback, conditioning, training, reinforcement by electrode or drug implant, and group therapeutic programs such as the 15-step programs for nonchemical compulsive disorders for those who choose the more conservative treatment.

The community sphere

The U.S. democracy is a mature political and social system, having lasted almost 250 years. Freshening winds blew through it periodically over the decades and centuries, giving it new life and viability, just when many people became jaded about the system's desirability and viability.

In the late 1990s and into this century, information technology brought the U.S. system to a new height of participation. People had the two things that ensured their role in the system: information about what was going on and access to the people responsible. Individually or in large and small groups, people were empowered by telecommunications technology.

Several other critical forces are reshaping community life:

- Greater population mobility nationally and globally—keeping the population fluid and dynamic and continually renewing and refreshing the processes of community formation and governance.

- Global telecommunications networks—bringing the power of participation to an international level and raising the viability of international community life and governance.

- Pressing new and newly important issues arising from human activity, including sustainability, environmental quality, and quality of life, which raised the stakes for building effective, self-regulating social mechanisms.

2025

It became clear as early as the 1990s that technology can lead to a distortion of democracy. U.S. society developed public access to information that was unprecedented anywhere in the world. By the 2010s, U.S. federal, state, and local governments gained the ability to keep track of what people want and think, through instant polling and continuous referenda. The capability ended up tying their hands, pressuring them to respond to short-term interests at the expense of greater leadership for desperately needed changes in society with long-term outcomes.

By 2005, it was clear that U.S. state and federal governments were not working well. Given the chance, and without more elaborate public discourse, people fall into the habit of letting parochial and pocketbook issues shape their voting and polling decisions. Under such conditions, the time horizon of relevance to leaders shrinks to as little as weeks. By the 2010s, analysts of government and some leaders were dismayed at the persistent short-sightedness of large and small governing bodies in the United States.

A key driver of the problem of distorted democracy is also the solution. Information technology is the tool for providing more and better information on public issues. Used effectively, it is possible to educate people to look at the consequences of a decision on a multiyear, decade, or lifetime horizon.

Modeling alternative futures

Futures modeling is increasingly a routine tool of public policy. With adequate training and using sophisticated computer software, legislators, lobbyists, and analysts can begin to do the work of the public policy futurist, looking out years and even decades at the implications of today's legislative actions. The only credible practice is to have "open construction" models, in which parties on every side can examine the assumptions, data, and workings of the model.

Information, combined with models of future outcomes can be constructed by government or advocates on either side and critically examined by the educated voter or legislator. Government today undertakes most of its actions as experiments and uses the public and other stakeholders to assess outcomes and fine-tune policy.

Instant polling and other immediate uses of public input are used more judiciously than in earlier decades and are used in conjunction with strategies for examining the longer view. The results are becoming clear as U.S. federal, state, and local governments get on top of problems that pervaded the public agenda until at least the 2010s.

Technology reshapes property ownership

Several forces in U.S. society reshaped property ownership and the use of community space in the early part of this century:

- The rise of condominium housing and business arrangements last century,

- Pressures on time,

- Higher population densities in U.S. cities, and

- Greater attention to strategies to manage resources, living space, and natural and human systems.

Property ownership mattered most when the United States was an agricultural society. Then, land ownership was a measure of success or at least of independence. That attitude lingered long after most people stopped farming. By the 2010s, owning land and dwellings had begun to lose its appeal. Residential developers had already moved to a mixed strategy of sales and leases by the end of the last century. So long as they could lease 20% to 40% of their properties up front, or show collateral for development financing loans, lenders were satisfied.

The rate of home ownership fell to 53% last year, down from a high of about 64% in 1980. Population mobility and changing values about land ownership were behind the dramatic drop. People found alternative, more liquid and reliable ways to save and invest their wealth.

The concurrent rise in rental and a trend toward more condominium ownership and rental means that more people live in designed communities with shared communal property. In a hard-won battle, such communities have gradually taught people to accept the responsibilities of shared property, using elaborate covenants and sometimes automated reporting where people fail to comply with the local rules. Accompanying shared property in such arrangements are amenities, including health facilities, child care and eldercare, and shopping and gourmet food services.

Coping with social problems

New problems of human interaction and human behavior seem to arise just when old ones are solved. The three big social problems of the 2020s are:

- Infopathologies. Obsessions and compulsions associated with the cybernetic world. Sometimes they include destructive rebellious behavior on the network. Claims of insanity provoked or triggered by information technology have been made for decades but have not been proven.

- Rejection and disruption of sustainability practices. This form of nihilism has resulted in episodes of terrorism against environmental protection installations and some ecoterrorism.

- Illegal genetic tampering, including the introduction of destructive organisms, deliberate damage to agricultural crops, and intellectual property theft related to licensed organisms.

Finding solutions to such problems therefore means fighting on a rolling front. Not all human behavior problems can be solved through genetic and drug therapies or social contracting and adjudication. Community policing will probably always be necessary.

Information technology can help people monitor problems and target the people involved. Expert systems help put solutions in the hands of parents, social workers, and disciplinary authorities.

Public affairs—the revolutionizing role of information technology

Where do people get their information, opinions, and received wisdom? Increasingly, people turn to public interest groups and affinity groups to inform them about issues. These groups act as news services, and compete with the mainstream press and broadcast media. The typical informed person in the United States will develop a personal news monitoring strategy that covers some of the traditional press but also gets some of its information from other networked sources.

Knowledge assistance, such as with knowbots, makes it possible for anyone to design their own news service. Probably 60% to 75% of adults do this at least some of the time. They may use knowbots, a simpler software system, or a service to filter their information down to what interests and affects them. A smaller group, perhaps 10% to 15% of adults, uses the global information network in an aggressive way to ferret out information from any source, and carefully tailor what they view, listen to, or read. Some build in alert mechanisms to inform them instantly of certain developments.

Government activities—promoting two-way information flows

People still make decisions, despite 30 years' evolution of expert systems and artificial intelligence. The role of government still is to gather and analyze information, listen to interested parties, and make judicious decisions about public goods, leadership, and law. Information technology helps leaders and bureaucrats manage the flow of information. It also helps them educate the public about its actions and their purposes. The art of communicating about government activity is still evolving.

Government is also a clearinghouse of information. Governments are data collectors, publishers of regulatory codes, and operators of large-scale systems to manage social and natural systems.

Finally, governments pay for public works, social programs, security, and defensive systems. Taxation is the key to public sector finance. Today's technology permits more equitable taxation. Efficiency in tax collection has lowered overall taxation in the United States since 2015. Electronic monitoring of transactions is the key taxation mechanism used today. It allows taxing financial transactions, property, and income. Similar mechanisms permit regulators and the owners of copyrights and trademarks to collect royalties, and government and others to collect user fees.

The justice grows less formal

Trends in criminal and civil justice over the past 30 years include two fundamental shifts:

- *Public-sector adjudication—private adjudication*—this trend is underway today. Perhaps 50% to 80% of commercial disputes, and many civil disputes could be mediated and solved through private voluntary or adjudicator-for-hire mechanisms.

- *After-the-fact solutions—prevention*—expert systems and better formal and informal mechanisms increasingly ensure that people avoid legal disputes in civil matters. Adjudication expert systems are well-known as a recourse for settling disputes quickly and inexpensively. Less well-known is their value in designing contracts and other legal arrangements so that there is little room for disputes to arise. Advocates of these systems expect they will have their greatest payoff when additional monitoring systems come about to allow expert systems to monitor compliance as well.

Crime, security, corrections, and how far the system should go

Society is capable of monitoring and tracking behavior to prevent and punish criminality and preserve security. The debate over how much behavior monitoring is acceptable continues. U.S. society will probably decide that the traditional legal principle of probable cause can govern how much intrusion authorities may make into individual behavior. Probable cause will also govern testing for criminal tendencies.

Civil libertarians still slow the spread of genetic testing for criminal tendencies, although psychological testing is more accepted. The problem with genetic tests is that there are combinations of genes responsible for criminal tendencies. Hence congenital tendencies do not always lead to crime.

Forensic specialists catch the bad guys

Crime detection is still more successful than crime prevention in the United States Forensic experts are among the most cross-disciplinary scientists today. Some forensic specialties are new since the turn of the century:

- Telecommunications forensics requires expertise in photonics and net-work systems.

- Detection of genetic tampering requires the standard techniques of the geneticist plus subtler capabilities in cross-comparing samples from an organism's lineage and using probability to determine plausible versus implausible genetic shifts.

- Intellectual property theft detection is particularly reliant on binary and fault bit code deciphering.

By the 1990s, the U.S. criminal justice system was overburdened with offenders, especially repeat offenders. Most were put, at least briefly, in jails or halfway houses. The dual purposes were to punish and to prevent further crimes. The prisons grew overcrowded, and in most U.S. jurisdictions new prison construction could not keep pace with the need for space.

Early solutions included electronic anklets that enforced house arrest for certain criminals. This was readily extended to house arrest/work furlough arrangements in which felons could be electronically monitored through continuous PCS transmission and tracking using implanted or worn monitoring devices.

Crimes of the 2020s

- Genetic tampering
- Electronic theft
- Radiomagnetic interference
- Pollution
- Traffic code violations
- Tax evasion

Nonphysical solutions for repeat offenders came about by the 2010s. Electronic, aversion, hormonal, and drug therapies made it possible for some criminals to have their propensity to violence or theft reduced or eliminated. By 2010, brain scientists had discovered physiological connections for some violent behavior and for compulsive behavior, including theft. Other 1990s crime categories, including illegal gambling and use of narcotics and other controlled substances, are not a problem today because of decriminalization.

Criminal behavior for which there are legally permitted treatments:

- compulsive larceny
- some forms of violence
- addictions

The Supreme Court case *Morris v. New York* in 2019 established that such treatments could not be compulsory except for certain repeat felons. Currently, any other convicted criminal must agree to the treatments. About 54% do so, because for most corrective treatments there is no evidence of side effects, undue risk, or a degrading of quality of life. Also, having the treatment allows reduced sentences.

The world of money

The 20th century concept of money held that every economic transaction had an assignable cash value. Behind that, in turn, was the notion that you could exchange currency for gold or some other precious substance, something that held value anywhere in society, and even anywhere on earth. Cash and currency are still with us, but few people today think of them as immutable substances, backed by gold as they once were when the U.S. Treasury Department stored gold in Fort Knox, Kentucky.

Today, an electronic or photonic coded binary signal takes the place of the unique serial number that once graced monetary instruments. This electronic currency is accepted globally. Value is exchanged more often through electronic transactions and rarely through physical means.

Regulators slow the freeing of the market system

Financing services, the sale of goods, and investment could be entirely fluid, dynamic, and continuous processes if regulators removed the last blockages and filters from the international exchange system. Instead, reporting requirements and built-in waiting periods make the system more viscous than it could be. The ideal system would make it possible for anyone in the world to borrow money from the cheapest sources at a given moment. They could with just a moment's effort, bundle together the credit they needed from a combination of sources. Goods and services exchanges could likewise be utterly open and free to happen across the world were it not for regulation, tariff and non-tariff trade rules, and other hindrances.

The traditional model of economics held that if supply and demand meet, then goods and services are exchanged at an equilibrium price. This could work because theoretically everyone involved had full access to the information they need to make an economic decision. Although free and open markets worked for most of the 20th century, and even dominated the world economy by the turn of the century, it took several decades saturation of advanced information technology to bring the system to its fullest development. Only with instant, universally accessible pricing, bid, and offer communications could prices truly reach equilibrium levels. Today we sometimes achieve that ideal, but there is still some inequality of access, especially timely access, to information. Also, the capability to assess the meaning of information is unequally distributed, especially in complex exchange situations.

The following two sections give comparison accounts for Worlds 2 and 3. They present summaries for the personal, intellectual, and community spheres, and give examples from around the world.

SOCIAL AND CULTURAL NEEDS IN WORLD 2 —Striving for the benefits of affluence

The middle-income societies typically fulfill their social, cultural, and human needs with little income to spare. Rarely can they afford the luxuries of World 1 in fine-tuning quality of life and steadily improving and redesigning social institutions. Their drive for quality of life is more basic. They want to extend the fruits of success to more of their people and keep climbing the economic ladder to affluence. Many societies have a tenuous hold on the ladder, and are just as likely to slide backward to destitution.

Their successes are emerging on an uneven front. Educational systems are furthest in front, while the amenities of personal life lag. Innovations in community organization and participation stabilize middle-income societies and reduce their potential for backsliding. The most advanced innovations in community life will remain less relevant until middle-income societies reach greater affluence.

The personal sphere

Successes on the personal front for the World 2 societies include better health and hygiene. Nearly universal access to indoor plumbing and health clinics ensures good health. Early detection of disease and, for growing numbers of people, genetic interventions are reducing suffering from congenital diseases and cancers. Mass inoculation campaigns are almost universally successful in the middle-income countries. Energy technology, including photovoltaics and other distributed power-generation technologies make universal electrification possible as well.

Child-rearing practices increasingly resemble those in World 1, though there is far less attention to perinatal intervention. Daycare systems are being established at workplaces and in neighborhoods. In several countries, including Turkey, Venezuela, and Thailand, the government had mandated that employers provide safe, affordable day care for children. The goal of these governments is to ensure that women have the option to earn incomes through work outside the home. Still, in World 2 about 60% of mothers care for their children at home until school age.

Elder-care outside the home is rare. In many World 2 countries, traditional reverence for the old is in conflict with the economic realities that compel people to work outside the home. Millions of people rely on cellular and wired telephony to keep track of elderly relatives while they are at work and if they live long distances away.

Dying and death are typically handled in traditional ways in the middle countries, though there is a steady increase in death counseling and euthanasia options. However, some traditional societies still ban assisted death.

The intellectual sphere

Certain information technologies are nearly universally accessible in World 2. Nearly every middle-country home has a telephone and a television. Interactive televideo with computing capabilities has reached about 34% saturation in homes throughout the middle-income world. In efforts to spur greater economic development through education, Thailand in 2019, and Greece in 2020 established the principle that every citizen must have access to the telecommunications network. This policy has extended the network to more remote villages and poorer suburbs of large cities. Other World 2 countries are exploring the possibility of implementing similar policies.

Education

Middle-income countries are chasing the technological heights reached in the education systems of the affluent world. Interactive video, virtual reality, and networked classrooms are increasingly available in the schoolrooms of World 2. Teacher training considerably lags the technology. Much of Southeast Asia and Latin America benefit from educational broadcast networks that supplement formal schooling.

The penetration of television to at least 82% of homes in Latin America and 88% in Southeast Asia helps supplement formal education. Starnet and TV America serve the two regions. Educational organizations in World 1 have a booming business in delivering educational technologies to the middle and destitute countries. Sterling Educational Systems, for example, opened 654 schools around the world between 2015 and the end of last year.

The community sphere

Public input into government is increasingly possible, though governments remain less responsive due to greater pressures on their basic performance. Labor actions including general strikes, for example, are readily called because of the near universality of telecommunications. Greater public input has been a boon to the promotion of democracy throughout World 2 societies. It played a central role in getting rid of authoritarian regimes in at least a dozen countries since 2000.

Most financial and exchange mechanisms mirror those in affluent societies. Global interconnectedness requires this for any international transaction. Nevertheless, the middle-income societies still use paper or plastic currency and coins as basic means of exchange.

The middle societies have worked hard to achieve their relative affluence. Their struggle is reflected in the continuing devotion people have to striving for property ownership. Just as the affluent societies evolve toward reduced property ownership, World 2 societies are still raising their numbers of homeowning families. In Turkey, for example, 38% of urban dwellers own their own homes, despite fast metropolitan growth and the difficulty of financing a mortgage.

Middle-income societies make criminal justice a priority. Internal and external security are integral to keeping their tenuous hold on the economic ladder. Many rely on international agreements to help track fugitives and to share forensic expertise. Interpol through its five regional components serves the world's need for communicating information on crimes and criminals.

World 2 societies also strive to make their criminal justice systems more fair and impartial and to remove politics from law enforcement. Criminal justice academies in Europe and North America have trained over 200,000 police officers for the middle and destitute countries since 2015.

Information channels guarantee greater public access to what is going on and have forced the hand of some despotic regimes. In China in 2016, for example, PCS systems made it possible for protesters to stage a general strike in 16 cities involving an estimated 130 million workers. The government was forced to free 1,200 political prisoners.

SOCIAL AND CULTURAL NEEDS IN WORLD 3—Striving for basic human needs

Destitute people struggle to get enough to eat, which saps their energy for social and cultural pursuits. The result is boredom, underdevelopment of their political and social systems, and stalled economic development. Their plight is little changed from 30 years ago because of economic failures, warfare, and unchecked population growth.

Meeting the food, shelter, energy, and health needs of World 3 societies is a continuing struggle. Most depend on international development assistance. Improvements to social institutions and governments are done on foreign models with foreign consultants offering expertise.

The personal sphere

Almost two-thirds of the people in destitute countries have poor sanitation services. Half live in communities with intermittent water, electrical, or waste-collection services. An estimated 80% of the bridges and roadways and 36% of dams in the destitute countries are in urgent need of repairs. Seventy percent of the people of World 3 are poorly nourished at least some of the time. The massive urban growth of the past 30 years has meant that governments cannot build and repair basic infrastructures fast enough. Warfare, anarchy, and natural disasters slow or prevent needed repairs to basic infrastructure.

The result is that a large part of every day for many people in the poorest countries is spent filling basic needs for food, shelter, water, energy, and waste disposal and traveling to and from workplaces. This reduces the time they might use for learning, socializing, and striving to improve daily life. Subsistence incomes mean that there is little or nothing left over for improvements in personal life. For others, there is not enough to do after survival needs are met, and boredom is a chronic problem. The spread of electrical power to more villages and the growth of television gives people something to do, and television has been embraced like few other technologies by the world's destitute.

The intellectual sphere

Poverty also sharply reduces any chance that people have to improve themselves intellectually. Hunger dulls the mind's ability to explore new ideas, and hunger afflicts schoolchildren across World 3.

Schools in most destitute countries are better than they were 30 years ago, thanks in large part to new educational techniques and, perhaps most importantly, to programs to feed children in school. Food programs have built attendance to 80% or 90% in some cases, compared with rates as low as 30% for communities without the programs. Better-fed children are more capable of concentrating and learning. Thousands of destitute-country schools today are established and operated by international agencies and commercial ventures.

In 2005, UNESCO designed a program for use throughout the poor countries to improve the prospects for school-age children. The program Bright Horizons has been adopted and maintained with some consistency in 34 of the poorest countries. It emphasizes intellectual stimulation and practical skills. Its educational programs mesh with programs to educate adults about the value of their children's educations and about community and business development.

Despite poverty, increasing numbers of the world's poor have access to learning through national and international educational programs for adults. Education, particularly the education of women in destitute societies has pay-offs in fewer low birth weights and higher family incomes.

Tools for educating impoverished adults include television, radio, class-room learning, and demonstrations. Playing on the popularity of television, UNESCO has placed 2.3 million televisions in villages, rural coffee houses, religious complexes, schoolhouses, and eating places in 43 countries. The UNESCO televisions are programmed to pick up the organization's channel 8 and channel 5 which mix educational programming and entertainment in a reportedly popular format.

Religion also caters to the intellectual needs of people in World 3. Religious complexes are often learning places for adults and children, and successful missionary work most often includes practical advice on how to develop home workshops and produce and trade goods. Religion has a much stronger influence in World 3 than in the World 1 or World 2. In many, it holds the place of other intellectual pursuits and serves as an avenue for values and political attitude-shaping information. Lutheran Missionary Services runs adult literacy programs throughout Africa and claims to have taught 22 million people to read.

The community sphere

Community life in the destitute countries is less participatory, more often confrontational, and less guided by contractual agreements and laws. Resolving problems often involves confrontation and violence. Cultural prohibitions and codes of conduct that once shaped behavior in traditional societies have been corrupted or dissolved by ethnic mixing, urbanization, and modernization. Many of the destitute societies have failed to put anything in their places.

Governance

Governments in World 3 are typically corrupt, inept, or both. Democracy, if it exists, is in a diluted form. The societies do not benefit from the access and participation that information technology can provide. Not enough people have access to information technology, and authoritarian governments limit its use.

On the other hand, more communications nationally and internationally have built up the role of political opposition and international pressures on governments. The result is often some improvement as checks on authoritarian power and reductions in civil injustice. Examples include live video monitoring of high-profile courtroom trials and elections. Zaire's 2024 elections were held in front of observers from 14 countries and 5 international agencies, largely through satellite broadcasts of video images.

Exchange

Money and finance in the destitute countries continues to rely on currency, with some incursions from optical valuecards. The abstraction of value measured and transferred electronically does not work if people are ignorant or suspicious of the medium of exchange. Internationally, World 3 businesses and governments use global networked electronic exchange.

As has long been traditional, barter and countertrade are central means of exchange. In local barter, verbal agreements enforced by social pressures make it possible for barter to take place over the network where traders have access. This capability has tripled the volume of regional and interregional trade in Africa and southern Asia. Development experts consider such trade one key to improving the economic prospects of destitute countries. Internationally, countertrade is supported by better access to market information and the international electronic goods auction.

Crime and criminal justice

Some of the tools of the affluent world's criminal justice system are becoming available to police forces in the destitute countries. As a priority of national governments, police systems are afforded more money than most parts of World 3 governments. Often the forces are paramilitary or incorporated with military forces. Pakistan's national police force merged with its national militia in 2012 to help put down increased ethnic conflict. That country's concerted efforts, along with the transfer of new terrorist interdiction technologies, are credited with preventing civil war and partition.

Keeping the peace is inarguably a necessity in the destitute countries, though the methods used continue to trouble organizations such as the UN High Commission on Civil Rights and Amnesty International.

Common tools for criminal justice in World 3 are genetic typing, enforcement databases, forensic specialties, and remote sensing. Bills of rights and other affluent-country standards of government conduct are not common in World 3.

Rebuilding national cultural identity

Nationalism is a fundamental driver of social and political developments in many of the destitute countries. Even some of the poorest countries, for example, pay for telecommunications satellites to be launched as a matter of national pride. More rigorous and probably more effective is the maintenance of national entertainment and newsmedia, even where those industries lose money. The poorest countries will at least arrange national programming on Starnet or another international satellite network. *Burundi Hour* and Bolivia's *Ahora* are examples.

Critical Developments 1990-2025

Year	Development	Effect
Late 1990s	First knowbot software commercialized.	Kicked off the development of personal-assistant software used to navigate increasingly complex information networks.
2005	UNESCO's Bright Horizons program established.	Launched the drive to improve education in the destitute countries as a route to economic development.
2010	Recognition of prenatal psychology as a scientific discipline.	Established and mainstreamed the practice of prenatal interventions for mental stimulation and to shape personality.
2010s	Rise of the QOL movement.	Put strong emphasis across society on broad scale and subtle improvements of everyday life and the aesthetics and amenities of the home and the community.
2010s	Discovery of clear physiological links for key categories of criminal behavior.	Led to therapies for treating people with criminal tendencies; reduced the numbers of people jailed for violent crime and other antisocial behavior.
2011	Virtual reality technologies become available for retail sales.	Changed the way people shopped. It became possible to try before you buy even in remote shopping, as with using virtual reality to try on clothes.
2017	Founding of Starnet's pan-Asia telecommunications and television network.	Brought new educational and cultural programming in 16 languages to over 2 billion; expected to spur development as it helps educate the Asian population.
2019	National Commission on Education (United States) sets guidelines for primary and secondary education.	Reconfigured the school day, establishing a redefined role for the school teacher as teacher-coach, and acknowledging the roles of self-learning and home education in education.
2019	*Morris v. New York* Supreme Court decision.	Restricted compulsory therapies for criminals to certain repeat felons.

Unrealized Hopes and Fears

Event	Potential effects
End to formal classroom education.	Institution of parent-assisted self-teaching based on the use of information technology; decline in the cohesion formerly gained in classroom socialization with other children.
Permanent treatments found for all forms of antisocial and criminal behavior based on testing for predisposition at or before birth.	Evolution of a conflict-free society, requiring no jails and obviating the need for criminal justice systems.
Instant democracy brought by continuous public opinion polling and referenda to guide all government decisions.	Government by lowest-common-denominator decision making; excessive focus on the near term and pocketbook issues.

16

BALANCING WORK AND LEISURE

Scientific and technological advances have driven a transition to an information- or knowledge-based economy that has reduced work hours and increased leisure and entertainment time, especially in the affluent nations. Information technology, genetics, materials, and energy technologies have shaped this social change, comparable in scope to the agricultural and industrial revolutions. The transition accelerated with the introduction of the personal computer in 1980 and continues today.

Less work and more leisure in the knowledge economy did not come immediately. It accrued as the superstructure of global communications, financial, transportation, and environmental monitoring networks took root and people adapted to using them.

Free time created an explosion in varieties, possibilities, and choices for those in World 1, and the promise or hope of the same for World 2 and World 3 nations. More leisure has translated into more fulfilling lives. Survey data indicate that people are more satisfied with their lives today than any time in the last 100 years. More and more people report that they are "doing what they want to do."

Machines working hard and smart

The historic pattern of the replacement of human labor with machine labor has continued over the last 30 years. The emphasis is shifting from machines replacing manual labor to replacing intellectual labor. Machines have already sharply reduced manual labor in the affluent nations. Advances in computer science and technology, especially AI, will further reduce the intellectual workload required of most people. Already, the less sophisticated intellectual functions are done by machines, and decision support programs are essential aids for more complex intellectual labor. In some cases, this enables people to focus on more challenging intellectual tasks. In others, people have lost their work and been unable to find anything comparable.

2025

Growing numbers of people in World 2 are meeting their basic needs and enjoying more discretionary income. Spending on leisure and entertainment is a mark of success. While entertainment is generally a positive experience in these World 2 and World 1 nations, it is still seen as a relief from tedium and boredom in most World 3 nations.

In the decades just before and after the turn of the century, governments in affluent nations struggled to reduce seemingly intractable unemployment. A short-term focus characteristic of political and business leaders kept them trying to maintain and create more of the jobs that were already disappearing from the economy. This was difficult even in affluent nations and impossible in middle and destitute ones. It has become apparent over the last 15 years that a different approach to work was required. Legislation to reduce the work-week was a first step.

Rising affluence and value shifts have changed what society and people consider to be work. For example, psychology, psychiatry, and other forms of therapy and counseling became occupations just in the last century. Before that, therapeutic needs were met informally and nonprofessionally by the family or church. Today's societies, particularly the more affluent ones, can afford to train professionals to meet these needs.

Legislation and rising incomes in World 1 have reduced the average workweek from 40 hours in the 1990s to 32 today. Advances in sleep management, including hormone and light therapies, have reduced sleep requirements from an average of 56 to 47 hours per week. As a result, leisure time has increased more than 50% from 40 hours in 1990 to 64 hours per week today. Discretionary income has doubled from $4,300 in 1990 to $9,000. These statistics have translated into hundreds of billions of dollars for leisure and entertainment industries.

How People Spend Their Time, Weekly, United States

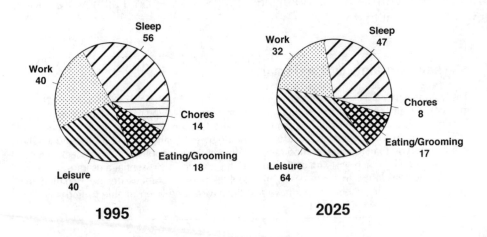

Sleep 56 · Work 40 · Chores 14 · Eating/Grooming 18 · Leisure 40

1995

Work 32 · Sleep 47 · Chores 8 · Eating/Grooming 17 · Leisure 64

2025

WORLD 1—Converging on best practices

Work practices have been converging in the affluent nations over the last 30 years. The North America, EC, and east Asia trade blocs have borrowed from one another. The strengths of one bloc have been benchmarks for others. For example, the EC is acknowledged as the leader in social welfare, Japan in building long-term market share, and the United States in managing diverse workforces.

Leisure and entertainment practices are also converging, but cultural differences show up more prominently than in work. For instance, each culture approaches time, or the pace of life, differently. Japan is the fastest paced and Europe the most relaxed. The cultures also vary in how they accommodate new practices or allow new people to join. Japan is the least accommodating, and the United States is the most.

Different cultural practices coexist alongside the mainstream global culture that has emerged over the last 50 years due to advances in communications technology and mass-media exposure. The United States leads the world in producing this mass culture. U.S. software underlies vids, films, nets, and virtual reality games. English is the global language of business and entertainment. The United States is often the trendsetter in foods, fashion, and fitness.

A CASE STUDY—U.S. work, leisure, and entertainment

The United States experience with work, leisure, and entertainment is similar to other World 1 countries. For growing numbers of workers, the distinction between work and leisure is blurred. When interests and jobs meet, it is hard to say when one is working or not. Is the professional gardener who taps into the gardening issue forum on the net when she gets home at night working or at leisure? The answer does not matter, unless one is doing a statistical analysis. What is important for the society, is that the transition to a knowledge economy enabled by advances in science and technology has improved the lives of people to where they are more often able to work at what they enjoy doing.

The transition has been rough in many areas. Millions of blue-collar workers and middle managers lost their jobs in seemingly endless downsizings in the 1990s and 2000s. They were often unable to find suitable replacements. Many had to do low-skill, low-wage service tasks that were not fulfilling. Rising affluence is improving social welfare programs so that people's basic needs are generally being met, but society is still wrestling with how to provide enough interesting work for everyone. Some argue, however, that finding interesting work is a task for individuals, not for government.

2025

Forces that shape work, leisure, and entertainment

Underlying values and beliefs about work, leisure, and entertainment are still evolving. The primary forces that shape work, leisure, and entertainment include:

- advances in science and technology, in particular:
  - information technology, which has already reshaped organizations and work, now has become a key leisure and entertainment tool
  - genetics and brain science, which are capable of assessing aptitudes and influencing human capabilities
  - materials, which increase safety and lead to new sports and games
- demographics: the aging of advanced nations and youth explosions in the rest of the world
- increasing discretionary income for most of the middle class
- values shifts
  - from materialism to self-actualization
  - from quantity to quality
  - from passivity to interactivity
- risk analysis, which plays an informative, advisory, and enhancing role
- government policy, by reducing work hours

Information technology has enabled distributed work to become common today. Workers use an array of technologies such as personal communicators, palmtop and laptop computers, networking and groupware, and videoconferencing. At the same time information technology has reconfigured organizations to flat, networked, and team-based structures. It has also opened the organization to the outside—customers, competitors, and regulators across the globe. It is a primary training tool as well.

Information technology has permeated leisure and entertainment. People have acquired new tastes in information as it became more fun. Since entertainment is information-intensive by nature, information technology has made leisure activities more interactive, except in cases where people just want to relax. Marshall McLuhan's "medium is the message" concept has played out in this area. The virtual reality medium, for example, is leading to applications and games beyond the imagination of its initial proponents.

Genetics and brain science have combined to enhance learning, creativity, and emotional awareness using genetic testing and alteration and mind- and

performance-enhancing drugs. Genetic testing and profiles point out aptitudes or areas to watch. Their use was prevalent first in sports. Brain science
advances in photic, acoustic, transmeditative techniques and imaging technologies such as MRI, PET, EEG (electroencephalography), MEG
(magnetoencephalography), and SPECT (single photon emission computerized tomography) pinpoint centers of brain activity. Language centers were
the first of many intellectual centers to be manipulated in 2009.

Materials science and technology advances have increased the safety of traditional leisure and entertainment activities—such as Pliantex^{tm} making lacrosse
safer—as well as contributing to new ones. Materials have led to new sports
and games such as piezoelectric rubber balls, photonic play jewelry, photonic racetracks, crazy cushions, electrotag shirts, virtual reality suits, and
Stickysand^{tm}. Artificial surfaces have been particularly important in extending
the seasonal and geographic ranges of sports—ice sailing, for example.

Demographic forces include aging in World 1 and the youth explosion in
Worlds 2 and 3. The composition of workforces and the types of leisure and
entertainment services that are required or do well differ considerably if one is
serving a young or old market. The United States is the only World 1 country
that does not have a flat or declining population. Denmark, for example, has
been losing population since the 1990s.

Per capita and discretionary income has increased throughout World 1, doubling in the United States. Many World 2 societies have had larger gains in
per capita income, but it has not necessarily led to increasing discretionary
income, because their populations' basic needs are still being met and improved on.

Value shifts in affluent nations include shifts from materialism to spirituality
and self-actualization, or needs from the neck up, and from passivity to
interactivity. Shifts in work values include workers' calls for socially significant work, acknowledgment of their work functions, implicit or explicit employment contracts, and fair compensation.

There has been a general move to improving the quality of life. The
QOL movement became firmly established by the 2010s. Personal values of
quality, reliability, and service, have become guiding principles of business.
QOL has led to personal change facilitators and consultants working to help
people self-actualize. Cosmetic surgery has been complemented by personality makeovers. Sustainability principles are firmly embedded in the values
of World 1. Any entertainment activity that harms the environment—off-road
and recreational vehicles, tourism, etc.—risks public opprobrium.

Risk analysis plays an informative, advisory, and enhancing role in risk-taking recreation in which people use risk to heighten competitive aspects. It is also a key tool for insurers, who have had to come up with locked-tight contracts for the retailer of more risky sports.

Government policy for reducing unemployment has led to mandates for shorter workweeks. The Reduced Time Act of 2000 reduced the full-time work week to 36 hours. Ten years later the act was amended to further reduced the work week to 34 hours. The shorter weeks led to more holidays and vacation time.

Working or playing?

People get more out of life as work, leisure, and entertainment meld. Here, work refers to making a product or performing a service for which one is compensated. Leisure is the mirror of what has to be done, or the time left over after essential activities like working, washing, or cooking are done. Entertainment is a subcategory of leisure in which people purchase a good or service, such as virtual reality and muds (multiuser dungeon) games, vids, interactive television services and programs, films, and music.

Are you working or playing, or not sure?

Responses to a February 2025 Gallroper On-line Survey on employment contracts and distinguishing work and leisure were tallied as follows:

Always do work beyond employment contract: 5%
Sometimes go beyond employment contract: 57%
Never go beyond employment contract: 38%

Always find it difficult to distinguish work and leisure: 37%
Sometimes find it difficult to distinguish between work and leisure activities: 40%
Never confuse work and leisure: 23%

Work, leisure, and entertainment are deeply intertwined. In industrial society, goods consumption had to absorb production capacity. People had to have leisure time to purchase the products and services they made at work. Sustainability as a contemporary principle is reshaping attitudes toward production and consumption. The assumption that more is better is widely disavowed. Quality is winning out over quantity—society is moving to higher-quality, more durable material goods production.

Work, leisure, and entertainment have become closer with the knowledge society. Learning, for example, falls between work and leisure. Some learning is a prerequisite for work, some is purely for personal enrichment; most is mixed.

Visiting Africa

Sara, age 8, races to the HWSC center, puts on the virtual reality headset, and plunks down on the carpet. She could hardly wait to get home and begin her zoology assignment. She can move to the next level if she can complete today's assignment successfully. She calls up the quiz on the wallscreen.

The program is tailored to provide students with the help they need while encouraging them to think independently. The new virtual reality programs engage the students with sight, sound, and touch.

Sara is virtually transported to Africa, where she successfully identifies elephants, lions, cheetahs, and monkeys. She also correctly points out which animals do not belong. She misses only one example, which is logged for retest at a later date.

Sara can move on to the next level. She runs for a snack, because her science lesson, a tour of Zyotix Laboratories by videoconference, will not begin for a half hour.

The home work-study center is another example of integration. There has been almost a tenfold increase in the dollar value of information technology in the home today since 1990. Homes are fully wired for connection to broadband fiber networks. The HWSCs are conduits to the world of information, be it for work, school, or play.

Productivity gains fuel leisure and entertainment

Productivity growth has been the key to higher incomes and more free time. Productivity growth has been above 3% annually since the turn of the century, using today's revised measures. Productivity growth also increased as people become more comfortable and capable with new technologies. A stronger commitment to training and retraining further boosted productivity.

Reducing hours, increasing productivity

In most cases, reducing work hours increases productivity. What shrinks is idle periods, such as breaks, water cooler talks, and the need to run errands. Concentration is also sustained better, because fatigue impairs effectiveness. Having a larger pool of workers working fewer hours adds to flexibility and the ability to meet a sudden upswing in workload.

Productivity problems of the 1990s and 2000s were partially definitional. Old manufacturing-based definitions based on input-output efficiency were extended to the information and services sector for lack of a better measure. They failed to capture the contrast between efficiency and effectiveness: gains in efficiency describe routine work being done faster and cheaper, and gains in effectiveness reflect improvements in the quality of work.

A new measure becoming more popular today is transformation, which is defined as innovation in how the work is done. Transformation would measure progress in improving processes and systems within the knowledge economy.

2025

Your money or your time?

Productivity growth presents people with the opportunity for more free time or more money. Before the turn of the century, most people chose more money. After 2000, however, people began choosing more time. They turned away from the powerful positive feedback loop of work-and-spend, and began to focus more on spiritual matters.

Working and playing around-the-clock

The United States and other World 1 countries moved to an around-the-clock, 24-hour society. Pharmaceuticals are used by some to work longer or alleviate jet lag or a change in shift work. Advances in chronobiology, such as locating the so-called biological clock, have been a key enabler of the 24-hour society.

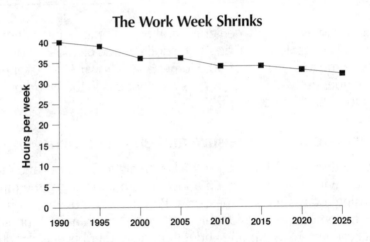

The Work Week Shrinks

People have regained sight of the fact that the point of economic success is a good life. That may come from increased work for some or increased leisure for others. The notion that work is a moral obligation to be pursued at the expense of leisure no longer dominates value systems.

The average workweek today is 32 hours. The first reductions came from the Reduced Time Acts of 2000 and 2010. These acts were intended to reduce unemployment, the rationale being that a shorter workweek would lead businesses to hire more people. And because health-care and pension benefits are provided for under a national plan, companies are not penalized by having to provide benefits for new hires.

Voluntary reductions often took the form of performance bonuses in free time rather than increased pay. It was simply computed by how many days off a pay raise would buy.

There has been a longstanding controversy over whether people really have more leisure time. Legitimate arguments could be made about how leisure time was measured. Some measures did not include the time adults spend going to school, taking part in clubs and other career-related organizational activities. Many people, especially two-income parents, report feeling busier and more hurried. They actually have more leisure time, but simply fill it up with activity. Although some high-powered executives and core workers who are exceptionally dedicated to their careers are losing free time, on balance most people are gaining it.

Work in the United States: continuity amidst change

Sixty-four percent of workers are classified as information workers, up from 60% in the 1990s. Fourteen percent are full-time information workers. Today, a smaller percentage of the workforce makes physical goods; more are involved in the abstract manipulation of symbols. Over 98% of all work involves some use of information technology.

Work is increasingly a means of meeting needs high up on the Maslovian scale and an important part of people's self-actualization. It is not sharply distinguished from leisure. Work is a love or hobby for more and more people today.

The Census Bureau and the Bureau of Labor Statistics adopted the system developed by I.M. Smith and F.M. Garcia in 1999. It was partially implemented in 2003 and became standard by 2007. Smith and Garcia received the Nobel prize in economics in 2018 for their work. The new structure of the workforce, primarily framed around categories relevant to the information era, benefited public-and private-sector planning, tax policy, education policy, and scores of other systems. The central feature of the new system is that it reflects the current dominance of information technology in the workforce. It goes beyond forced choice categories and uses the fuzzy logic techniques developed in the 1980s in the United States to allow a worker to participate partially in several work categories. Furthermore, the basis of employment, whether it is full-time, part-time, or contingent (i.e., temporary or term employment) is again coded into the work situation, as is the location, whether the work occurs in a factory, mill, central place office, partially off-site, or at home. The system has added a great degree of graininess to the understanding of the ebb and flow of work, labor rates, income, and so on. The two fundamental categories—information workers and service workers—are described as follows:

Information workers are

1.   Those who create knowledge (primarily scientists and other researchers).
2.   Workers who are 95%-100% involved with the use of infor mation (doctors, lawyers, professors, clergy).

3.   Those who collect information (survey workers, reporters, census workers).

4.   Those who primarily use information tools (managerial and supervisory workers).

Service workers are

1.   Those who do physical labor (janitors, cleaners, maids).

2.   Those who do heavy work (loggers, miners, farmers, port workers).

3.   Those who do light-duty work (primarily maintenance and repair workers).

4.   Those who perform productive functions (such as short-order cooks).

5.   Those who are engaged in the arts and crafts.

The vast restructuring of the economy implicit in the new categories is highlighted when compared with the old standard industrial classifications. The old categories fail to convey the degree of economic restructuring.

Comparing the industrial and knowledge economies using 1990 Standard Industrial Classifications shares of GDP, 1990 and 2025

Industry	% of GDP (1990)	% of GDP (2025)
Agriculture, forestry, and fisheries	2.2	1.9
Mining	1.5	0.7
Construction	4.8	3.2
Manufacturing	18.7	20.0
Transportation and public utilities	8.9	10.3
Wholesale trade	6.6	2.6
Retail trade	9.4	10.3
Finance, insurance, and real estate	17.4	17.2
Services	18.8	21.3
Government and government enterprises	12.0	13.5

The power imbalance between management and labor redressed

The power balance between labor and management is in rough balance today after a bitter struggle. For two decades before and after the turn of the century, management clearly had the upper hand and often used it perniciously.

Unionization was declining to all-time lows as employment in manufacturing, the traditional bastion of labor unions, fell below 10% of the workforce.

Waves of downsizing, smaller pay raises, squeezes on benefits, job exports, and rising workloads in the name of international competitiveness began to wear thin by the turn of the century. Workers came to believe that temporary calls for belt-tightening were becoming institutionalized, so they began to organize. They learned from the many failures of past unions, such as insisting on obsolete work rules to protect jobs. They embraced the principle of merit- rather than seniority-based pay.

Knowledge and service worker unions also redressed the imbalance. These new unions were more sophisticated than their predecessors as well. Job actions, lawsuits, strikes, and seeking legislative relief were done efficiently and effectively based on the unions' first-rate organizing skills.

The government intervened with mandated reductions in work hours. New social programs to deal with health insurance and pension benefits removed the issues from the management-labor dispute. Another key legislation was the Bradshaw Act of 2004, which shifted fiscal and tax policy to encourage business to invest in the long-term future.

Some redress arose naturally in two-income households, as a move away from materialism and towards spirituality led many to use their independence to leave bad job situations. Businesses adopted friendlier policies in order to attract the best people—unfortunately, in many cases it only happened reluctantly.

Distributed work continues to grow

Forty percent of the workforce today works at least twenty hours per week off-site, often from the home, up from 2% to 3% in 1990. Distributed work provides flexibility for organizations and workers. Organizations reduce overhead costs and improve customer service by reducing the amount of time and space necessary to house its workforce. Workers spend more time at customer, vendor, or regulator sites and keep in close contact with the home office through information technology. Vehicles are mobile offices, equipped with plug-in modules to meet wide ranging information technology needs. Information technology enables more personal service. Workers balance family and lifestyle needs more effectively.

Growth of Distributed Work

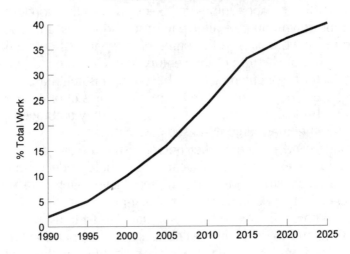

The transition to distributed work has had its glitches. Abuses have taken place on both sides: Home workers have loafed or been unproductive and management has passed over distributed workers for promotions. There were tough legal issues of liability as the lines between work and nonwork time blurred. Corporate culture often suffered. The cohesion vital to nonhierarchal, team-based work was difficult to instill with conflicting schedules and far-apart team members. Greater comfort with information technology solved many of these problems. People eventually became used to meeting with people by videoconference, voice conference, and in person at the same time.

The flexible workplace: core and contingent workers

Flexible core workers supplemented by contingents is a standard business structure today. Today's common work formats for the core and contingent workforce include flexitime, part-time, job sharing, leaves of absence, telecommuting, job rotation, apprenticeships, temporaries, consultants, contractors, leasing, and internships.

More than half of all U.S. businesses operate with some form of core/contingent workforce. The core/contingent structure has evolved from the 1990s model when contingents accounted for about 30% of the workforce. Contingent work then was probably 90% involuntary. It was a means for many businesses to reduce payrolls and cut benefits, as well as the more positive goal of aligning the workforce with workloads.

This issue was central to management-worker struggles from the 1980s to the 2010s. Contingent work became more voluntary. Legislation to stop the practice of outsourcing and hiring workers back as consultants or contract workers was passed in 2005. Another trend was the professionalizing of the contingent worker. More highly skilled people sought contingent work, and more organizations sought to hire highly skilled workers on a contingent basis. Many workers wanted the flexibility that contingent work provided, whereas others preferred the greater security and commitment of being core workers.

Networked organizations redirect power flows

Today's networked organizations are more open—there are open spaces in which the amount of restricted, confidential, proprietary, and secret information is small compared with that of similar organizations last century. Organizations have greater contact, often electronically, with customers. Practically every employee meets with some customers. Training in interpersonal skills has improved over the last few decades but could be improved further.

Hierarchies and Networks

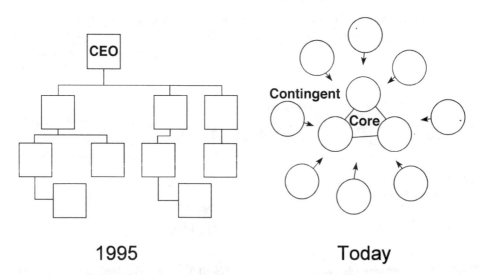

CEO

Contingent

Core

1995   Today

Work and family issues are balanced

Work and family issues are in balance today after a long and perhaps unnecessary struggle. Women in the workforce brought family issues such as child care and elder care to the forefront. The inability of management and workers to agree on work and family issues led to legislation. Parental leave and health-care reform set the stage in the 1990s. The bargaining power of dual-income couples was also a critical factor.

Women and minorities diversify corporate culture and practices

The predominant issue regarding women and minorities since the turn of the century has been their advance, or lack of advance, into upper management. Rates of 6% of women and 3% of minorities in upper management in the 1990s have been boosted to 35% and 29% respectively. Pay disparities have lessened to under 10% between men and women in comparable jobs.

Corporate cultures and practices have opened organizations to diversity. There has also been a shift towards more collegial and cooperative management styles. The composition and expectations of workforces have changed and required unprecedented degrees of flexibility in virtually every dimension of work.

The scientific and technical workforce expands worldwide

Science and technology developments draw people from all over the world to U.S. centers of excellence in corporations and universities. This globalizing of science and technology has opened up new opportunities for business collaboration.

Alliances become standard practice

Alliances have been one of the most significant developments in business practices over the last 30 years. Alliances are legal, semilegal, handshake, well- understood, or cooperative arrangements between two or more organizations. Lean, flat organizations are well suited to forming alliances to meet needs as they come along. They are a means of spreading risk as well. Alliances provide the missing components for a project. Virtual organizations are temporary groups that form to meet a single task and disband upon completion. Managing workers of different organizations and managing the flow of workers across companies in the alliances are still challenges today. As workers learn more about what competitors offer, they press for parity. Benchmarking reinforces this trend.

Education is the key to corporate and career success

Quaternary—postcollege and professional school—education and new kinds of skills are necessary to compete in the knowledge-based global economy. Businesses' increased requirements for educated and skilled workers were competing against a backdrop of failing K-12 education. Business has been forced to pick up the slack by extensive investment in training and retraining. The top percent of workers have had acceptable skills, but the lower half have been difficult to reach.

The long term: a work-free society?

Some futurists and visionaries forecast a paid work-free society as early as 50 years from now. This is somewhat a question of semantics, since people will still carry out activities that today are labeled as work. The primary difference is that people will be guaranteed a stipend that will take care of basic material needs. People will not be forced to "work," but will do so either because they enjoy it or they want to earn extra income. Detractors raise objections about creating a society of loafers, but proponents feel that only a small percentage will loaf, and the rest will find what they want to do.

As the relationship between educational attainment and economic success became common knowledge, learning became more prominent and practiced. Plans to raise the national average CPQ by 4 points between 2027 and 2037 are intended to take advantage of benefits gained from advances in genetics and brain and social science.

U.S. leisure and entertainment: participating, learning, and escaping

Choices in leisure and entertainment reflect social and cultural values. People seek high-quality experiences. Because basic social and material needs are generally met, leisure pursuits are often directly tied to self-actualization. Of course, not all people are consciously seeking personal fulfillment in leisure. Some are looking to get away from it all and recharge their physical and intellectual batteries.

People are working more with their minds than their hands. As a result, many seek physical exertion with their leisure. And many seek to avoid thinking-types of recreation. On the other hand, others have become so accustomed to cerebral work, that they pursue it in leisure as well.

Americans have more time for leisure—64 hours per week today compared with 40 in 1990. Although the amount of leisure has changed, the distribution has been relatively stable. The big gainer within leisure has been entertainment, dominated by information-technology-based multimedia games and interactive video. Socializing has resurged after holding steady and even declining in the last part of the 20th and early 21st century. Socializing was often sacrificed when people felt pressed for time, and it regained popularity with gains in free time.

Weekly Use of Leisure*

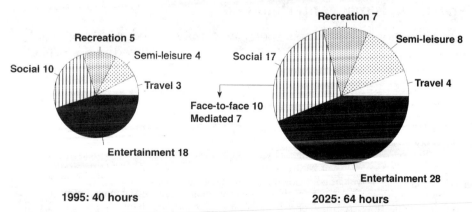

1995: 40 hours 2025: 64 hours

* definitions of categories are for *semi-leisure*: adult education, religious activities, organizational activities; *social*: visiting and conversing; *recreation*: sports and outdoors, hobbies and games; *entertainment*: TV, reading, music, relaxing; *travel*: to and from leisure; tourism not included.

Spending on leisure and entertainment ($ billions)		
Type	1985	2025
tourism*	49	183.5
toys/sports supplies	20	69.4
magazines/newspapers	13	38.7**
flowers, seeds, and potted plants	5.5	16.3
boats	5.9	7.1
spectator sports	2.8	14.5
cable TV	8.6	53***
virtual reality	0	35.2

*   entertainment spending only
**   includes electronic delivery
*** includes interactive programming and information services

The average household spends $5,625 a year on entertainment, up from $1,500 a year in 1990. The single largest entertainment expense today is for interactive television and network access. Broadcast and cable television were the leading expenses at the end of last century.

Two strong trends and a countertrend have influenced leisure and entertainment over the last 35 years:

• Leisure and entertainment have become more active, participatory, and interactive,

• Information technology is enhancing the leisure and entertainment experience, and

• Escapism has become a prominent countertrend to the above two.

Sports leads mass entertainment

Traditional professional sports are supplemented by new sports. Technology has had a strong influence, leading to enhanced performance and safety and creating the boom industry of interactive sports.

Crime as entertainment? or training?

Crime has been a popular topic for newspapers, television programs, and movies for decades. It has dominated news headlines. Crime reenactment shows became popular about 35 years ago. Their offspring today are interactive crimebusters shows. They have become a national fad. Participants enter the programs as detectives or police officers and try to stop criminals. Equally popular, as well as controversial, are programs in which people try to commit and get away with crimes. Opponents of these programs claim that these games are teaching and training people how to commit crime. Supporters claim they are channeling base instincts into harmless play.

Genetic testing identifies proclivities to superior performance as well as susceptibility to injury. Athletes take advantage of biofeedback and sometimes drugs, hormones, or even blood transfusions to enhance their performance. The enhancement movement that has its roots in sports has spread to business, self-help, education, and health fields. The traditional Olympics are supplemented by an enhanced Olympics that allows athletes to push performance frontiers by any means. On average, enhanced athletes perform 10% better. In events like distance running, enhanced women can now defeat nonenhanced men.

Sports have become safer, thanks to new technologies and the growing popularity of less-violent sports. Soccer, for example, is gaining adherents at the expense of football. The appalling levels of injuries and permanent maiming in sports like football led to innovations that did not reduce the violence but did reduce injury. Body armor developed with extra support for vulnerable spots like the knees was phased in and did not change the appearance of players to fans, who might have resisted tampering with their pastimes.

Virtual reality games use footage from live sports and allow game players to get inside of their hero, or enemy, and experience what it is like. It also enables the hardier souls to test their mettle against the pros (which often leads to unsubstantiated claims of prowess).

New games follow the model of the ancient games of Rome, the Middle Ages, and even television shows of last century like *American Gladiators*. People can experience being thrown in with lions, jousting with Sir Lancelot, or battling against their friends. Some use these games as ways to resolve disputes, or at least blow off steam. Robotic competitions that stake company, university, or even national pride are a growth area.

Some chores move into leisure and entertainment

The once-onerous tasks of shopping, eating out, gardening, and cooking have become pleasurable leisure pursuits for many. As technology reduced the monotony of the tasks, it enabled people to focus on the parts they like. In cooking, developments like the tunable microwave and robot chef in 2019 and the automated pantry in 2020, took care of the burden of daily meal preparation and let the amateur chef concentrate on trying to replicate great-grandma's apple pie.

Top five leisure activities: 1990, 2025

1990*	**2025**
1. [TV] viewing**	1. viewing
2. socializing	2. interactive Net games
3. reading	3. network socializing
4. do-it-yourself projects	4. outdoors
5. shopping	5. volunteering

* These five activities account for half of the weekly total.
** Consumes 1/3 of Americans' free time during the week and 1/4 on weekends

---

Health and fitness, including bicycling, racquetball, running, ice-skating, rollerblading, swimming, and tennis, rose and fell in popularity in direct proportion to the numbers of young people. Interest was flat in the 1990s, surged around 2000, and receded again by 2020. There is a hard-core group of people fanatically dedicated to fitness. They typically resist any enhancement technologies, preferring to grunt and sweat their way to good health.

Learning as leisure has been an enormous growth area. It blends in with work and formal and quaternary education. The growth of quaternary education has led to an explosion in leisure learning. People use AI-based personal tutors to guide them into microworlds to satisfy intellectual curiosity. Finally, napping is still a favorite leisure activity. Many take advantage of technology to induce sleep, anytime and anyplace.

The art and practice of socializing renaissances

Socializing has made a comeback over the last 20 years. Contacts with family and friends often suffered as two-income households felt pressed for time. Socializing was gradually declining from about the 1970s to the 2000s.

Computer-assisted socializing is a service for helping people find friends. The stigma of its roots in personal dating services had largely passed by the turn of the century. More mobile societies made it harder to keep friends and continually presented the challenge of trying to make new ones. Programs mix preferences with similarities and complementary characteristics to improve compatibility.

Changes in indoor activities

Information technology is a key tool to making indoor activities come alive. Art galleries, museums, and historical sites are not simply spectacles to be viewed and quickly forgotten, but experiences to learn from. At historical sites, for example, reenactments have become standard features. Whether participating physically with other people, remotely by video, or by virtual reality, the visitor comes away with a feel for how it was.

> ## Escape from technology
>
> The growing science and technology basis of society has led many to seek escape from activities related to these areas in their free time. For some, anti-technology has become a cause. But for most, there is simply the occasional need to get away from it. This escapism comes out in a return to traditional, and often labor-intensive, activities. There has been a significant rebirth in crafts and handiwork. The once-forgotten arts of coopering and tinsmithing have resurged. Similarly, people often seek manual labor for the purpose of exercise or to experience how things used to be done.

The arts, such as painting, acting, holography, singing and dancing, have also made mild renaissances. The trend to participation is strong. People still go to museums and the theater, but they also appreciate creating art as amateurs. Seeing the professionals enhances their own experience. Many, especially older folks with less mobility, make video visits to these sites. They can take interactive guided tours, or simply browse.

The primary development in music is in its delivery. The distressing alphabet soup changes in formats, from LP to CD to DCC to DAT, became obsolete as the industry began piping recordings directly to home entertainment centers. People did not have to go to stores. Music stores have evolved into information centers where patrons sample the latest music, videos, and learn about performers and upcoming concerts.

Global affinity groups for collectibles such coins, sports cards, photo CDs, and sea shells evolved over the nets. Hobby auctions are accessible around the world for remote participation.

Smart bars are an increasingly popular form of indoor entertainment. People can order generic smart drugs or use their health smart cards for tailored pharmaceuticals to boost cognitive capacity. People can also experience pharmaceuticals with effects similar to illegal drugs of the past. This practice is closely regulated. Although many still resist legalized intoxicating experiences, it is surely an improvement over the drunkenness and debauchery common to many bars and nightclubs last century. Alcohol and illegal drug use are way down from record levels back in the 1960s, 1970s, and 1980s.

Changes in outdoor activities

Sustainability principles have pervaded people's approach to outdoor recreation. Any activity that is detrimental to the environment has come under scrutiny and has typically been revised or eliminated. Hunting animals such as duck or deer has been banned for almost 10 years. Speedboating and snowmobiling have met similar fates. Less disruptive activities like sailing and cross-country skiing have surged in place of them. In many cases, outdoor activities are designed to improve nature along with simply enjoying it. Eco-restoration vacations have become popular.

Information technology has enhanced the outdoor experience much as it has the indoor one. Remote and in situ sensing networks provide ready information about parks, woods, hiking, canoeing, or fishing conditions.

Anthropology or archaeology expeditions resemble Outward Bound expeditions. Participants can expect to get their hands dirty or swim with the whales. These day or weekend trips have been popular since the turn of the century.

People's vital health functions are monitored while they participate in activities with heavy physical exertion. Participants can plug their health smart cards into kiosks to check whether the day's planned activities fall within their health limits.

Information technology dominates entertainment

People still watch television today, but they demand far greater input into programming. Interactive television, in which viewers become participants in the program, has become the norm. Flat screens, which became widely available around 2010, led to today's video walls, or for the very affluent, four-wall screens.

There are fewer megafilms, but a proliferation of lower-budget niche films. The United States remains the unparalleled leader in exporting films. Theaters to suit a wide range of tastes exist. A key development has been theater owners' attention to the social side of going to the movies. Sixty-two percent show interactive films (on large 3-D IMAX screens) in which members of the audience influence the plot and choose the ending. Thirty-eight percent show integrated, single stories. A certain snob appeal keeps theaters showing two-dimensional small screen films. Three-dimension virtual reality films are the choice of the young.

Editors have come back in vogue as indicators of quality for people sorting through infoglut in their reading. Hypertext stories break the bounds of linear plots. The reader directs the story line, and the text is annotated with images. Of course, many still prefer to be led by the author.

Entertainment under the sea

Undersea recreation has been a booming industry over the last 15 years. Undersea tours are popular. Thousands of undersea tours in glass-bottom boats, submarines, and semisubmersibles, which have deep hulls configured to look like submarines, go on each day. The Great Barrier Reef in Australia is the world's most popular underwater tourist attraction.

Aquariums and marine theme parks are proliferating, along the lines of, but much more sophisticated than, the old Sea Worlds. They provide displays to accurately depict marine creatures in their natural environments and in some cases allow customers right into the exhibit. In other cases, virtual reality supplements the live exhibits.

Floating marine entertainment structures began operating at the turn of the century in the Bahamas, the Virgin Islands, and Israel. They are free-floating structures with exhibits above and below the water. Underwater habitats and hotels, long on the drawing board, are now under construction in the Caribbean and Japan.

Images are integrated into computer information systems, mingling video, graphics, text, sound, and voice. They are essential ingredients in entertainment such as holography, Virtual Reality, and interactive video.

Games are piped in over the Net for the individual or as part of local national, or global interactive games. There have been nostalgic, almost cultish movements springing up around the early games like Pong, Space Invaders, and Mutant Death Racers, similar to the nostalgia clubs for old radio programs.

The Nintendo generation has been succeeded by the virtual generation. Virtual reality games invaded the arcades in the 2000s and the home by the 2010s. The spillover or technology transfer from military applications has spawned many games. Virtual War is the number one game in the world today.

Hide and seek via satellite

Global positioning games are the base of many modern hide and seek and tracking games. Entertainment companies lease transponders and provide access to the eager hunters who compete in leagues. They use mobile video technology, specially equipped indoor or outdoor sites, and virtual reality to enhance the chase.

Cybersuits for heightened sensation still cost too much for most families. Some parents and other critics feel that virtual reality is no more than a new hallucinogenic drug, as it puts users into trances, or intoxication. Survey data supports their contention that children exiting a virtual reality experience have a difficult time distinguishing what is real.

Changes in vacations

Tourism is the number one vacation activity. Growing affluence has translated into growing travel. Tourists are more sophisticated, seeking to go beyond pre-cooked tours. Video tourism has spurred, rather than supplanted, travel. Interacting with sites on video has motivated people to physically visit them. It improves trip satisfaction, because people can try before they buy. In eco- and anthro-tourism, for example, people participate in the culture, rather than visiting predesignated sites. Vacations to the Pacific Northwest for people to work for a week or two as forest rangers have long been popular. Video tours are popular with older folks.

Participating [not seeing] is believing

People increasingly want to experience what life is like in the cultures they visit. This has led to a niche tourist market, in which the traveler poses as a resident for a day, week, or month. The travel agent equips the traveler with the appropriate dress, hairstyle, cosmetics, and may even chemically alter physical appearance. The traveler is also equipped with language translation hearing aids, or even speech synthesis devices virtually indistinguishable from human speech. This has led to a few scary situations, when tourists were jailed in misunderstandings, but the vast majority of experiences have been safe and valuable for learning.

Travelplan^{tm} provides access to databases for any kind of transportation mode. It enables one to communicate by IVHS on rail, ship, or train. It is the essential piece of equipment for the cross-country, cross-continental, or trans-global traveler. Access to the Network enables the traveler to learn more about interesting sites and provides information about critical questions like where food, lodging, or the next virtual reality club are located.

Daily or weekend excursions often include new and expanded forms of gambling. Casino gaming is now a $50 billion a year industry.

Amusement and theme parks are showcases of technological prowess. They have the volume of business to support heavy R&D and the competition for the latest advances is fierce. Virtual reality simulations, animation, imaging, and holography combine with actors and sets to create or recreate almost any scene one can dream up. Dinoland has been a huge hit, especially with children. Science fiction sets, such as *Star Trek: This Generation*, have been a favorite of older people. Simulated space exploration, which uses actual footage, is also a big hit.

The growth of leisure industries and communities

Leisure has become the primary economic base of many cities, similar to the retirement community model of last century. Once-moribund economies turned around to become spectacular successes. Gaming locations have been among the most successful of these communities, although the competition has often been ruthless and many have not survived.

U.S. companies have had great success in tapping huge mass markets in World 2. Foreign markets are critical to the entertainment industry. In film, for example, U.S. distribution has not been enough to cover production costs since last century. The ripple effects from selling to overseas markets spill over into equipment providers and information and service industries.

Leisure and entertainment consulting is a growth business. The biggest market is among the estimated two million people who are having difficulty making the transition from less work to more leisure. They often felt adrift and directionless. Leisure and entertainment consultants help their clients balance goals, time, and money.

WORK, LEISURE, AND ENTERTAINMENT IN WORLD 2—Gathering the fruits of growing affluence

The middle-income countries have mixed economic and social success. Some are ascendant, with growing economies and industries and stable governments. Others are fighting to stay in the middle ranks and stand in danger of falling into destitution. The mixed story extends to work, leisure, and entertainment, varying according to how far each World 2 society developed.

Worklife bridges the old and the new

In some World 2 societies, work is much as it was 30 or 50 years ago, with traditional industries and crafts done by low-skilled workers for large state or commercial employers. There workers are often an expendable resource, with limited power and rights.

In other World 2 countries, global commercial interests and more modern national corporations have brought the latest workplace technologies and strategies and a growing emphasis on workers' rights and responsibilities. There industrial workers toil in modern plants assisted by automation, and service and professional workers are at work in modern offices providing financial and other services to national and international markets.

Leisure and entertainment in middle-income countries vary from traditional to global popular culture

Leisure and entertainment are in part a global industry and in part a local one. The share of local content depends on the society. Some more-traditional societies prefer and even aggressively promote local origination

491

television programming, sporting events, music, art, and theater. Others, a growing majority, are dominated by global popular culture, fueled in large part by arts and entertainment imported from the United States or modeled on U.S. popular culture.

English is the lingua franca of international television, and more and more people around the world speak English in part because of television, radio, and movie entertainment. However, other languages can be instantly dubbed today.

People in World 2 countries typically can afford entertainment in the home including television (color, noninteractive), radio, and electronic games. Fewer households, about one quarter, can afford interactive television and other electronic entertainment devices for home use.

About a fifth of middle-country people can also travel for leisure, at least every few years. Resort vacations at beaches, religious shrines, cultural landmarks, and visits to family members overseas are all affordable.

A CASE STUDY: Turkey works to build a modern, western society

Turkey is a World 2 country with good chances of making it at least to the bottom of the World 1 ranks within several decades. Turkey is strategically located for lucrative trade—where the largely fundamentalist Muslim Middle East meets secular Eastern and Western Europe. The country is uniquely capable of serving the two markets.

Turkey is among the most developed and affluent countries in the Middle East, and dominates the Turkic-speaking Muslim countries of central Asia economically and culturally. Though most of those central Asian countries are destitute, since the early 1990s they have been a growing market for Turkey as consumers of food, water, raw materials, cement, energy, and entertainment.

Turkey: on the path to success

Turkey's economy has grown an average 3.2% over the past 30 years, while its population has averaged under 2% growth. That has brought it from an uncertain status as a middle-income country to a country high in the economic ranks of the middle countries.

Western Europe and the United States encouraged Turkey's development through the end of the last century and into the first 20 years of this century. Their purpose was in part to hold it out as an example to its less stable, less peaceful, and less successful Middle Eastern neighbors.

No other Middle Eastern country, nor any majority Muslim country has the successful, heterogeneous economy that Turkey has. That has made Turkey the industrial center of its Muslim world, while it also serves markets to the north, northeast, and west.

Though Turkey failed to meet EC requirements for associate membership through the early part of this century, it has a better chance within the next 10 years, if it keeps its relations with Greece peaceful, and avoids unstable government.

In Turkey's favor, it has a relatively stable population and low birth rates. Though the country grew from 56 million in 1990 to 88 million today, it has made the demographic transition from a high-birth-rate country to a low-birth-rate country, and so stabilized its population at what most believe are sustainable levels. Today, the only demographic threats the country faces are international refugees from war-torn Muslim central Asian republics to the northeast, and periodic waves of migrants and refugees from elsewhere in the Middle East.

Turkey benefits from being a nexus of global trade between several distinct regions. That status drives Turkish society to be open to diversity, cosmopolitan, and highly mobile. This goes on in a context of basic Muslim values and a 17% share of the population being fundamentalist Muslim. The Turkish government, however, has remained secular and committed to modern development.

Turkish work life grows steadily more European

Turkey used to be a state-industry-dominated economy. In the 2000s and 2010s, the country privatized one after another of its state enterprises and loosened the grip of the state on the private sector. The same pattern happened in dozens of other World 2 countries, with varying success. Privatization led to national and international commercial interests setting up factories and service industries in Turkey. As a result, today the majority of Turks work at private-sector jobs.

Turkey supplied guest workers to labor-short northern Europe in the 1980s and 1990s. By 2000 or so, Turkey could provide good jobs at home for many who might formerly have traveled abroad for work. Working overseas is now a choice of only about 0.2% or about 300,000 Turks. Greater educational achievement in Turkey means that a greater share, nearly 40%, of those overseas workers are professionals and experts, whereas in 1990, most were low-skilled menial workers.

The slow pace of modernization and globalization in the Turkish workplace

Turkey has a workplace power imbalance reminiscent of that in the United States in the 1990s. Employers, especially in manufacturing, hire and fire workers according to demand for their firms' products. This practice was illegal during the Social Democratic regimes of the '00s, and '10s. Today's probusiness government has deregulated the workplace, and industrial development takes place on the backs of the working population.

Global corporations are slowly changing work in Turkey. Working directly for global corporations is a coveted status. But more people work for the domestic suppliers and contractors that support the Turkish operations of global corporations.

Increasingly, the local operations of international organizations come under international agreements that govern workplace conditions, working hours, compensation, hiring, and firing. For example, the International Compact on Workplace Rights (2017), is voluntarily supported by most *Cumhuriyei Merkez 100* businesses in Turkey.

Work life in Turkey approximates that of World 1 30 years ago. Workers go to workplaces away from home in proportions much higher than workers in World 1 do today. With few exceptions, work-at-home and distributed work are rare, though the information technologies required are becoming available to more Turkish service and professional workers.

Technological literacy is the key to the modernization of Turkish industry. The Turkish education system is trying to respond to greater demands placed on it by employers. Turkish businesses and global interests in Turkey established a network of community colleges in 2018-2022, which people can attend in person or electronically. The curriculum covers automation, network management, quality, and other essential topics for business. Interactive learning consoles in workplaces give workers after-hours access to the training.

The Turkish education system, as has been true of systems in World 1, is evolving to serve lifelong learning. The emphasis in Turkey is on training and retraining workers for the shifting demands of the workplace.

Leisure and entertainment for an increasingly middle-class population

Turks fill their leisure time with a mix of traditional and imported activities. Like any increasingly urban, middle-class society, Turkey is undergoing a flowering of leisure and entertainment activities, fueled by migration, growing affluence, the globalization of world culture, and the long reach of the telecommunications network.

Five most popular leisure activities in Turkey

- Television and interactive television

- Watching professional soccer

- Going to the beach

- Playing soccer, speed tag, basketball

- Shopping

More Turks today than ever before can afford to travel. Travel patterns for the population show the ironies of modern Turkey. Fundamentalist and conservative Muslims save to travel to Mecca and other shrines in the Muslim world. More secular Muslim and non-Muslim Turks typically travel west to Europe or the United States or take holidays on the Mediterranean or Black Sea. Disney Mediterranean, which opened on Cyprus in 2019, attracts 1.8 million visitors a year, and hundreds of thousands of them are from Turkey. The trip is a two-hour hovercraft ride from Adana, which is in turn linked to Ankara and Istanbul by bullet train.

Part of Turkey's growing interest in leisure travel stems from its success as an international tourist destination. Celebrating its tourist resources—ancient and Byzantine heritage, Black Sea beaches, rugged mountains, and even the semi-mythical status of Turkey as the resting place of Noah's Ark—has made the country a global tourist destination. Today, tourism is 22% of the country's GDP.

More time spent in leisure involves the professional sports that have pervaded Turkish life in the past 15 years. Istanbul hosted the 2012 Olympics, spurring heightened interest in televised sports competition and amateur sports participation. Fueled by the growth of televised sports in Turkey, organized amateur and professional sports took off as entertainment events. Wrestling, weight lifting, competitive martial arts, soccer, speed tag, and basketball dominate sports on television. The biggest celebrity in Turkey today is Sulayman Ugur, nicknamed Sulayman the Invincible for his soccer prowess. Virtual-reality-enhanced martial arts and fencing competitions are also runaway hits on television.

Manipulating the mind for leisure and escape

Turks turn to drugs for leisure and to escape boredom and poverty. The growing use of cannabis-derived recreational narcotics exemplifies this practice. Turkey historically produced and traded hashish and other narcotics, so their use fits with traditional customs while it also suits the modern stressful urban lifestyle. At first, the Turkish government reacted to the rise in narcotics use as a problem. But by 2018, the practice was legal and accommodated by social innovations to prevent it from affecting workplace safety and productivity. But though they are legal, most drugs are sold in the informal sector and evade government regulations and taxes.

Peasant migrants from rural Turkey are swelling the populations of the largest cities. The arriving migrants bring traditional values and thus leisure and entertainment customs to the centers of population. Commercial entertainment in Ankara, Istanbul, and other large cities has begun to serve these migrants more effectively. *Anatolia Hour* on Turkish television, and traditional dance and music shows and radio and television entertainment are growing increasingly popular.

Consumption as entertainment

Today's grand bazaars in Istanbul and Ankara, and the bazaars of other large cities, are thriving indoor shopping malls, the center of leisure for millions of Turks. They combine shopping with 3-D movie entertainment, virtual reality games, hashish parlors, fine dining and fast food, and socializing. Traditional ways of barter and bargaining still linger in some shops, while others conform to the standards of internationally franchised retailing, with set prices and standard products.

The arts

Ethnicity and nationalism have grown important to the increasingly Europeanized Turkey. This raises interest in traditional Islamic art and music and modern interpretations of them. Many among the wealthy have become avid art collectors, while the average Turk can afford to see musical events live and on television or order concerts for home viewing. Affordable, portable music synthesizers make traditional and modern Turkish or globally popular music available in almost any household. Wealthier urban Turks go to classes to learn traditional ethnic crafts such as rug weaving.

Food as entertainment

Turkish cuisine has long suffered from competition by more refined ethnic foods from Europe, Africa, and Asia. People who were affluent enough to enjoy eating out during the past 30 years could rarely find gourmet Turkish food to enjoy. They usually opted instead for French and other European foods, or international fast food from franchises. By 2000, the ethnic food available in the largest cities included choices from all over the world. Over the decades since then, the food choices and habits of Turks have grown more diverse. At the same time, new approaches to traditional Turkish cuisine and the rediscovery of regional dishes from the countryside have led to a renaissance in food.

Home entertainment

The ever-popular computer games played in many Turkish households are driving more families to get television units with which they can play more games and interact remotely with other players. Most Turks cannot afford the flat screens and video walls popular in the United States, Japan, and Europe.

Television dominates most Turkish homes. Ninety-eight percent of homes have a television, though most are noninteractive sets. Only 46% have interactive TV capabilities. Half of programming is Turkish-produced. The Turkish programming is broadcast to the Turkic-speaking countries of Central Asia, further cementing their growing bonds to Turkey as a regional leader. Turkish TV entertainment competes with translated programming from Greece, Europe, the United States, and occasionally, other Muslim countries.

The poorest urban Turks, living in the *gecekondu* squatter settlements, fill their leisure time with coffeehouse entertainment, television, radio, or visiting a nearby mosque. These settlements are not the ramshackle, bleak slums of Brazil or North Africa. They are made up of sturdy if modest houses in orderly neighborhoods. Social Democratic governments have made the support of these settlements a policy. In most cases, people are given title to their houses and the property they are on.

WORK, LEISURE, AND ENTERTAINMENT IN WORLD 3—The struggle of the destitute makes leisure and entertainment secondary

Destitute families struggle to feed, clothe, shelter, and warm themselves. For some, that struggle means long hours of work in factory or menial service jobs. For others, it means days of scavenging for food, fuel, and goods. When destitute people are not working at their survival, most have little energy and almost no money for leisure and entertainment.

Liquor, drugs, and television soak up the intense boredom of the world's poorest people. Many of the traditional cultural outlets for leisure and entertainment, sports, music, theater, games, and dance, have been lost in the migrations and urbanization that have characterized the development of the world's poor. Mass popular culture from Europe and the United States takes its place mostly in television, movies, and radio, and does not serve everyone well.

The world of work

Destitute people often suffer poor conditions at their workplaces. Most large industries are raw materials producers, make second-rate merchandise for regional consumption, or make components for goods assembled elsewhere. Workers are poorly skilled, poorly paid, and are not very productive.

Smaller-scale industries and home workshops offer varying conditions for workers. Skill levels may be much higher for these craft-driven enterprises, but pay and safety conditions vary. In a World 3 version of work-at-home, millions in the poor countries make their living as independent craft and service people, serving their surrounding neighborhoods.

Technology plays only a limited role in work in World 3. Where big business has built factories, there may be high levels of automation supported by some skilled workers. Telecommunications plays a role for even some of the smallest enterprises—those that trade goods across cities or regions, or that offer a transportation or message carrying service. They are usually equipped with cellular telephones or PCS units. Bartering done through such communications devices extends and improves the market for small-time traders just as it does for national and international enterprises.

Women work in industry and services in higher proportions than they did 20 years ago in the industrializing poor countries. Many have long been displaced from traditional work in crafts, agriculture, and trade. For example, West African women once produced goods and sold them in their own trading businesses, largely out of their homes and small shops. Today, 55% of industry workers in Nigeria are women, compared with far fewer in the 1990s. Women have benefitted from intensive programs to improve their education and financial prospects in Latin America, Africa, and southern Asia since the 1990s.

Leisure and entertainment

The destitute do not have the wide choices in leisure and entertainment that the affluent societies and even the middle-income societies offer. Most destitute people must settle for escape through liquor and other drugs, mass broadcast radio and television entertainment, religious activities, sports, and socializing.

Through national and international efforts, there is a movement today to make entertainment inWorld 3 countries a vehicle for education for people of all ages. National and international television broadcasters, such as those supported by UNESCO, are altering programming to instruct people on things like agricultural techniques, health practices, automobile repair, and sanitation. Programs also include literacy and civic lessons. The education programming is sometimes built into entertainment shows directly, such as with the long-popular soap opera saga *Jambo Jambo* shown in West Africa, which promote better health and sanitation practices in urban areas.

The prospects for improved life in World 3, and with it greater choice in leisure and entertainment, are poor. Fast population growth and poor government policymaking pervade the destitute world. Few countries have much chance to raise themselves to World 2 in the next few decades.

Critical Developments, 1990-2025

Year	Development	Effect
1990	Work and leisure time roughly equal at 40 hours a week.	People consistently report feeling rushed.
1999	Productivity growth under 2%.	Last year of flat productivity growth.
2000	Reduced Time Act passed.	Mandates full-time workweek of 36 hours.
2001	Virtual reality industry surpasses $2 billion in annual sales.	Combination of entertainment, military, simulation, and business training sales.
2003	Productivity measures changed to accommodate knowledge economy.	Productivity found to be above 3% per year.
2005	Census Bureau and Bureau of Labor Statistics adopt Smith-Garcia work categories.	Reflects shift to knowledge economy.
2005	IRS issues revised rules on contract work.	Practice of shifting workers to contract work involuntarily is prohibited.
2007	50% of homes have access to net communications and games.	Penetration of information technology continues.
2010	Reduced Time Act Amendments passed.	Mandatory full-time work week reduced to 34 hours.
2018	Smith and Garcia receive Nobel Prize in economics.	Recognizes their earlier work in reorganizing work categories.
2019	75% of workers now categorized as Class I knowledge workers in the Smith-Garcia classification.	Knowledge economy firmly established.
2020	37% of workers engaged in distributed work part of the week.	Growing comfort with information technology established distributed work as a preferred format.
2025	Executive boards are now 35% women and 29% minorities.	Slow but steady progress through glass ceilings.

Unrealized Hopes and Fears

Event	Potential Effects
Unemployment at depression levels above 12% in World 1.	General strikes or violent political revolution.
Machines continue to put people out of work; low-skill, low-wage service jobs are predominant.	Antitechnology campaigns to stop new machine technology; rise of voluntary simplicity movements.
Lessons from entertainment industry in capturing people's interest are adopted by educators and trainers and lead to more educated and skilled workforce.	Workers meet and exceed requirements of knowledge economy, leading to productivity boom of rates above 5% annually.

Index

Y

Z

About the Authors

Joseph F. Coates—President of Coates & Jarratt, Inc., a think tank and policy research company that consults with organizations on key factors shaping their futures. Coates has worked with at least 45 of the Fortune 100 companies, numerous smaller firms, all levels of government, and many trade, professional, and public interest groups. A frequent lecturer on trends and future developments, he has published more than 300 articles and has co-authored several books including *Future Work* and *What Futurists Believe.*

John B. Mahaffie—An associate with Coates & Jarratt, Inc. Mahaffie has authored over three dozen futures studies for corporations, government agencies, and nonprofit groups. He is an author and speaker on the future of science and technology, health and medicine, work and worklife, and other futures topics. He co-authored *Future Work,* with colleagues Jarratt and Coates.

Andy Hines—Formerly an associate with Coates & Jarratt, Inc. for six years, Hines had been a principal analyst and author on several dozen futures research projects. He frequently speaks at conferences and has appeared on television, radio, and on-line programs about the future. He is president of the Metropolitan Washington D.C. Chapter of the World Future Society.